Recurrence in Topological Dynamics

Furstenberg Families and Ellis Actions

THE UNIVERSITY SERIES IN MATHEMATICS

Recent volumes in the series:

THE CLASSIFICATION OF FINITE SIMPLE GROUPS: Volume 1: Groups of Noncharacteristic 2 Type
Daniel Gorenstein

COMPLEX ANALYSIS AND GEOMETRY
Edited by Vincenzo Ancona and Alessandro Silva

ELLIPTIC DIFFERENTIAL EQUATIONS AND OBSTACLE PROBLEMS
Giovanni Maria Troianiello

FINITE SIMPLE GROUPS: An Introduction to Their Classification
Daniel Gorenstein

AN INTRODUCTION TO ALGEBRAIC NUMBER THEORY
Takashi Ono

MATRIX THEORY: A Second Course
James M. Ortega

PROBABILITY MEASURES ON SEMIGROUPS: Convolution Products, Random Walks, and Random Matrices
Göran Högnäs and Arunava Mukherjea

RECURRENCE IN TOPOLOGICAL DYNAMICS: Furstenberg Families and Ellis Actions
Ethan Akin

A SCRAPBOOK OF COMPLEX CURVE THEORY
C. Herbert Clemens

TOPICS IN NUMBER THEORY
J. S. Chahal

VARIATIONS ON A THEME OF EULER: Quadratic Forms, Elliptic Curves, and Hopf Maps
Takashi Ono

Recurrence in Topological Dynamics

Furstenberg Families and Ellis Actions

Ethan Akin
The City College
New York, New York

Plenum Press • New York and London

Library of Congress Cataloging-in-Publication Data

Akin, Ethan, 1946-
Recurrence in topological dynamics : Furstenberg families and Ellis actions / Ethan Akin.
p. cm. -- (The university series in mathematics)
Includes bibliographical references and index.
ISBN 0-306-45550-1
1. Topological dynamics. 2. Point mappings (Mathematics) I. Title. II. Series: University series in mathematics (Plenum Press)
QA611.5.A38 1997
514'.322--dc21 97-24363
CIP

ISBN 0-306-45550-1

A Division of Plenum Publishing Corporation
233 Spring Street, New York, N. Y. 10013

http://www.plenum.com

10 9 8 7 6 5 4 3 2 1

Printed in the United States of America

For Dr. Karl Gruber, Ph.D.
In memory of a warm friend.

Preface

In the long run of a dynamical system, after transient phenomena have passed away, what remains is recurrence. An orbit is recurrent when it returns repeatedly to each neighborhood of its initial position. We can sharpen the concept by insisting that the returns occur with at least some prescribed frequency. For example, an orbit lies in some minimal subset if and only if it returns almost periodically to each neighborhood of the initial point. That is, each return time set is a so-called syndetic subset of $T =$ the positive reals (continuous time system) or $T =$ the positive integers (discrete time system). This is a prototype for many of the results in this book. In particular, frequency is measured by membership in a family of subsets of the space modeling time, in this case the family of syndetic subsets of T. In applying dynamics to combinatorial number theory, Furstenberg introduced a large number of such families.

Our first task is to describe explicitly the calculus of families implicit in Furstenberg's original work and in the results which have proliferated since. There are general constructions on families, e.g., the dual of a family and the product of families. Other natural constructions arise from a topology or group action on the underlying set. The foundations are laid, in perhaps tedious detail, in Chapter 2. The family machinery is then applied in Chapters 3 and 4 to describe family versions of recurrence, topological transitivity, distality and rigidity. Many different notions are unified by this family viewpoint. In particular, various ideas of mixing appear as family versions of transitivity. In Chapter 5 we use Gelfand theory to describe how such properties are carried over to compactifications of the original system.

For compact systems Robert Ellis developed something completely different. He defined the enveloping semigroup of a system by taking the pointwise closure of the set of maps of the flow. The result is an important example of a class of compact semigroups which we call Ellis semigroups. The idempotents play an important role in the resulting theory. For example, a point is recurrent exactly when it is fixed by certain idempotents of the enveloping semigroup. We review the theory of Ellis semigroups and their idempotents in Chapter 6.

The two viewpoints are brought together in Chapter 7. As a specific example consider the discrete case with T the set of positive integers. Addition on T extends to define a nonabelian composition on the Stone–Čech compactification βT, giving this space the structure of an Ellis semigroup. A compact, discrete dynamical system is an action of T on a compact space and such an action extends to an action of the semigroup. On the other hand, the closed subsets of βT correspond to filters on T with the points associated to ultrafilters, i.e., maximal filters. Filters and their duals are families and the family constructions for them have natural semigroup interpretations. Furthermore, the semigroup structure yields special results for such families. On the other hand, there are important families which are neither filters nor filterduals. Such families can be used to describe aspects of βT.

Finally, in Chapter 8 we apply all the earlier work to study equicontinuous and related systems.

For help with this work I would like to thank my fellow travelers in topological dynamics, particularly Joseph Auslander, Ken Berg and Eli Glasner with whom I have shared many conversations both real and virtual. Especially fruitful were my meetings with the two artists who inspired all this: Hillel Furstenberg and Robert Ellis. At the production end, Kate March once again translated my pen scratchings into a manuscript, and at Plenum it was a joy to meet and work with Sy Marchand (it was a pleasure even to lose an editorial argument with him). To my wife Jean: Thank you for your continuing support of my mathematical enthusiasms, not to mention your patience when I used math as an excuse to avoid other work.

Contents

Introduction

If $f : X \to X$ is a continuous map and $x \in X$, we say that y is a limit point for the associated dynamical system with initial value x, or y is an ω limit point of x, when y is a limit point of the orbit sequence $\{f^n(x) : n \in T\}$ where T is the set of nonnegative integers. This means that the sequence enters every neighborhood of y infinitely often. That is, for any open set U containing y, the entrance time set $N(x,U) = \{n \in T : f^n(x) \in U\}$ is infinite.

It is often useful to keep track of just how frequently these entrance times occur. In *Recurrence in Ergodic Theory and Combinatorial Number Theory*, Hillel Furstenberg used families of subsets of T to keep track of the frequencies. A family $\mathcal{F}$ for T is a collection of subsets, i.e., a subset of the power set $\mathcal{P}$ of T, which is hereditary upward. That is, if $F_1 \subset F_2$ and $F_1 \in \mathcal{F}$ then $F_2 \in \mathcal{F}$. A family is proper if it is a proper subset of $\mathcal{P}$, i.e., $\mathcal{F} \neq \emptyset, \mathcal{P}$. In view of heredity, this says that $\mathcal{F}$ is proper when $T \in \mathcal{F}$ and $\emptyset \notin \mathcal{F}$. For a proper family $\mathcal{F}$, we say that y is an $\mathcal{F}$ ω limit point of x if $N(x,U) \in \mathcal{F}$ for all neighborhoods U of y. In Chapter 2 we present the elementary theory implicit in Furstenberg's work for such families; in Chapters 3 and 4, we apply it to dynamical systems in general and to topologically transitive systems in particular.

This family approach goes back at least to Gottschalk and Hedlund (1955), who introduced *admissible subset* collections to unify several notions of recurrence — exactly our purpose.

When the state space X is compact, the semigroup theory of Robert Ellis can be applied. For our purposes, this is best described here as an extension of the action of T on X to an action of βT, the Stone–Čech compactification of T, on X. For $x \in X$, we can define the orbit map $\varphi_x : T \to X$ by $\varphi_x(n) = f^n(x)$. Since T is discrete, this is a continuous map which therefore extends to a map $\Phi_x : \beta T \to X$. For $p \in \beta T$, we write $p(x) = \Phi_x(p)$, thus regarding elements of βT as functions on X. Ellis observed that βT has a natural semigroup structure satisfying $(pq)(x) = p(q(x))$. However this hybrid between the algebra and the topology on βT appears at first to be rather mulish. Right translation is continuous, but left translation is not; i.e., $p \mapsto pq$ is continuous, $q \mapsto pq$ is not. Similarly $p \mapsto p(x)$ is continuous, but p itself, $x \mapsto p(x)$, is usually not. Perhaps unsurprisingly the composition in βT, while it extends addition on the dense subset T, is not commutative. Despite these infelicities, the semigroup structure on βT has proved very fruitful in studying dynamical systems.

The reader should be aware that Ellis uses right rather than left actions, so his semigroup structure on βT is the reverse of ours. His left translations are continuous, and his compatibility equation reads $(xq)p = x(qp)$. Thus the view in Chapters 6 and 7 is the mirror image of the Ellis way.

We can regard the points of βT as ultrafilters on T. A filter is a proper family that is closed under the operation of intersection. An ultrafilter is a maximal filter. Thus the two approaches meet. The Furstenberg theory applies to ultrafilters as it does to all families; some of the Ellis constructions are nicely expressed in a family way. Furthermore the family approach is more general. Various family constructions do not preserve the filter property. On the other hand, the semigroup structure with its associative law and collections of idempotents reveal properties about certain special families that would not be otherwise apparent.

An outline of our book follows.

1. Monoid Actions

Once you abandon compactness assumptions, you discover that various dynamic notions like equicontinuity and chain recurrence are really uniform space notions. The very definitions require a uniform structure. The associated topological spaces, i.e., completely regular spaces, appear in full generality when you take arbitrary products and subsets. Each dynamical system in this book is a *uniform action* of an abelian *uniform monoid* on a uniform space, written $\varphi : T \times X \to X$. This chapter presents set-up work that enables the reader to make sense of these phrases.

An abelian topological group has a unique translation invariant uniform structure obtained from neighborhoods of the identity. An abelian uniform monoid is a submonoid T of an abelian topological group; i.e., T is closed under addition and zero is in T, but inverses may not be. Also T satisfies a mild technical condition, the *Interior Condition*, on the set of tails. For $t \in T$, the associated tail T_t is the image of T under translation by t, $T_t = \{s+t : s \in T\}$. Any discrete abelian monoid, e.g., the nonnegative integers, $\mathbf{Z}_+$, and any abelian topological group; e.g., $\mathbf{R}$, is uniform monoid. The nonnegative reals $\mathbf{R}_+$ under addition is also uniform.

Note: After Chapter 1 all monoids are assumed to be abelian unless otherwise mentioned; all topologies are assumed to be Hausdorff.

An action is a function $\varphi : T \times X \to X$ such that time t maps defined by $f^t(x) = \varphi(t,x)$ satisfy the composition property $f^{t_1} \circ f^{t_2} = f^{t_1+t_2}$. For φ to be a uniform action, we first assume that each $f^t : X \to X$ is a uniformly continuous map of the uniform space X. Hence the adjoint associate $\varphi^\#$ of φ that associates $t \mapsto f^t$ is a homomorphism from T to $C^u(X;X)$, the space of uniformly continuous maps on X. The second condition of a uniform action is that the homomorphism $\varphi^\#$ is continuous, and hence uniformly continuous, when $C^u(X;X)$ is given the uniformity of uniform convergence. Thus φ is a uniform action if each f^t is uniformly continuous, and $t_i \to 0$ in T implies that $f^{t_i} \to 1_X$ uniformly on X. It then follows that φ is a continuous map (though not uniformly continuous), i.e., φ is a topological action. When T is discrete, the second condition is trivial. In particular a uniform action of $T = \mathbf{Z}_+$ is just given by the iterates of a uniformly continuous map $f = f^1$ on X. If T is uniform and X is compact, then any topological action $\varphi : T \times X \to X$ is uniform. Recall that a compact space has a unique uniformity consisting of all neighborhoods of the diagonal.

The use of monoids allows us to apply the theory to semiflows and noninvertible maps. More importantly even for a homeomorphism f on a compact space X, it is useful to distinguish between the $\mathbf{Z}_+$ action using f, the reverse action that is the $\mathbf{Z}_+$ action using f^{-1}, and the extended $\mathbf{Z}$ action that includes both. The limit point set for x associated with these are, respectively, the ω limit set, the α limit set, and the closure of the entire orbit of x. For a monoid we move toward infinity using the tails T_t. If T is a group, then $T_t = T$ for all t.

While our theory is motivated by the cases $T = \mathbf{Z}_+, \mathbf{R}_+, \mathbf{Z}$, and $\mathbf{R}$, there is at least one other example worth mentioning here, namely, $T = \mathbf{Z}^*$, the positive integers under multiplication (discrete uniformity). If X is a compact topological group, e.g., the unit circle in $\mathbf{C}$, then $\mathbf{Z}^*$ acts on X via exponentiation. That the $\mathbf{Z}^*$ action on the circle is strongly mixing proves useful once definitions are in place to make sense of the statement.

2. Furstenberg Families

For a uniform monoid T, a *family* $\mathcal{F}$ is a subset of $\mathcal{P}$, the power set of T, which is hereditary upward. $\mathcal{F}$ is a proper family when $\emptyset \notin \mathcal{F}$ and $T \in \mathcal{F}$. The *dual* $k\mathcal{F}$ is $\{F : F \text{ meets } F_1 \text{ for every } F_1 \in \mathcal{F}\}$ or equivalently $F \in k\mathcal{F}$ iff $T \setminus F \notin \mathcal{F}$. For any family $\mathcal{F}$, $k\mathcal{F}$ is a family and $kk\mathcal{F} = \mathcal{F}$. Clearly $k\mathcal{P} = \emptyset$, so $k\mathcal{F}$ is proper iff $\mathcal{F}$ is. The largest proper family is $\mathcal{P}_+ = \mathcal{P} \setminus \{\emptyset\}$, whose dual $k\mathcal{P}_+$ is $\{T\}$.

A *filter* is a proper family that is closed under intersection. A *filterdual* is a family whose dual is a filter. A proper family $\mathcal{F}$ is a filterdual iff it satisfies what Furstenberg calls the *Ramsey Property*: $F_1 \cup F_2 \in \mathcal{F} \Rightarrow F_1 \in \mathcal{F}$ or $F_2 \in \mathcal{F}$. An *ultrafilter* is a maximal filter or equivalently a self-dual filter.

Using the action of T on itself by translation, we define $g^t : T \to T$ to be the translation map by t whose image is the tail T_t. Call a family $\mathcal{F}$ *translation invariant* if for all $t \in T, F \in \mathcal{F}$ iff $g^{-t}(F) \in \mathcal{F}$, where $g^{-t}(F)$ denotes the preimage

$(g^t)^{-1}(F)$. $\mathcal{F}$ is *thick* if $F \in \mathcal{F}$ and $t_1,\dots,t_k \in T$ imply $\cap_{i=1}^k g^{-t_i}(F) \in \mathcal{F}$. For any family $\mathcal{F}$, $\gamma\mathcal{F}$ denotes the smallest translation invariant family containing $\mathcal{F}$ and $\tilde{\gamma}\mathcal{F}$ the largest translation invariant family contained in $\mathcal{F}$, so that $k\gamma\mathcal{F} = \tilde{\gamma}k\mathcal{F}$. Define $\tau\mathcal{F}$ to be the largest thick family contained in $\mathcal{F}$. Observe that a translation invariant filter is automatically thick.

The family $\tilde{\gamma}\mathcal{P}_+$, denoted $\mathcal{B}_T$, is the largest translation invariant proper family. $F \in \mathcal{B}_T$ iff $g^{-t}(F) \neq \emptyset$ for all $t \in T$. Its dual, $k\mathcal{B}_T = \gamma k\mathcal{P}_+$ is the family generated by the tails. $F \in k\mathcal{B}_T$ iff $g^{-t}(F) = T$ for some $t \in T$. $k\mathcal{B}_T$ is the smallest translation invariant proper family. It is a filter, so $\mathcal{B}_T$ is a filterdual. Notice that if T is a group, then $\mathcal{B}_T = \mathcal{P}_+$ and $k\mathcal{B}_T = k\mathcal{P}_+ = \{T\}$.

In the case $T = \mathbf{Z}_+$, $\mathcal{B}_T$ is the family of infinite subsets and the dual $k\mathcal{B}_T$ is the family of cofinite subsets. The family $\tau\mathcal{B}_T$ is called the family of thick sets of $\mathbf{Z}_+$. $F \in \tau\mathcal{B}_T$ iff F has arbitrarily long runs; i.e., for every $N \in \mathbf{Z}_+$ there exists $t \in \mathbf{Z}_+$ such that $t, t+1, \dots, t+N \in F$. The dual $k\tau\mathcal{B}_T$ consists of the syndetic or relatively dense sets. That is, $F \in k\tau\mathbf{B}_T$ iff there exists N such that every interval of length N meets F; i.e., for every $t \in \mathbf{Z}_+$, $\{t, t+1, \dots, t+N\} \cap F \neq \emptyset$. The family $\tau k\tau\mathcal{B}_T$ consists of what we call *replete* sets. $F \in \tau k\tau\mathcal{B}_T$ if for every N the positions where length N runs begin form a syndetic set. All of these families are translation invariant, and $\tau k\tau\mathcal{B}_T$ is a filter. Translation invariant filters are quite useful. In general if $\mathcal{F}$ is a filter, then $\mathcal{F}$ is contained in some translation invariant filter iff $\mathcal{F} \subset \tau\mathcal{B}_T$.

Using the uniform structure on T, we call $\mathcal{F}$ an *open family* if every $F \in \mathcal{F}$ is a uniform neighborhood of some other element of $\mathcal{F}$; i.e., there exists $F_1 \in \mathcal{F}$ and V in the uniformity $\mathcal{U}_T$ such that $F \supset V(F_1)$. One of the purposes of the Interior Condition on a uniform monoid is to ensure that the filter $k\mathcal{B}_T$ generated by the tails is an open family. Of course if T is discrete, so the diagonal $1_T \in \mathcal{U}_T$, then every family is open.

3. Recurrence

For $\varphi : T \times X \to X$, a uniform action and subsets A, B of X we define the *meeting time set* $N(A,B) = \{t \in T : f^t(A) \cap B \neq \emptyset\}$. If we identify each map f^t with its graph, we can define the relation f^F for $F \subset T$ to be $\cup\{f^t : t \in F\}$. Thus F meets $N(A,B)$ iff $f^t(A) \cap B \neq \emptyset$ for some t in F and so iff $f^F(A) \cap B \neq \emptyset$.

Define $\mathcal{N}(A, u[B])$ to be the family of subsets of T generated by all $N(A,U)$ where U is a uniform neighborhood of B and $\mathcal{N}(u[A], u[B])$ the family generated by all $N(W,U)$ where W and U are uniform neighborhoods of A and B, respectively. When proper each of these families is an open filter.

Given a family $\mathcal{F}$ and a nonempty subset A of X, we define $\omega_{\mathcal{F}}\varphi[A] = \cap\{\overline{f^F(A)} : F \in k\mathcal{F}\}$. A point $y \in \omega_{\mathcal{F}}\varphi[A]$ iff $N(A,U) \in \mathcal{F}$ for every neighborhood

U of y, i.e., iff $\mathcal{N}(A,u[y]) \subset \mathcal{F}$. In particular we define the relation $\omega_{\mathcal{F}}\varphi \subset X \times X$ by $\omega_{\mathcal{F}}\varphi(x) = \omega_{\mathcal{F}}\varphi[x] = \cap\{\overline{f^F(x)} : F \in k\mathcal{F}\}$. When X is compact and $\mathcal{F}$ is a filterdual, then $\omega_{\mathcal{F}}\varphi[A]$ is nonempty; if U is an open set containing $\omega_{\mathcal{F}}\varphi[A]$, then U contains $\overline{f^F(A)}$ for some $F \in k\mathcal{F}$.

We define the closed relation $\Omega_{\mathcal{F}}\varphi$ to be $\cap\{\overline{f^F} : F \in k\mathcal{F}\}$, taking the closure in $X \times X$. Two points x,y satisfy $y \in \Omega_{\mathcal{F}}\varphi(x)$ iff $N(W,U) \in \mathcal{F}$ for every neighborhood W and U of x and y, respectively, i.e., iff $\mathcal{N}(u[x],u[y]) \subset \mathcal{F}$. Equivalently $\Omega_{\mathcal{F}}\varphi(x) = \cap\omega_{\mathcal{F}}\varphi[W]$, intersecting over all neighborhoods W of x.

We say that x $\mathcal{F}$ *adheres* to a set B if $\mathcal{N}(x,u[B]) \subset \mathcal{F}$, i.e., $N(x,U) \in \mathcal{F}$ for every uniform neighborhood U of B. If B is compact, then the sufficient condition $\overline{f^F(x)} \cap B \neq \emptyset$ for all $F \in k\mathcal{F}$ is necessary as well. So if B is compact and $\mathcal{F}$ is a filterdual then x $\mathcal{F}$ adheres to B iff $\omega_{\mathcal{F}}\varphi(x) \cap B \neq \emptyset$. For any family $\mathcal{F}$, $\omega_{\mathcal{F}}\varphi(x) = \{y : x\ \mathcal{F} \text{ adheres to } y\}$. In particular $x \in \omega_{\mathcal{F}}\varphi(x)$ iff x $\mathcal{F}$ adheres to x. We call such a point $\mathcal{F}$ *recurrent*.

Note: The usual notions of $\omega\varphi$ and $\Omega\varphi$ [cf. Akin (1993)] correspond to $\mathcal{F} = \mathcal{B}_T$, so we drop the subscript in that case.

To describe the meaning of these concepts in certain important cases, recall that an action φ is called *minimal* if for B a nonempty, closed subset of X, $f^t(B) \subset B$ for all $t \in T$ (i.e., B is + invariant) implies $B = X$. A closed, nonempty, + invariant subset is called a *minimal subset* if the restriction of the action to B is a minimal action. Every compact, nonempty, + invariant subset of X contains a minimal subset. The closure of the union of all minimal subsets of X is called the *mincenter* of X.

We call x a *fixed point* for φ if $f^t(x) = x$ for all $t \in T$ or equivalently if $\{x\}$ is + invariant and hence a minimal subset.

Now assume that $\varphi : T \times X \to X$ is a uniform action, with X compact. Let $x \in X$ and B be a closed subset of X. If $\mathcal{F}$ is a filterdual, then x $\mathcal{F}$ adheres to B, iff $\omega_{\mathcal{F}}\varphi(x)$ meets B and x $k\mathcal{F}$ adheres to B iff $\omega_{\mathcal{F}}\varphi(x) \subset B$. In particular x $\mathcal{B}_T$ adheres to B iff $\omega\varphi(x) \cap B \neq \emptyset$ and x $k\mathcal{B}_T$ adheres to B iff $\omega\varphi(x) \subset B$. The family $k\tau k\tau\mathcal{B}_T$ is a filterdual, and $\omega_{k\tau k\tau\mathcal{B}_T}\varphi(x)$ is the mincenter of $\omega\varphi(x)$. So x $k\tau k\tau\mathcal{B}_T$ adheres to B iff B meets the mincenter of $\omega\varphi(x)$, and x $\tau k\tau\mathcal{B}_T$ adheres to B iff B contains the mincenter of $\omega\varphi(x)$.

Furthermore under this compactness hypothesis, x $k\tau\mathcal{B}_T$ adheres to B iff B meets every minimal subset of $\omega\varphi(x)$, while x $\tau\mathcal{B}_T$ adheres to B iff B contains some minimal subset of $\omega\varphi(x)$. In particular x is $k\tau\mathcal{B}_T$ recurrent iff x is contained in some minimal subset of X [which is then necessarily $\omega\varphi(x)$], and x is $k\tau k\tau\mathcal{B}_T$ recurrent iff the minimal subsets of $\omega\varphi(x)$ are dense in $\omega\varphi(x)$.

4. Transitive and Central Systems

A uniform action $\varphi : T \times X \to X$ is called $\mathcal{F}$ *central* for a family $\mathcal{F}$ if $1_X \subset \Omega_{\mathcal{F}}\varphi$, i.e., $\mathcal{N}(u[x], u[x]) \subset \mathcal{F}$ for all $x \in X$. This means that for every nonempty open set U, the return time set $N(U,U)$ is in $\mathcal{F}$. The action is called $\mathcal{F}$ *transitive* if $X \times X = \Omega_{\mathcal{F}}\varphi$, i.e., $\mathcal{N}(u[x], u[y]) \subset \mathcal{F}$ for all $x, y \in X$. This means that for every pair U, W of nonempty open sets $N(W,U) \in \mathcal{F}$. The action φ is called *central* (or *transitive*) when it is $\mathcal{B}_T$ central (resp. $\mathcal{B}_T$ transitive). When φ is central, it is a dense action; that is, $f^t(X)$ is a dense subset of X for every $t \in T$. In the compact case this means of course that $f^t(X) = X$ for all t.

Suppose X is a complete, separable metric space, e.g., compact metric space, and that φ is a uniform action, with T separable as well. If φ is central, then the set of recurrent points $|\omega\varphi| = \{x : x \in \omega\varphi(x)\}$ is a residual subset of X. If φ is transitive, then the set of transitive points $\text{Trans}_\varphi = \{x : \omega\varphi(x) = X\}$ is a residual subset of X. For a central, uniform action φ with a separable T on any compact space X, the set of recurrent points is always dense, but the analogous result for transitivity is false. If T is separable and the continuous image $f^T(x)$ is dense in X, then X is separable, and it has cardinality at most 2^c, that of $\beta\mathbf{Z}_+$; however transitive actions occur on spaces of arbitrarily large cardinality.

The action φ is called *weak mixing* when the product action $\varphi \times \varphi$ on $X \times X$ is transitive. Furstenberg's beautiful *Intersection Lemma* shows that a uniform action φ is weak mixing exactly when it is $\tau\mathcal{B}_T$ transitive. For a translation invariant family $\mathcal{F}$, we call φ $\mathcal{F}$ mixing when it satisfies the following equivalent conditions: (1) $\varphi \times \varphi$ is $\mathcal{F}$ transitive, (2) φ is $\tau\mathcal{F}$ transitive, (3) φ is $\mathcal{F}$ transitive and weak mixing, (4) φ is $\mathcal{F}$ central and weak mixing, (5) φ is $\mathcal{F}_1$ transitive for some translation invariant filter $\mathcal{F}_1 \subset \mathcal{F}$. It follows that if φ is an $\mathcal{F}$ mixing action on X, then the product action induced on an arbitrary product X^I is also $\mathcal{F}$ mixing, and so it is a fortiori transitive.

We call φ *strong mixing* when it is $k\mathcal{B}_T$ transitive (= $k\mathcal{B}_T$ mixing since $k\mathcal{B}_T$ is a filter), topologically ergodic or just *ergodic* when it is $k\tau\mathcal{B}_T$ transitive, and *ergodic mixing* when it is both ergodic and weak mixing, which is equivalent to $\tau k\tau\mathcal{B}_T$ mixing. Many transitive systems are in fact ergodic, and the gap between the two notions consists of peculiar systems.

A uniform action $\tilde{\varphi} : T \times \tilde{X} \to \tilde{X}$ is called an *eversion* if $\tilde{X}$ is compact, the action is surjective $[\tilde{f}^t(\tilde{X}) = \tilde{X}$ for all $t]$, and there is a fixed point $e \in \tilde{X}$ such that for every neighborhood U of e, the set of times $\{t : f^t(X \setminus U) \subset U\} \in \mathcal{B}_T$. If $\varphi : T \times X \to X$ is a surjective uniform action with X compact and $x \in X$ such that $x \notin \Omega_{k\tau\mathcal{B}_T}\varphi(x)$ (and so φ is not $k\tau\mathcal{B}_T$ central), then there is an eversion $\tilde{\varphi}$ with fixed point e and a continuous map h from X onto $\tilde{X}$ relating the actions such that $x \notin h^{-1}(e)$. It follows that a transitive but nonergodic system on a compact space has a transitive but nontrivial eversion as a factor.

One application of this machinery is an extension of a theorem of Kronecker: Let $\varphi : T \times X \to X$ be a uniform action with X compact metric and T separable. If φ is weak mixing, then there exists a Cantor subset A of X, i.e., a compact, perfect, zero-dimensional subset such that $\{f^t|A : t \in T\}$ is dense in $C(A;X)$. Thus every continuous map from A into X can be uniformly approximated by the special maps $f^t|A : A \to X$. The existence of so-called Kronecker subsets of the circle arise from the application of this result to the strongly mixing action of $T = \mathbf{Z}^*$ on the circle by $(n,z) \mapsto z^n$.

5. Compactifications

For a uniform space X, let $\mathcal{B}(X)$ denote the Banach algebra of bounded, real-valued continuous functions on X with the sup norm. Let $\mathcal{B}^u(X)$ denote the closed subalgebra of uniformly continuous functions in $\mathcal{B}(X)$.

For any closed subalgebra E of $\mathcal{B}(X)$, let j_E denote the map from X to the dual space E^* associating to $x \in X$ evaluation at x, i.e., $j_E(x)(u) = u(x)$ for $u \in E$. The map $j_E : X \to E^*$ is continuous when E^* is given the weak* topology. Let X_E denote the closure of the image $j_E(X)$. The set X_E consists of all algebra maps from E to $\mathbf{R}$ and X_E is compact with the topology induced from E^*. Furthermore the induced map $j_E^* : \mathcal{B}(X_E) \to \mathcal{B}(X)$ is a B algebra isometry with image E. This Gelfand theory classifies the compactifications of X, i.e., continuous maps from X to a compact space, via the closed subalgebras of $\mathcal{B}(X)$. Associated with $\mathcal{B}(X)$ is the Stone–Čech compactification βX and with $\mathcal{B}^u(X)$ is the uniform Stone–Čech compactification denoted $\beta_u(X)$; the latter is especially important for our purposes. The map $j_u : X \to \beta_u(X)$ $[j_u \equiv j_{\mathcal{B}^u(X)}]$ is uniformly continuous and a topological *embedding*, i.e., a homeomorphism of X onto its image in $\beta_u(X)$. It is however a uniform isomorphism onto its image only when X is totally bounded. The points of $\beta_u X$ can be identified with the maximal open filters on X.

The space X_E is metrizable iff the algebra E is separable. If X is a separable metric space with bounded metric d, then a special metrizable compactification, the *Gromov compactification*, is obtained by using the algebra E_d generated by the functions $\{d(x) : x \in X\}$ where $d(x)(x_1) = d(x,x_1)$.

If φ acts uniformly on X and E is a closed subalgebra of $\mathcal{B}(X)$, then each $f^t : X \to X$ factors through j_E exactly when the algebra maps $f^{t*} : \mathcal{B}(X) \to \mathcal{B}(X)$ all preserve E; i.e., when E is φ + invariant. If in addition $E \subset \mathcal{B}^u(X)$ then φ extends to a uniform action $\varphi_E : T \times X_E \to X_E$ such that $j_E : X \to X_E$ maps the action φ to φ_E.

If φ is $\mathcal{F}$ central (or $\mathcal{F}$ transitive) and $E \subset \mathcal{B}^u(X)$ is φ + invariant then the compactified flow φ_E on X_E is $\mathcal{F}$ central (resp. $\mathcal{F}$ transitive). Conversely if φ_E is $\mathcal{F}$ central (or $\mathcal{F}$ transitive) for any compactification with $j_E : X \to X_E$ an embedding, or when T is separable, for all compactifications with E separable (and hence with X_E metrizable), then φ is $\mathcal{F}$ central (resp. $\mathcal{F}$ transitive).

6. Ellis Semigroups and Ellis Actions

Addition, $T \times T \to T$, can be regarded as the translation action of T on itself. For a uniform monoid, this translation action is uniform, so it extends to a uniform action $T \times \beta_u T \to \beta_u T$, where $\beta_u T$ is the uniform Stone–Čech compactification. Fixing the second coordinate yields a uniformly continuous map from T to $\beta_u T$ that extends to $\beta_u T$. The result is an associative composition $\beta_u T \times \beta_u T \to \beta_u T$, $(p,q) \to pq$; however only the right translation $\Phi_q(p) = pq$ is continuous on $\beta_u T$.

For a uniform action $T \times X \to X$ with X compact, fixing the second coordinate yields a uniformly continuous map of T to X that extends to $\beta_u T$. The result is an action of the semigroup $\beta_u T$ on the space X, but by not necessarily continuous maps. The map $\Phi_x : \beta_u T \to X$ defined by $\Phi_x(p) = px$ is continuous for each x and $(pq)(x) = p(qx)$, extending the associative law on $\beta_u T$.

Starting with the preceding example, Ellis studied compact semigroups. An *Ellis semigroup* S is a usually nonabelian semigroup with a compact topology such that each right translation is continuous on S. For any compact space X, the function space X^X is an Ellis semigroup under composition of maps with the product topology. An *Ellis action* of an Ellis semigroup S on a compact space X is a not usually continuous function $\varphi : S \times X \to X$ such that the adjoint associate $\varphi^\#$ is a continuous semigroup homomorphism from S to X^X. Its image, denoted S_φ, is an Ellis subsemigroup of X^X called the *enveloping semigroup* of φ.

An element e of S is called *idempotent* when $e^2 = e$. Namakura's Lemma says that any compact semigroup contains idempotents. For example if φ is an Ellis action on X and $x \in X$, then $\mathrm{Iso}_x = \{p \in S : px = x\}$ is a closed subsemigroup if it is nonempty. It then follows that idempotents fixing x exist. A point x is called S *recurrent* if the isotropy set Iso_x is nonempty.

7. Semigroups and Families

In $\beta_u T$ the semigroup structure can be described using family ideas. For $p \in \beta_u T$ we use the embedding $j_u : T \to \beta_u T$ to pull back the filter of neighborhoods of p. We obtain $\mathcal{F}_p$, a maximal open filter of subsets of T. The point px is the point in the singleton set $\omega_{\mathcal{F}_p}\varphi(x) = \omega_{k\mathcal{F}_p}\varphi(x)$.

In general for any open filter $\mathcal{F}$ of subsets of T, the hull $H(\mathcal{F}) = \{p \in \beta_u T : \mathcal{F} \subset \mathcal{F}_p\}$ is a closed subset of $\beta_u T$ and the compact subset $\{px : p \in H(\mathcal{F})\}$ is $\omega_{k\mathcal{F}}\varphi(x)$.

Recall that a filter $\mathcal{F}$ is thick if $F \in \mathcal{F}$ and $t \in T$ imply $g^{-t}(F) \in \mathcal{F}$. Define $F \in \mathcal{F}$ to be $\mathcal{F}$ *semiadditive* for a filter $\mathcal{F}$ if $t \in F$ implies $g^{-t}(F) \in \mathcal{F}$. A filter $\mathcal{F}$ is called semiadditive if it is generated by $\mathcal{F}$ semiadditive sets, i.e., $F \in \mathcal{F}$ implies $F \supset F_1$ with $F_1 \in \mathcal{F}$ an $\mathcal{F}$ semiadditive set.

For any open filter $\mathcal{F}$, let H be the hull of $\mathcal{F}$. The filter $\mathcal{F}$ is semiadditive iff $\omega_{k\mathcal{F}}\mu[H] \subset H$ where μ is the translation action of T on $\beta_u T$. Notice that this implies $\omega_{k\mathcal{F}}\mu(H) = \{pq : p,q \in H\}$ is contained in H, so H is a closed subsemigroup. Note that $\omega_{k\mathcal{F}}\mu(H) = \omega_{k\mathcal{F}}\mu[H]$ in the particular case where H is a singleton. As a corollary we see that the maximal open filter $\mathcal{F}_p$ is semiadditive iff p is an idempotent.

Assume the hull $H(\mathcal{F})$ is a subsemigroup, e.g., $\mathcal{F}$ is semiadditive or translation invariant. If $\varphi : T \times X \to X$ is a uniform action with X compact and $x \in X$, then x is recurrent for the filterdual $k\mathcal{F}$, i.e., $x \in \omega_{k\mathcal{F}}\varphi(x)$, iff x is $H(\mathcal{F})$ recurrent; i.e., the isotropy set $\mathrm{Iso}_x \subset \beta_u T$ meets $H(\mathcal{F})$. There then exists an idempotent e such that $\mathcal{F} \subset \mathcal{F}_e \subset k\mathcal{F}$.

The family $\cup\{\mathcal{F}_e : e \in \beta_u T \backslash T \text{ and } e^2 = e\}$ is the filterdual generated by the so-called IP sets.

8. Equicontinuity

For a uniform action $\varphi : T \times X \to X$, Lyapunov stability of a point $x \in X$, or of its orbit, is equicontinuity at x of the family of functions $\{f^t : t \in T\}$. To obtain the version associated with a family $\mathcal{F}$, we define for $V \in \mathcal{U}_X$ and $F \subset T$:

$$V_\varphi^F = \cap_{t \in F}(f^t \times f^t)^{-1}(V) \qquad Eq_{V,\varphi}^F = \{x \in X : (x,x) \in \mathrm{Int}\, V_\varphi^F\}$$

Then let $Eq_{V,\varphi}^{\mathcal{F}} = \cup_{F \in \mathcal{F}} Eq_{V,\varphi}^F$ and $Eq_\varphi^{\mathcal{F}} = \cap_{V \in \mathcal{U}_X} Eq_{V,\varphi}^{\mathcal{F}}$.

We call x an $\mathcal{F}, V$ *equicontinuity point* if x is in the open set $Eq_{V,\varphi}^{\mathcal{F}}$ and an $\mathcal{F}$ *equicontinuity point* if it is in $Eq_\varphi^{\mathcal{F}}$, i.e., if it is an $\mathcal{F}, V$ equicontinuity point for all V in $\mathcal{U}_X$. Thus, $x \in Eq_\varphi^{\mathcal{F}}$ if for every $V \in \mathcal{U}_X$ there is a neighborhood U of x and $F \in \mathcal{F}$ such that $(f^t(x_1), f^t(x_2)) \in V$ for all $(t, x_1, x_2) \in F \times U \times U$.

The action is called $\mathcal{F}$ *equicontinuous* if $X = Eq_\varphi^{\mathcal{F}}$, i.e., $X = Eq_{V,\varphi}^{\mathcal{F}}$ for all $V \in \mathcal{U}_X$. The action is $\mathcal{F}$ *almost equicontinuous* if for each $V \in \mathcal{U}_X$ the open set $Eq_{V,\varphi}^{\mathcal{F}}$ is dense in X. So for an $\mathcal{F}$ almost equicontinuous action on a complete, metrizable space X, the set $Eq_\varphi^{\mathcal{F}}$ of $\mathcal{F}$ equicontinuity points is a dense G_δ. For the case $\mathcal{F} = k\mathcal{B}_T$, the filter generated by the tails, we drop the superscript $\mathcal{F}$ and refer simply to equicontinuity point, almost equicontinuity etc.

For a translation invariant family $\mathcal{F}$, if φ is $k\mathcal{F}$ transitive and $\mathcal{F}$ almost equicontinuous, then φ satisfies the a priori stronger condition of almost equicontinuity.

In fact for every $V \in \mathcal{U}_X$, $Eq^{\mathcal{T}}_{V,\varphi}$ is open and dense, and $Eq^{\mathcal{T}}_{\varphi} = Eq_{\varphi} = \text{Trans}_{\varphi} = \{x : \omega\varphi(x) = X\}$.

If φ is an almost equicontinuous action on a compact space with $\text{Trans}_{\varphi} \neq \emptyset$ but that is not minimal and therefore not equicontinuous, then φ is not topologically ergodic, thus it has as a factor a topologically transitive, nontrivial eversion. The eversion factor can be chosen almost equicontinuous as well. Such peculiar actions do in fact exist.

1

Monoid Actions

All of our spaces are assumed to be completely regular and Hausdorff. This class is closed under the taking of subspaces and arbitrary products; it is in fact the class of subspaces of compact Hausdorff spaces. It is also the class of spaces whose topology can be associated with some Hausdorff uniformity. We follow Kelley (1955) in using uniformities, distinguished collections of neighborhoods of the diagonal, to define uniform spaces and uniformly continuous maps. The *gage*, the set of uniformly continuous pseudometrics, provides an equivalent characterization. Recall that a uniformity is metrizable iff it has a countable base. Also a compact space has a unique uniformity consisting of all neighborhoods of the diagonal.

We use the relation notation of Akin (1993), identifying a function $f: X_1 \to X_2$ with its graph $\{(x, f(x)) : x \in X_1\} \subset X_1 \times X_2$. For example the identity map 1_X is the diagonal subset of $X \times X$. On the other hand any subset F of $X_1 \times X_2$ can be regarded as a relation from X_1 to X_2, so that for $A \subset X_1$:

$$F(A) = \{y \in X_2 : (x,y) \in F \text{ for some } x \in A\}$$

The inverse relation F^{-1} is $\{(y,x) : (x,y) \in F\} \subset X_2 \times X_1$. With d a pseudometric on X, we denote by V_ε^d (or $\bar{V}_\varepsilon^d$) the relation on X $\{(x_1,x_2) : d(x_1,x_2) < \varepsilon\}$ [resp. $\{(x_1,x_2) : d(x_1,x_2) \leq \varepsilon\}$]. Thus for $x \in X$, $V_\varepsilon^d(x)$ and $\bar{V}_\varepsilon^d(x)$ are the open and closed balls centered at x with radius ε.

We regard the discrete topology as the default topology; that is, any set not otherwise topologized is regarded as a discrete metric space with the zero/one metric. The discrete uniformity on X consists of all subsets of $X \times X$ containing the diagonal 1_X.

For spaces X and Y, the set of maps from X to Y is denoted by Y^X and equipped with the product topology, i.e., the topology of pointwise convergence. $C(X;Y)$ is the subspace of continuous maps. If Y has a uniformity $\mathcal{U}_Y$, then for $V \in \mathcal{U}_Y$, we define the set

$$V^X = \{(f_1, f_2) : (f_1(x), f_2(x)) \in V \text{ for all } x \in X\} \subset (Y^X) \times (Y^X)$$

These sets generate a uniformity on the set Y^X with a finer topology, that of uniform convergence. If $\mathcal{U}_Y$ comes from a metric d on Y, then the uniformity on the function space is associated with the sup metric:

$$d(f_1, f_2) \equiv \sup\{d(f_1(x), f_2(x)) : x \in X\} \tag{1.1}$$

Here we assume that d is bounded, replacing it otherwise by the uniformly equivalent metric $\text{Min}(d, 1)$. Let $C(X;Y)$ denote the set of continuous functions equipped with this uniformity or metric and the associated topology of uniform convergence.

For $F \subset X$ we obtain the restriction map $Y^X \to Y^F$. Collecting the preimages of the V^Fs for V varying over $\mathcal{U}_Y$ and F over the finite subsets of X, we generate the pointwise uniformity with the product topology on Y^X. When Y has a uniformity, we assume that Y^X and $C(X;Y)$ are so equipped. By letting F vary over compact subsets of X, we generate an intermediate uniformity, and we let $C_c(X;Y)$ denote the uniform space of continuous functions with this uniform convergence on compacta. The associated topology is the compact open topology with subbase the set of maps of pairs:

$$C((X,F);(Y,U)) = \{f \in C(X;Y) : f(F) \subset U\}$$

for F compact and U open.

Observe that the topologies on Y^X, $C(X;Y)$ and $C_c(X;Y)$ depend only on the topologies of X and Y. In particular we let $C_c(X;Y)$ denote $C(X;Y)$ equipped with the compact open topology even when Y does not carry a chosen uniformity. On the other hand, the topology on $C(X;Y)$ depends in general on the choice of uniformity on Y, not just on its topology.

When X as well as Y has a uniform structure, we denote by $C^u(X;Y)$, $C^u(X;Y)$, and $C^u_c(X;Y)$ the subset of uniformly continuous maps with uniform structure induced from $C(X;Y)$, $C(X;Y)$, and $C_c(X;Y)$, respectively.

A *monoid* is a set T with an associative composition that has an identity element 0. A *group* is a monoid such that every element has an inverse. A *topological monoid* T is a monoid equipped with a completely regular Hausdorff topology such that composition is a continuous map from $T \times T$ to T. A *topological group* is a group that is a topological monoid such that the inversion map is continuous on T as well.

Suppose that a uniformity $\mathcal{U}_T$ is given for a monoid T. An element $V \in \mathcal{U}_T$ is called *right + invariant* if for all $s_1, s_2, t \in T$:

$$(s_1, s_2) \in V \Rightarrow (s_1 + t, s_2 + t) \in V \tag{1.2}$$

V is called *right invariant* if for all $s_1, s_2, t \in T$:

$$(s_1, s_2) \in V \Leftrightarrow (s_1 + t, s_2 + t) \in V \tag{1.3}$$

If we let $g^t, \tilde{g}^t : T \to T$ denote right and left translation by t, respectively:

$$g^t(s) = s + t \qquad \tilde{g}^t(s) = t + s \tag{1.4}$$

then (1.2) says $g^t \times g^t(V) \subset V$, and (1.3) says $(g^t \times g^t)^{-1}(V) = V$ for all $t \in T$.

A *right + invariant uniformity* (or a *right invariant uniformity*) on a monoid T is a Hausdorff uniformity generated by right + invariant (resp. right invariant) elements. For example if $\mathcal{U}_T$ is the uniformity associated with a right invariant metric, a metric d_T on T satisfying:

$$d_T(s_1 + t, s_2 + t) = d_T(s_1, s_2) \tag{1.5}$$

for all $s_1, s_2, t \in T$, then $\mathcal{U}_T$ is a right invariant uniformity.

Because the uniformity is assumed Hausdorff, a right invariant uniformity can only exist when each right translation map g^t is injective. This means that T allows cancellation on the right; i.e., $s_1 + t = s_2 + t$ implies $s_1 = s_2$.

Beginning with Chapter 2, all of our monoids are assumed to be abelian. We then speak of a + invariant or invariant uniformity, dropping the adjective right.

We say that a right invariant uniformity $\mathcal{U}_T$ satisfies the *Interior Condition* when:

For every $V \in \mathcal{U}_T$, there exists $W \in \mathcal{U}_T$ such that for every $t \in T$ there exist $t_1, t_2 \in V(t)$, so that $W(g^{t_1}(T)) \subset g^t(T)$, and $W(g^t(T)) \subset g^{t_2}(T)$. (1.6)

A *homomorphism* h between monoids T and T_1 is a map $h : T \to T_1$ such that $h(0) = 0$ and $h(s + t) = h(s) + h(t)$ for all $s, t \in T$.

LEMMA 1.1. *Let T, T_1 be monoids with right + invariant uniformities $\mathcal{U}_T$, $\mathcal{U}_{T_1}$.*

a. If every left translation map $\tilde{g}^t : T \to T$ is continuous with respect to the $\mathcal{U}_T$ topology, then T is a topological monoid with respect to the $\mathcal{U}_T$ topology; i.e., composition is a continuous map from $T \times T$ to T.

b. If $\mathcal{U}_T$ is a right invariant uniformity satisfying the Interior Condition and a particular left translation map $\tilde{g}^t : T \to T$ is continuous at 0, then $\tilde{g}^t$ is uniformly continuous with respect to $\mathcal{U}_T$.

c. If $\mathcal{U}_T$ is a right invariant uniformity satisfying the Interior Condition and a homomorphism $h : T \to T_1$ is continuous at 0 with respect to the $\mathcal{U}_T$ and $\mathcal{U}_{T_1}$ topologies, then h is uniformly continuous with respect to $\mathcal{U}_T$ and $\mathcal{U}_{T_1}$.

PROOF. (a) Given $V \in \mathcal{U}_T$, there exists V_1 a right + invariant member of $\mathcal{U}_T$ such that $V_1 \circ V_1 \subset V$. Given $(t,s) \in T \times T$, continuity of g^t implies there exists a neighborhood U of s such that $s_1 \in U$ implies $g^t(s_1) \in V_1(g^t(s))$. If $t_1 \in V_1(t)$ and $s_1 \in U$, $(t+s, t+s_1) \in V_1$, and by right + invariance, $(t+s_1, t_1+s_1) \in V_1$. Thus $(t+s, t_1+s_1) \in V_1 \circ V_1$, and $t_1 + s_1 \in V(t+s)$.

(c) Given $\tilde{V} \in \mathcal{U}_{T_1}$, choose $\tilde{V}_1 \in \mathcal{U}_{T_1}$ right + invariant and symmetric (i.e., $\tilde{V}_1^{-1} = \tilde{V}_1$) such that $\tilde{V}_1 \circ \tilde{V}_1 \subset \tilde{V}$. By continuity of h at 0, we can choose $V_1 \in \mathcal{U}_T$ so that $h(V_1(0)) \subset \tilde{V}_1(0)$. Now choose $V \in \mathcal{U}_T$, right invariant and symmetric with $V \circ V \subset V_1$ and choose $W \subset V$ and right invariant to satisfy the Interior Condition (1.6) with respect to V. We show that $h \times h(W) \subset \tilde{V}$.

Let $s \in W(t)$. Then $s,t \in W((g^t(T)) \subset g^{t_2}(T)$ for some $t_2 \in V(t)$. Thus there exist $s', t' \in T$ so that $s = s' + t_2$ and $t = t' + t_2$. Of course, $t_2 = 0 + t_2$. $(t,s) \in W$ and $(t,t_2) \in V$ so by right invariance of W and V, $(t',s') \in W$ and $(t',0) \in V$. By symmetry, $(0,t')$ and $(0,s') \in W \circ V \subset V_1$. Because $h(V_1(0)) \subset \tilde{V}_1(0)$, we have $(0,h(t'))$ and $(0,h(s')) \in \tilde{V}_1$. Since h is a homomorphism:

$$h(s) = h(s') + h(t_2) \qquad h(t) = h(t') + h(t_2) \qquad h(t_2) = 0 + h(t_2)$$

Because $\tilde{V}_1$ is right + invariant $(h(t_2),h(t))$ and $(h(t_2),h(s)) \in \tilde{V}_1$. By symmetry $(h(t),h(s)) \in \tilde{V}_1 \circ \tilde{V}_1 \subset \tilde{V}$, as required.

(b) The proof is analogous to that for (c): V_1 is chosen so that $g^{\tilde{t}}(V_1(0)) \subset \tilde{V}_1(\tilde{t})$. The homomorphism condition is replaced by:

$$g^{\tilde{t}}(s) = g^{\tilde{t}}(s') + t_2 \qquad g^{\tilde{t}}(t) = g^{\tilde{t}}(t') + t_2 \qquad g^{\tilde{t}}(t_2) = \tilde{t} + t_2$$

■

COROLLARY 1.1. *Let $\mathcal{U}_T$ and $\mathcal{U}_{T_1}$ be right invariant uniformities on a monoid T that satisfy the Interior Condition. If they are topologically equivalent (i.e., induce the same topology on T), then $\mathcal{U}_T = \mathcal{U}_{T_1}$. If in addition the left translation maps are each continuous at 0 then they are all $\mathcal{U}_T$ uniformly continuous, and T is a topological monoid with respect to the $\mathcal{U}_T$ topology.*

PROOF. Apply (c) to the identity maps from $\mathcal{U}_T$ to $\mathcal{U}_{T_1}$ and back to obtain $\mathcal{U}_T = \mathcal{U}_{T_1}$. Then apply (b) followed by (a). ■

A *uniform monoid* T is a monoid together with a right invariant uniformity $\mathcal{U}_T$ satisfying the Interior Condition and such that each left translation is uniformly continuous. Of course right translations are uniformly continuous injections as well by right invariance. Corollary 1.1 shows that a uniform monoid is a topological monoid and the uniformity is determined by the topology. A *metric monoid* T is a monoid with a right invariant metric with respect to which T is a uniform monoid.

Examples

(1) If T is a topological group, then neighborhoods of the identity determine a right invariant uniformity. Associated to V_0 a neighborhood of 0 is $\{(t_1,t_2) : t_2 - t_1 \in V_0\} = V$, the corresponding right invariant member of $\mathcal{U}_T$ such that $V(0) = V_0$. Since each translation map g^t is surjective condition (1.6) is satisfied with $t_1 = t_2 = t$ for any $W \in \mathcal{U}_T$. Thus $\mathcal{U}_T$ satisfies the Interior Condition. Each left translation is continuous and so it is uniformly continuous by Lemma 1.1b. So a topological group is a uniform monoid. If neighborhoods of 0 have a countable base then by applying the construction in the proof of the uniformity metrization theorem [see Kelley (1955) Theorem 6.13] to a sequence of invariant elements, we obtain a right invariant metric, and T becomes a metric group.

(2) If T is a discrete monoid satisfying cancellation on the right, then the zero/one metric on T is an invariant metric. Using $W = V_1 = 1_T$ and $t_1 = t_2 = t$, we obtain the Interior Condition (1.6), so T is a metric monoid.

(3) The set of nonnegative reals $\mathbf{R}_+$ is an abelian monoid under addition. The usual metric is invariant. Given $\varepsilon > 0$ such that $V_\varepsilon \subset V$ we can choose $W = V_{\varepsilon/2}$ and $t_1 = t + (\varepsilon/2)$, $t_2 = \max(0, t - (\varepsilon/2))$ to satisfy the Interior Condition.

(4) For a uniform space X, the uniformity of uniform convergence restricts to a right + invariant uniformity on $C^u(X;X)$, the set of uniformly continuous maps on X, with composition of maps providing the monoid structure. Uniform continuity of $f : X \to X$ implies that the left translation by f is uniformly continuous on $C^u(X;X)$. So by Lemma 1.1a, $C^u(X;X)$ is a topological monoid with respect to the topology of uniform convergence.

(5) For a uniform space X, let $C^u is(X;X)$ denote the subset of invertible elements of $C^u(X;X)$; i.e., $C^u is$ is the set of uniform isomorphisms on X. $C^u is$ is a submonoid of the topological monoid C^u, and it is a group. In fact it is a topological group. To show that inversion is continuous on the subset $C^u is$ of C^u, it suffices to prove continuity at 1_X, since translations by members of $C^u is$ are homeomorphisms on C^u and $f_1^{-1} = f^{-1}[f_1 f^{-1}]^{-1}$. So f_1 is close to f iff $f_1 f^{-1}$ is close to 1_X and f_1^{-1} is close to f^{-1} iff $[f_1 f^{-1}]^{-1}$ is close to 1_X. Now for $V \in \mathcal{U}_X$, $(1_X, f) \in V^X$ iff $(x, f(x)) \in V$ for all x iff $(f^{-1}(z), z) \in V$ for all z [use $x = f^{-1}(z)$ as a change of variable] iff $(f^{-1}, 1_X) \in V^X$. Continuity of inversion at 1_X follows. In fact if $V = V^{-1}$, the neighborhood $V^X(1_X)$ is invariant under the inversion map.

(6) For any set X, the set X^X is a monoid with respect to composition. If X is a topological space, then the set $C(X;X)$ is a submonoid of X^X. As in Example (5), the set of bijections, the invertible members of X^X, is a subgroup of X^X, and the set of homeomorphisms, the invertible members of $C(X;X)$, is a subgroup of $C(X;X)$.

Now suppose that T is an abelian monoid satisfying the cancellation property; i.e., each translation map g^t is injective. T can be identified as a submonoid of the canonically defined *group of quotients* G_T. On the product monoid $T \times T$ define the equivalence relation $\approx_T$ by:

$$\begin{aligned}(t_1,s_1) \approx_T (t_2,s_2) &\Leftrightarrow t_1+s_2=t_2+s_1 \\ &\Leftrightarrow \text{For some } u,v \in T \quad (t_1+u,s_1+u)=(t_2+v,s_2+v)\end{aligned} \tag{1.7}$$

Clearly the first condition implies the second, with $u = s_2$ and $v = s_1$. The second condition implies $t_1+s_2+u+v = t_2+s_1+u+v$, so it implies the first by cancellation. Let $[t/s]$ denote the $\approx_T$ equivalence class of (t,s). Notice that $[t_1/0] = [t_2/0]$ iff $t_1 = t_2$. Composition on the set G_T of $\approx_T$ equivalence classes is well-defined by the condition that the quotient map $T \times T \longrightarrow G_T$ is a homomorphism. The composition $T \xrightarrow{1\times 0} T \times T \longrightarrow G_T$ defines the canonical injection of T as a submonoid of G_T. The monoid G_T is a group because:

$$[t/s]+[s/t] = [t+s/t+s] = [0/0]$$

for all $s,t \in T$.

PROPOSITION 1.1. *Let T be an abelian uniform monoid. The group of quotients G_T is naturally an abelian topological group such that the canonical inclusion of T into G_T is a uniform isomorphism; i.e., the restriction of the invariant uniformity $\mathcal{U}_{G_T}$ to the subset T is exactly the invariant uniformity $\mathcal{U}_T$.*

If G is a group and $h: T \to G$ is a homomorphism, then h extends uniquely to a homomorphism $h'': G_T \to G$. If G is a topological group and h is continuous then h'' is continuous.

PROOF. Since T is uniform, it satisfies the cancellation property, so G_T can be defined as stated.

Given V a right invariant element of $\mathcal{U}_T$, define $V' \subset T \times T \times T \times T$:

$$V' = \{(t_1,s_1,t_2,s_2) : (t_1+s_2,t_2+s_1) \in V\} \tag{1.8}$$

Since V is invariant, $(t_1,s_1,t_2,s_2) \in V'$ iff $(t_1+u,s_1+u,t_2+v,s_2+v) \in V'$ for any $u,v \in T$. Thus V' depends only on the equivalence classes $[t_1/s_1]$ and $[t_2/s_2]$ and it is the preimage of a subset V'' of $G_T \times G_T$. The set V'' contains 1_{G_T} because

V contains 1_T; V'' is clearly invariant. If W is invariant and $W \circ W \subset V$ then $W'' \circ W'' \subset V''$:

$$(t_1+s_2,t_2+s_1),\ (t_2+s_3,t_3+s_2) \in W$$

implies

$$(t_1+s_3+s_2,t_2+s_1+s_3),\ (t_2+s_1+s_3,t_3+s_1+s_2) \in W$$

Thus $(t_1+s_3+s_2,t_3+s_1+s_2) \in V$, which implies $(t_1+s_3,t_3+s_1) \in V$. Thus, such invariant V''s generate an invariant uniformity on G_T. As in Example (1), the Interior Condition is trivial because each translation map is a bijection on G_T. Thus G_T is a uniform monoid. Since $([t_1/s_1],[t_2/s_2]) \in V''$ iff $([s_2/t_2],[s_1/t_1]) \in V''$, we see that inversion is uniformly continuous.

Now $([t_1/0],[t_2/0]) \in V''$ iff $(t_1,t_2) \in V$, so V'' restricts to V on T. On the other hand, suppose W is an invariant subset of $G_T \times G_T$ and W restricts to V on T:

$$\begin{aligned}([t_1/s_1],[t_2/s_2]) &= ([t_2+s_2/s_1+s_2],[t_2+s_1/s_1+s_2]) \\ &= ([t_1+s_2/0]+[0/s_1+s_2],[t_2+s_1/0]+[0/s_1+s_2])\end{aligned}$$

implies $([t_1/s_1],[t_2/s_2]) \in W$ iff $([t_1+s_2/0],[t_2+s_1/0]) \in W$ and iff $(t_1+s_2,t_2+s_1) \in V$. Thus $W = V''$.

If $h : T \to G$ is a homomorphism, then $h(T)$ is a commutative subset of G so the subgroup generated by $h(T)$ is abelian. Hence $h' : T \times T \to G$ defined by $h'(t,s) = h(t) - h(s)$ is a well-defined homomorphism of the product, which factors to define $h'' : G_T \to G$, extending h uniquely.

Now suppose G is a topological group and h is continuous. If W_0 is a neighborhood of 0 in G, then $W = \{(u,v) : u - v \in W_0\}$ is the associated right invariant element of $\mathcal{U}_G$. By Lemma 1.1c, h is uniformly continuous, so there exists $V \in \mathcal{U}_T$ invariant and such that $h \times h(V) \subset W$. This says exactly that $(t,s) \in V$ implies $h'(t,s) \in W_0$. Hence $[t/s] \in V''([0/0])$ implies $h''([t/s]) \in W_0$. Thus h'' is continuous at 0, and it is uniformly continuous on G_T by Lemma 1.1c again. ■

For the monoids $\mathbf{Z}_+$ and $\mathbf{R}_+$, the associated groups are $\mathbf{Z}$ and $\mathbf{R}$, respectively.

An *action* of a monoid T on a set X is a map $\varphi : T \times X \to X$ such that:

$$\begin{gathered} f^{t_1} \circ f^{t_2} = f^{t_1+t_2} \qquad f^0 = 1_X \\ \text{where } f^t(x) = \varphi(t,x) \end{gathered} \tag{1.9}$$

Thus if $\varphi^\# : T \to X^X$ is defined by $\varphi^\#(t) = f^t$, then condition (1.9) states that $\varphi^\#$ is a homomorphism.

A *topological action* is an action of a topological monoid on a topological space such that φ is continuous on the product $T \times X$. It follows that $\varphi^\#$ maps T into $C(X;X)$ and $\varphi^\# : T \to C(X;X)$ is continuous with respect to the topology of pointwise convergence. In general however continuity of this homomorphism does not suffice for continuity of φ on the product (i.e., for joint continuity).

A *uniform action* is an action of a uniform monoid on a uniform space which satisfies the following conditions:

For each $t \in T$, $f^t : X \to X$ is uniformly continuous with respect to the uniformity $\mathcal{U}_X$ on X. (1.10)

For every $V \in \mathcal{U}_X$, there exists $W \in \mathcal{U}_T$ so that $(t_1, t_2) \in W$ implies $(f^{t_1}(x), f^{t_2}(x)) \in V$ for all $x \in X$; i.e., $(t_1, t_2) \in W$ implies $(f^{t_1}, f^{t_2}) \in V^X$. (1.11)

Clearly, (1.10) says that $\varphi^\#$ maps T into the set $C^u(X;X)$, and (1.11) says that the homomorphism $\varphi^\# : T \to C^u(X;X)$ is uniformly continuous with respect to the uniformity of uniform convergence on $C^u(X;X)$. By Lemma 1.1c uniform continuity of $\varphi^\#$ follows from continuity at 0, because T is a uniform monoid and $C^u(X;X)$ has a right + invariant uniformity. Thus we can replace (1.11) by the apparently weaker condition:

For every $V \in \mathcal{U}_X$, there exists a neighborhood W_0 of 0 in T such that $t \in W_0$ implies $(x, f^t(x)) \in V$ for all $x \in X$; i.e., $t \in W_0$ implies $(1_X, f^t) \in V^X$. (1.12)

LEMMA 1.2. *A uniform action is a topological action. A topological action of a uniform monoid T on a compact space X is a uniform action.*

PROOF. It is easy to check that evaluation Ev: $C^u(X;X) \times X \to X$ is continuous, i.e., it is a topological action of $C^u(X;X)$ on X. So a uniform action φ is continuous because it is the composition $\mathrm{Ev} \circ (\varphi^\# \times 1_X) : T \times X \to X$.

If X is compact then (1.10) is clear for a topological action. Given V a neighborhood of 1_X in $X \times X$, $\{(t,x) : (x, \varphi(t,x)) \in V\}$ is a subset of $T \times X$ containing $0 \times X$ in its interior. By compactness it contains $W_0 \times X$ for some neighborhood W_0 of 0 in T. Thus (1.12) holds, and the action is uniform. ■

On any monoid T, composition $T \times T \to T$ is an action, the *left action* of T on T; i.e., the map associated to $t \in T$ is the left translation g^t. If T is a topological (or uniform) monoid then the left action is a topological (resp. uniform) action.

For any action φ of T on X we define the *orbit map* $\varphi_\# : X \to X^T$ by:

$$\varphi_\#(x)(t) = \varphi(x,t) \tag{1.13}$$

If φ is a topological action, then $\varphi_\#$ maps X to $C(T;X)$, and it is easy to check that $\varphi_\# : X \to C(T;X)$ is continuous with respect to the topology of pointwise convergence; in fact a stronger result is true.

PROPOSITION 1.2. *If φ is a topological action then $\varphi_\# : X \to C_c(T;X)$ is a continuous map (i.e., compact open topology on the function space). If φ is a uniform action, then $\varphi_\# : X \to C^u_c(T;X)$ is a uniformly continuous map.*

PROOF. Let F be a compact subset of T and $V \in \mathcal{U}_X$. For $x \in X$ $\{(t,x_1) : (\varphi(t,x), \varphi(t,x_1)) \in V\}$ contains $T \times x$ in its interior. By compactness there is a neighborhood U of x such that this set contains $F \times U$. For $x_1 \in U$, $(\varphi_\#(x), \varphi_\#(x_1))$ restricts to an element of V^F.

Now assume that φ is a uniform action. Choose V_1 symmetric in $\mathcal{U}_X$ so that $V_1^3 \subset V$. Apply (1.11) to choose W in $\mathcal{U}_T$ so that $(t_1,t_2) \in W$ implies $(f^{t_1}, f^{t_2}) \in V_1^X$. By compactness we can choose a finite subset F_0 of F so that $\{W(t) : t \in F_0\}$ covers F. Finally, by (1.10) we can choose $V_2 \in \mathcal{U}_X$ so that $(f^t \times f^t)(V_2) \subset V_1$ for all $t \in F_0$. Now suppose $(x_1,x_2) \in V_2$ and $t \in F$. Choose $\tilde{t} \in F_0$ so that $t \in W(\tilde{t})$. Then

$$
\begin{aligned}
(f^t(x_1), f^{\tilde{t}}(x_1)), (f^t(x_2), f^{\tilde{t}}(x_2)) &\in V_1 \\
(f^{\tilde{t}}(x_1), f^{\tilde{t}}(x_2)) &\in V_1
\end{aligned}
$$

by choice of $\tilde{t}$, W, and V_2. So $(f^t(x_1), f^t(x_2)) \in V_1^3 \subset V$ for all $t \in F$. Hence $(x_1,x_2) \in V_2$ implies $(\varphi_\#(x_1), \varphi_\#(x_2))$ restricts to a member of V^F. ∎

Continuity of $\varphi_\# : X \to C^u(T;X)$, topology of uniform convergence, is a much stronger condition that rarely holds. It is a form of equicontinuity that we study in Chapter 8.

If φ_1 and φ_2 are actions of a monoid T on sets X_1 and X_2, then a map of actions, or *action map*, $h : \varphi_1 \to \varphi_2$ is a map $h : X_1 \to X_2$ such that the following diagram commutes:

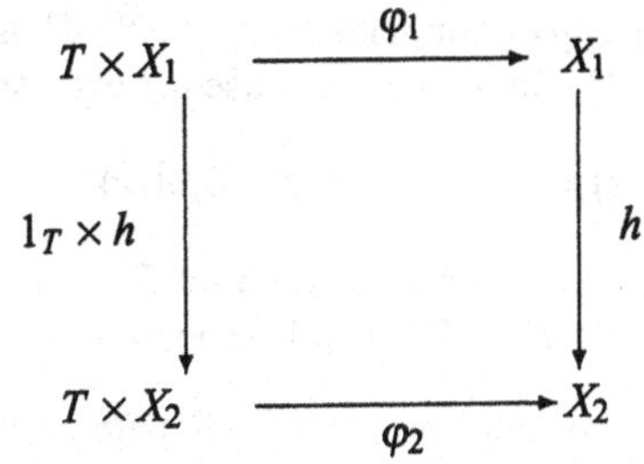

$$
\begin{array}{ccc}
T \times X_1 & \xrightarrow{\varphi_1} & X_1 \\
\downarrow{\scriptstyle 1_T \times h} & & \downarrow{\scriptstyle h} \\
T \times X_2 & \xrightarrow[\varphi_2]{} & X_2
\end{array}
\tag{1.14}
$$

Equivalently $h : X_1 \to X_2$ satisfies

$$
f_2^t \circ h = h \circ f_1^t \tag{1.15}
$$

for all $t \in T$. The action map h is called a continuous action map if φ_1 and φ_2 are topological actions and h is continuous. The action φ_1 is called a *subsystem* of φ_2 if h is an embedding, that is, a homeomorphism of X_1 onto its image in X_2. The action φ_2 is called a *factor* of φ_1 if h is a quotient map, that is, a surjection on X_2 inducing the quotient topology from X_1.

We can also express diagram (1.14) by using the orbit map:

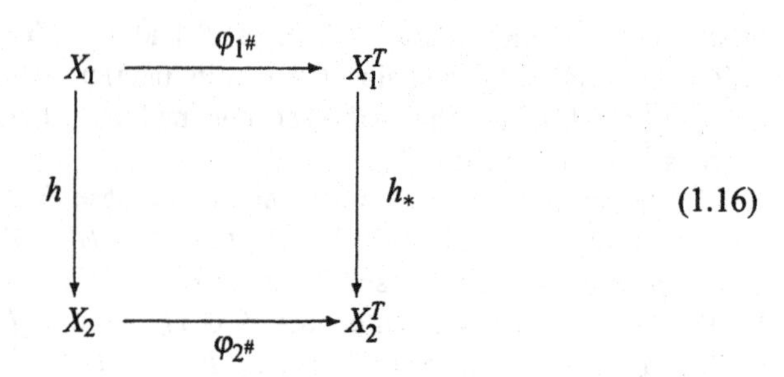

(1.16)

where $h_*(u) = h \circ u$.

For an action φ of an abelian monoid, each $f^t : X \to X$ is an action map on φ.

If T_1 is a submonoid of T, then the restriction of an action $\varphi : T \times X \to X$ to $T_1 \times X$ is the *restriction* of φ to the corresponding T_1 action. On the other hand, we say that a T_1 action φ_1 *extends* to a T action φ if φ_1 is the restriction to T_1 of φ.

For an abelian monoid T, we call an action $\varphi : T \times X \to X$ *reversible* if each $f^t : X \to X$ is a bijection. In that case, the *reverse action* $\overline{\varphi} : T \times X \to X$ is defined by:

$$\overline{\varphi}(t,x) = (f^t)^{-1}(x) \tag{1.17}$$

Thus φ is reversible iff the homomorphism $\varphi^\#$ maps T to the group of invertible elements in X^X. This is equivalent to assuming that the homomorphism $\varphi^\#$ extends to a homomorphism of the group of quotients G_T into X^X. Thus, φ is reversible iff it extends to an action φ'' of G_T on X in which case the extension is unique and:

$$\overline{\varphi}(t,x) = \varphi''(-t,x) = \varphi''([0/t],x) \tag{1.18}$$

A uniform action of an abelian uniform monoid T on a uniform space X is *uniformly reversible* if each $f^t : X \to X$ is a uniform isomorphism.

PROPOSITION 1.3. *For a uniform action $\varphi : T \times X \to X$ with T abelian the following are equivalent.*

(1) The action φ is uniformly reversible, i.e., $f^t : X \to X$ is a uniform isomorphism for each $t \in T$.

(2) The action φ is reversible, and the reverse action $\overline{\varphi}$ is a uniform action.

(3) The action φ extends to a uniform action φ'' of the group of quotients G_T on X.

PROOF. (3) $\Rightarrow$ (2) by (1.18) and (2) $\Rightarrow$ (1) is obvious. But (1) says precisely that the continuous homomorphism $\varphi^{\#}: T \to C^u(X;X)$ maps into the group $C^u is(X;X)$. By Proposition 1.1 this implies $\varphi^{\#}$ extends to a continuous homomorphism of G_T into $C^u is(X;X)$. Equivalently φ extends to a uniform action of G_T on X.

If for some indexed family of sets $\{X_\alpha\}$, each $\varphi_\alpha : T \times X_\alpha \to X_\alpha$ is an action, then the *product action* $\varphi_\pi : T \times \prod_\alpha X_\alpha \to \prod_\alpha X_\alpha$ is defined by $\varphi_\pi(t,x)_\alpha = \varphi_\alpha(t,x_\alpha)$. If each φ_α is a topological (or uniform) action, then φ_π is a topological (resp. uniform) action. In particular for a uniform space Y and a topological (or uniform) action $\varphi : T \times X \to X$, we obtain a topological (resp. uniform) action φ_* on X^Y by $(f^t)_*(u) = f^t \circ u$ for $u \in X^Y$ and $t \in T$. The projections $\pi_\alpha : \varphi_\pi \to \varphi_\alpha$ or evaluations $ev_y : \varphi_* \to \varphi$ are uniformly continuous action maps.

For a uniform action φ, the function space action φ_* is a uniform action on $C(Y;X)$ and $C_c(Y;X)$ as well. When T is abelian, we can also define the T action φ^* on $C(X;Y)$ by $(f^t)^*(u) = u \circ f^t$. Each $(f^t)^*$ is uniformly continuous, and on the $+$ invariant subset $C^u(X;Y)$ we obtain a topological action if φ is a uniform action. On $C_c(X;Y)$, φ^* is a topological action when φ is.

When the index set is $\{1,\ldots,n\}$ we write $\varphi_1 \times \ldots \times \varphi_n$ for the product action on $X_1 \times \ldots \times X_n$.

For a property P of maps on a set X, we say that P is *right canceling* (or *left canceling*) if $f \circ g$ satisfies P implies f satisfies P (resp. g satisfies P). Thus f is surjective is right canceling and f is injective is left canceling.

LEMMA 1.3. *Let φ be an action of an abelian monoid on a set X and let P be a right canceling or left canceling property of maps on X. If $\{s : f^s$ satisfies $P\}$ meets $g^t(T)$ for every $t \in T$ then f^t satisfies P for all $t \in T$.*

PROOF. For $t \in T$, there exists t_1 so that f^s satisfies P with:

$$s = t + t_1 = t_1 + t$$

As $f^s = f^{t_1} \circ f^t = f^t \circ f^{t_1}$, it follows that f^t satisfies P. ∎

REMARK. If P is the conjunction of a right canceling and a left canceling property (e.g., f is bijective), then applying the result to each component property, we see that it holds for P as well. ∎

We say that an action is P or satisfies P if f^t satisfies P for all $t \in T$. Thus we can speak of injective, surjective, or bijective (= reversible) actions. For a topological action we can thus speak of open, dense, homeomorphic, and almost open actions. Recall that a map $f : X_1 \to X_2$ is dense when $f(X_1)$ is dense in X_2, open when U open in X_1 implies $f(U)$ open in X_2, almost open when U open and nonempty in X_1 implies $f(U)$ has a nonempty interior in X_2 (see Appendix). Lemma 1.3 applies to all of these properties.

Convention. From now on all monoids are assumed abelian. So if T satisfies cancellation, it can be regarded as a submonoid of the abelian group G_T. In particular a uniform monoid is a submonoid of the abelian topological group G_T.

2

Furstenberg Families

On a nonempty set S, we denote by $\mathcal{P}$ the power set of S, the collection of all subsets of S, and by $\mathcal{P}_+$ the collection of nonempty subsets, i.e., $\mathcal{P}_+ = \{F \in \mathcal{P} : F \neq \emptyset\}$. Many of the following constructions would work equally well if $\mathcal{P}$ denoted instead some fixed algebra of subsets of S like the Borel sets or clopen sets of a topological space S.

A *family* $\mathcal{F}$ for S is a collection of subsets that is *hereditary upward*; that is, $F_0 \subset F_1$ and $F_0 \in \mathcal{F}$ imply $F_1 \in \mathcal{F}$. The family $\mathcal{F}$ is a *proper family* when it is a proper subset of $\mathcal{P}$, neither empty nor all of $\mathcal{P}$. Since $\mathcal{F}$ is hereditary, it is proper iff $S \in \mathcal{F}$ and $\emptyset \notin \mathcal{F}$. The collection $\mathcal{F}$ is a *filter* when it is a proper family closed under intersection, i.e.:

$$F_0, F_1 \in \mathcal{F} \Rightarrow F_0 \cap F_1 \in \mathcal{F} \tag{2.1}$$

Any subset $\mathcal{A}$ of $\mathcal{P}$ generates a family:

$$[\mathcal{A}] = \{F \in \mathcal{P} : F \supset A \text{ for some } A \in \mathcal{A}\} \tag{2.2}$$

If $\mathcal{A}$ consists of a single set A then we write $[A]$ for the associated family:

$$[A] = \{F \in \mathcal{P} : A \subset F\} \tag{2.3}$$

The family $[\mathcal{A}]$ is proper exactly when $\mathcal{A}$ is a nonempty subset of $\mathcal{P}_+$. $\mathcal{A}$ is called a *filterbase* if $[\mathcal{A}]$ is a filter. Thus $\mathcal{A}$ is a filterbase if $\mathcal{A} \subset \mathcal{P}_+$, $\mathcal{A} \neq \emptyset$ and:

$$A_0, A_1 \in \mathcal{A} \Rightarrow A_0 \cap A_1 \supset A_2 \text{ for some } A_2 \in \mathcal{A} \tag{2.4}$$

For example, if $A \neq \emptyset$, $[A]$ is a filter.

For a family $\mathcal{F}$, we define the *dual family*:

$$\begin{aligned} k\mathcal{F} &= \{F \in \mathcal{P} : S \setminus F \notin \mathcal{F}\} \\ &= \{F \in \mathcal{P} : F \cap F_1 \neq \emptyset \text{ for all } F_1 \in \mathcal{F}\} \end{aligned} \tag{2.5}$$

Notice that if $F_1 = S \backslash F \in \mathcal{F}$ then $F \cap F_1 = \emptyset$ with $F_1 \in \mathcal{F}$. Conversely if $F \cap F_1 = \emptyset$ for some F_1 in $\mathcal{F}$, then $F_1 \subset S \backslash F$ and so $S \backslash F \in \mathcal{F}$ by heredity.

Clearly $k\mathcal{P} = \emptyset$, while for $\mathcal{P}_+$, the largest proper family, $k\mathcal{P}_+$ is $\{S\}$, the smallest proper family.

For families $\mathcal{F}_1, \mathcal{F}_2$, define

$$\mathcal{F}_1 \cdot \mathcal{F}_2 = \{F_1 \cap F_2 : F_1 \in \mathcal{F}_1 \text{ and } F_2 \in \mathcal{F}_2\} \tag{2.6}$$

We now collect the elementary consequences of these definitions.

PROPOSITION 2.1. *Let $\mathcal{F}$, $\mathcal{F}_1$, $\mathcal{F}_2$ be families for S.*

a. $k\mathcal{F}$ is a family; it is proper if $\mathcal{F}$ is. The operator k on families is involutive and reverses inclusions:

$$\mathcal{F} = kk\mathcal{F} \qquad \mathcal{F}_1 \subset \mathcal{F}_2 \Rightarrow k\mathcal{F}_2 \subset k\mathcal{F}_1$$

b. For a collection $\{\mathcal{F}_\alpha\}$ of proper families, the intersection $\cap_\alpha \mathcal{F}_\alpha$ is a proper family, and $F \in \cap_\alpha \mathcal{F}_\alpha$ iff $F = \cup_\alpha F_\alpha$ for some choice of elements $F_\alpha \in \mathcal{F}_\alpha$. Furthermore the union $\cup_\alpha \mathcal{F}_\alpha$ is a proper family and:

$$\begin{aligned} k(\cup_\alpha \mathcal{F}_\alpha) &= \cap_\alpha k\mathcal{F}_\alpha \\ k(\cap_\alpha \mathcal{F}_\alpha) &= \cup_\alpha k\mathcal{F}_\alpha \end{aligned} \tag{2.7}$$

c. The collection $\mathcal{F}_1 \cdot \mathcal{F}_2$ is a family with:

$$\mathcal{F}_1 \cup \mathcal{F}_2 \subset \mathcal{F}_1 \cdot \mathcal{F}_2 \tag{2.8}$$

The family $\mathcal{F}_1 \cdot \mathcal{F}_2$ is proper iff $\mathcal{F}_2 \subset k\mathcal{F}_1$. More generally:

$$\mathcal{F}_1 \cdot \mathcal{F}_2 \subset \mathcal{F} \Leftrightarrow \mathcal{F}_1 \cdot k\mathcal{F} \subset k\mathcal{F}_2 \tag{2.9}$$

The operation $\cdot$ is commutative and associative and preserves inclusion:

$$\mathcal{F}_1 \subset \mathcal{F} \Rightarrow \mathcal{F}_1 \cdot \mathcal{F}_2 \subset \mathcal{F} \cdot \mathcal{F}_2$$

d. The family $\mathcal{F}$ is a filter iff $\mathcal{F}$ is a proper family and $\mathcal{F} \cdot \mathcal{F} \subset \mathcal{F}$. If $\mathcal{F}$ is a filter, then

$$\mathcal{F} = \mathcal{F} \cdot \mathcal{F} \subset \mathcal{F} \cdot k\mathcal{F} = k\mathcal{F} \tag{2.10}$$

If $\mathcal{F}_1$ and $\mathcal{F}_2$ are filters, then $\mathcal{F}_1 \cup \mathcal{F}_2$ is contained in some filter iff $\mathcal{F}_1 \cdot \mathcal{F}_2$ is proper. In that case, $\mathcal{F}_1 \cdot \mathcal{F}_2$ is the smallest filter containing $\mathcal{F}_1 \cup \mathcal{F}_2$. For a collection $\{\mathcal{F}_\alpha\}$ of filters $\cap_\alpha \mathcal{F}_\alpha$ is a filter.

e. For any proper family $\mathcal{F}$, $k(\mathcal{F} \cdot k\mathcal{F})$ is a filter contained in $\mathcal{F} \cap k\mathcal{F}$. This filter is the largest family $\mathcal{F}_1$ satisfying $\mathcal{F}_1 \cdot \mathcal{F} \subset \mathcal{F}$.

PROOF. (a) $S \in k\mathcal{F}$ and $\emptyset \notin k\mathcal{F}$ follow from $\emptyset \notin \mathcal{F}$ and $S \in \mathcal{F}$, respectively. If $F_0 \subset F_1$ and $S\backslash F_0 \notin \mathcal{F}$, then $S\backslash F_0 \supset S\backslash F_1$ implies $S\backslash F_1 \notin \mathcal{F}$. Thus $k\mathcal{F}$ is a proper family. From (2.5) we have for $F \in \mathcal{P}$:

$$F \in \mathcal{F} \Leftrightarrow S\backslash F \notin k\mathcal{F} \tag{2.11}$$

Thus $\mathcal{F} = kk\mathcal{F}$. Inclusion reversal is obvious.

(b) If $F = \cup_\alpha F_\alpha$, then $F \in \mathcal{F}_\alpha$; for all α by heredity and so $F \in \cap_\alpha \mathcal{F}_\alpha$; if $F \in \mathcal{F}_\alpha$ for all α, then we can choose $F = F_\alpha$ for all α to obtain $F = \cup_\alpha F_\alpha$. The intersection clearly contains S and omits $\emptyset$. Heredity is obvious.

$S\backslash F \notin \cap_\alpha \mathcal{F}_\alpha \Leftrightarrow S\backslash F \notin \mathcal{F}_\alpha$ for some α iff $F \in k\mathcal{F}_\alpha$ for some α iff $F \in \cup_\alpha (k\mathcal{F}_\alpha)$. So $k(\cap_\alpha \mathcal{F}_\alpha) = \cup_\alpha (k\mathcal{F}_\alpha)$. Applying this equation to the family $\{k\mathcal{F}_\alpha\}$ we obtain the other equation of (2.7).

(c) If $F_1 \in \mathcal{F}_1$, $F_2 \in \mathcal{F}_2$ and $F_1 \cap F_2 \subset F$, then $F = (F_1 \cup F) \cap (F_2 \cup F)$ and so $F \in \mathcal{F}_1 \cdot \mathcal{F}_2$ by heredity. Thus $\mathcal{F}_1 \cdot \mathcal{F}_2$ is a family. Clearly $S \in \mathcal{F}_1 \cdot \mathcal{F}_2$. So $\mathcal{F}_1 \cdot \mathcal{F}_2$ is proper iff $\emptyset \notin \mathcal{F}_1 \cdot \mathcal{F}_2$ i.e., $F_1 \in \mathcal{F}_1$ and $F_2 \in \mathcal{F}_2$ imply $F_1 \cap F_2 \neq \emptyset$. Thus:

$$\mathcal{F}_1 \cdot \mathcal{F}_2 \subset \mathcal{P}_+ \Leftrightarrow \mathcal{F}_2 \subset k\mathcal{F}_1 \tag{2.12}$$

The operation properties are obvious. Hence $\mathcal{F}_1 \cdot \mathcal{F}_2 \subset \mathcal{F}$ implies

$$(\mathcal{F}_1 \cdot k\mathcal{F}) \cdot \mathcal{F}_2 = (\mathcal{F}_1 \cdot \mathcal{F}_2) \cdot k\mathcal{F} \subset \mathcal{F} \cdot k\mathcal{F} \subset \mathcal{P}_+$$

By (2.12) $(\mathcal{F}_1 \cdot k\mathcal{F}) \subset k\mathcal{F}_2$. The reverse implication then follows from $kk\mathcal{F} = \mathcal{F}$.

The family $k\mathcal{P}_+ = \{S\}$ acts as an identity for $\cdot$, i.e.,

$$\mathcal{F} \cdot k\mathcal{P}_+ = \mathcal{F} \tag{2.13}$$

So $\mathcal{F}_1 \subset \mathcal{F}_1 \cdot \mathcal{F}_2$ follows from $k\mathcal{P}_+ \subset \mathcal{F}_2$. This establishes (2.8).

(d) A proper family $\mathcal{F}$ is clearly a filter iff $\mathcal{F} \cdot \mathcal{F} \subset \mathcal{F}$ from which equality follows from (2.8). Then $\mathcal{F} \cdot k\mathcal{F} \subset k\mathcal{F}$ follows from (2.9) with equality from (2.8) again. Finally, $\mathcal{F} \cdot \mathcal{F} = \mathcal{F} \subset \mathcal{P}_+$ implies $\mathcal{F} \subset k\mathcal{F}$ by (2.12).

If $\mathcal{F}_1$ and $\mathcal{F}_2$ are filters, then:

$$(\mathcal{F}_1 \cdot \mathcal{F}_2) \cdot (\mathcal{F}_1 \cdot \mathcal{F}_2) = (\mathcal{F}_1 \cdot \mathcal{F}_1) \cdot (\mathcal{F}_2 \cdot \mathcal{F}_2) = \mathcal{F}_1 \cdot \mathcal{F}_2$$

So $\mathcal{F}_1 \cdot \mathcal{F}_2$ is a filter when it is proper. Observe that any filter containing $\mathcal{F}_1 \cup \mathcal{F}_2$ contains $\mathcal{F}_1 \cdot \mathcal{F}_2$. Clearly the intersection of filters is a filter.

(e) Let $\tilde{\mathcal{F}} = k(\mathcal{F} \cdot k\mathcal{F})$ so that $\mathcal{F} \cdot k\mathcal{F} = k\tilde{\mathcal{F}}$ implies $\mathcal{F} \cdot \tilde{\mathcal{F}} \subset \mathcal{F}$ by (2.9). In general if $\mathcal{F}_1 \cdot \mathcal{F} \subset \mathcal{F}$, then $\mathcal{F}_1 \cdot (\mathcal{F} \cdot k\mathcal{F}) \subset \mathcal{F} \cdot k\mathcal{F}$. By (2.9) again this implies $\mathcal{F}_1 \cdot \tilde{\mathcal{F}} \subset \tilde{\mathcal{F}}$. So by (2.8) $\mathcal{F}_1 \subset \tilde{\mathcal{F}}$. With $\mathcal{F}_1 = \tilde{\mathcal{F}}$, we have $\tilde{\mathcal{F}} \cdot \tilde{\mathcal{F}} \subset \tilde{\mathcal{F}}$, so $\tilde{\mathcal{F}}$ is a filter because it is proper.

Finally $\tilde{\mathcal{F}} \subset \mathcal{F} \cap k\mathcal{F}$ by (2.7) and (2.8). ■

REMARK. For a family $\mathcal{F}$, we repeatedly use:

$$\mathcal{F} \text{ is proper } \Leftrightarrow k\mathcal{P}_+ \subset \mathcal{F} \subset \mathcal{P}_+ \tag{2.14}$$

■

A family $\mathcal{F}$ is called a *filterdual* when its dual $k\mathcal{F}$ is a filter. A proper family $\mathcal{F}$ is a filterdual iff it satisfies the *Ramsey Property* dual to (2.1):

$$F_0 \cup F_1 \in \mathcal{F} \Rightarrow F_0 \in \mathcal{F} \text{ or } F_1 \in \mathcal{F} \tag{2.15}$$

For any proper family $\mathcal{F}$, $\mathcal{F} \cdot k\mathcal{F}$ is a filterdual containing $\mathcal{F} \cup k\mathcal{F}$ (cf. Proposition 2.1e).

For any filter $\mathcal{F}$, we define congruence mod $\mathcal{F}$ on the subsets of S by

$$F_1 \equiv F_2 \ (\text{mod } \mathcal{F}) \Leftrightarrow F_1 \cap F = F_2 \cap F \text{ for some } F \in \mathcal{F} \tag{2.16}$$

Symmetry and reflexivity are obvious, while the filter property easily yields transitivity. Congruence preserves the set operations:

$$\begin{aligned} F_1 \equiv F_2 \ (\text{mod } \mathcal{F}) &\Rightarrow F_3 \backslash F_1 \equiv F_3 \backslash F_2 \\ F_3 \cap F_1 \equiv F_3 \cap F_2 \qquad & F_3 \cup F_1 \equiv F_3 \cup F_2 \ (\text{mod } \mathcal{F}) \end{aligned} \tag{2.17}$$

for all $F_3 \in \mathcal{P}$.

In most of our examples, there is a distinguished filterdual family $\mathcal{B}$ referred to as the *base family*. In applications it comes from a class of subsets $\mathcal{N}$, which we think of negligible or *null*. The collection $\mathcal{N}$ is assumed to be a proper subset of $\mathcal{P}$ that is *hereditary downward*, so that $F_0 \subset F_1$ and $F_1 \in \mathcal{N}$ imply $F_0 \in \mathcal{N}$. Thus $\emptyset \in \mathcal{N}$ and $S \notin \mathcal{N}$. Finally $\mathcal{N}$ is assumed closed under finite, and sometimes countable, unions. The associated base family $\mathcal{B}$ is the collection of non-null sets, i.e., $\mathcal{B} = \{F \in \mathcal{P} : F \notin \mathcal{N}\}$. The dual $k\mathcal{B}$ is the collection of sets with null complements, so that $k\mathcal{B} = \{F \in \mathcal{P} : S \backslash F \in \mathcal{N}\}$. It is clear that $k\mathcal{B}$ is a filter, so $\mathcal{B}$ is a filterdual.

Table 2.1 lists some common examples.

Table 2.1

	S	$\mathcal{N}$	$\mathcal{B}$	$k\mathcal{B}$
1.	Nonempty set	$\{\emptyset\}$	$\mathcal{P}_+$	$\{S\}$
2.	Infinite set	Finite subsets	Infinite subsets	Cofinite subsets
3.	Baire space	First category subsets	Subsets not of first category	Residual subsets
4.	Noncompact space	Bounded subsets[a]	Unbounded subsets	Cobounded subsets
5.	Measure space	Subsets of meas. zero	Subsets of positive meas.	Subsets of full meas.

[a] Subsets with compact closure.

Now with $\mathcal{B}$ a proper family, we define, by analogy with (2.5), the $\mathcal{B}$-dual of a family $\mathcal{F}$:

$$k_{\mathcal{B}}\mathcal{F} = \{F : F \cap F_1 \in \mathcal{B} \text{ for all } F_1 \in \mathcal{F}\} \tag{2.18}$$

Because $\mathcal{B}$ is proper, $F \cap F_1 \in \mathcal{B}$ implies $F \cap F_1 \neq \emptyset$, and so we have

$$k_{\mathcal{B}}\mathcal{F} \subset k\mathcal{F} \tag{2.19}$$

with equality in the special case $\mathcal{B} = \mathcal{P}_+$. Suppose now that $\mathcal{B}$ is a filterdual fixed to define a base family. Equality in (2.19) still holds for an important class of families $\mathcal{F}$. For example applying (2.10) to the filter $\mathcal{F} = k\mathcal{B}$, we see that $k\mathcal{B} \cdot \mathcal{B} = \mathcal{B}$. Furthermore it is easy to check the analogue of (2.12)

$$\mathcal{F}_1 \cdot \mathcal{F}_2 \subset \mathcal{B} \Leftrightarrow \mathcal{F}_2 \subset k_{\mathcal{B}}\mathcal{F}_1 \tag{2.20}$$

So we obtain $k\mathcal{B} \subset k_{\mathcal{B}}\mathcal{B}$. With (2.19) this implies

$$k_{\mathcal{B}}\mathcal{B} = k\mathcal{B} \tag{2.21}$$

LEMMA 2.1. *Assume $\mathcal{B}$ is a filterdual. For any family $\mathcal{F}$, we have*

$$\begin{aligned} &\mathcal{F} \subset k\mathcal{B} \cdot \mathcal{F} = \\ &\{F \in \mathcal{P} : F \equiv F_1 (\mathit{mod}\ k\mathcal{B}) \text{ for some } F_1 \in \mathcal{F}\} \\ &= k_{\mathcal{B}}k_{\mathcal{B}}\mathcal{F} = kk_{\mathcal{B}}\mathcal{F}. \end{aligned}$$

PROOF. We first prove that each set in the above list is contained in its successor. Since $S \in k\mathcal{B}$, the first inclusion is clear. Every element of $k\mathcal{B} \cdot \mathcal{F}$ is congruent mod $k\mathcal{B}$ to an element of $\mathcal{F}$, proving the next inclusion. If $F \cap Z = F_1 \cap Z$ for some $Z \in k\mathcal{B}$ and if $A \in k_{\mathcal{B}}\mathcal{F}$, then:

$$F \cap A \supset (F_1 \cap A) \cap Z \in \mathcal{B} \cdot k\mathcal{B} = \mathcal{B}$$

Thus F congruent to an element of $\mathcal{F}$ implies $F \in k_{\mathcal{B}}k_{\mathcal{B}}\mathcal{F}$. The last inclusion follows from (2.19).

To complete the proof, we show that $kk_{\mathcal{B}}\mathcal{F} \subset k\mathcal{B} \cdot \mathcal{F}$, or equivalently that $k(k\mathcal{B} \cdot \mathcal{F}) \subset k_{\mathcal{B}}\mathcal{F}$. Suppose $F \in k(k\mathcal{B} \cdot \mathcal{F})$ and $F_1 \in \mathcal{F}$. We must show $F \cap F_1 \in \mathcal{B}$. If not, then there exists $Z \in k\mathcal{B}$ such that $(F \cap F_1) \cap Z = \emptyset$. Then $F \cap (Z \cap F_1) = \emptyset$, contradicting $F \in k(k\mathcal{B} \cdot \mathcal{F})$. ∎

PROPOSITION 2.2. *Assume that $\mathcal{B}$ is a filterdual family of subsets of S.*

a. A family $\mathcal{F}$ is called full *(with respect to $\mathcal{B}$) if it is proper and satisfies the following conditions, which are equivalent for a proper family:*

(1) $k\mathcal{F} = k_{\mathcal{B}}\mathcal{F}$

(2) $\mathcal{F} = k_{\mathcal{B}}\mathcal{F}_1$ *for some family* $\mathcal{F}_1$

(3) $k\mathcal{B} \cdot \mathcal{F} \subset \mathcal{F}$

(4) $\mathcal{F} \cdot k\mathcal{F} \subset \mathcal{B}$

(5) $\mathcal{F}$ *is closed under congruence mod* $k\mathcal{B}$, *i.e.,* $F \equiv F_1$ *(mod* $k\mathcal{B}$*) and* $F_1 \in \mathcal{F}$ *imply* $F \in \mathcal{F}$.

b. A full family $\mathcal{F}$ *satisfies*

$$k\mathcal{B} \subset k\mathcal{B} \cdot \mathcal{F} = \mathcal{F} \subset \mathcal{F} \cdot k\mathcal{F} \subset \mathcal{B} \tag{2.22}$$

If $\mathcal{F}$ *is a filter, then* $k\mathcal{B} \subset \mathcal{F}$ *implies* $\mathcal{F}$ *is full.*

c. The families $\mathcal{B}$ *and* $k\mathcal{B}$ *are full. In general* $\mathcal{F}$ *is full iff its dual* $k\mathcal{F}$ *is full.*

d. A nonempty family $\mathcal{F}$ *is contained in some full family iff* $\mathcal{F} \subset \mathcal{B}$, *in which case* $k\mathcal{B} \cdot \mathcal{F}$ *is the smallest full family containing* $\mathcal{F}$. *If* $\mathcal{F}$ *is a filter with* $\mathcal{F} \subset \mathcal{B}$, *then* $k\mathcal{B} \cdot \mathcal{F}$ *is a full filter.*

PROOF. (a) (1) $\Rightarrow$ (2). By Lemma 2.1 and then (1), $k_{\mathcal{B}}k_{\mathcal{B}}\mathcal{F} = kk_{\mathcal{B}}\mathcal{F} = kk\mathcal{F} = \mathcal{F}$, so $\mathcal{F} = k_{\mathcal{B}}\mathcal{F}_1$ with $\mathcal{F}_1 = k\mathcal{F} = k_{\mathcal{B}}\mathcal{F}$.

(2) $\Rightarrow$ (3). By (2.21) and (2):

$$\mathcal{F}_1 \cdot (k\mathcal{B} \cdot \mathcal{F}) = k_{\mathcal{B}}\mathcal{B} \cdot (\mathcal{F}_1 \cdot k_{\mathcal{B}}\mathcal{F}_1) \subset k_{\mathcal{B}}\mathcal{B} \cdot \mathcal{B} \subset \mathcal{B}$$

So by (2.20) $k\mathcal{B} \cdot \mathcal{F} \subset k_{\mathcal{B}}\mathcal{F}_1 = \mathcal{F}$.

(3) $\Leftrightarrow$ (4). From (2.9).

(3) $\Leftrightarrow$ (5). Lemma 2.1.

(5) $\Rightarrow$ (1). By Lemma 2.1 and (5) $\mathcal{F} = kk_{\mathcal{B}}\mathcal{F}$; apply k.

(b) For a proper family $\mathcal{F}$, (3) and (4) and (2.8) yield (2.22). If $\mathcal{F}$ is a filter, then $k\mathcal{B} \subset \mathcal{F}$ implies $k\mathcal{B} \cdot \mathcal{F} \subset \mathcal{F} \cdot \mathcal{F} = \mathcal{F}$, which is condition (3).

(c) Condition (4) is symmetric in $\mathcal{F}$ and $k\mathcal{F}$. So $\mathcal{F}$ is full iff $k\mathcal{F}$ is. $\mathcal{B}$ is full by (2.21) and so $k\mathcal{B}$ is.

(d) By Lemma 2.1, $k\mathcal{B} \cdot \mathcal{F}$ contains $\mathcal{F}$ and satisfies condition (2). So it is a full family provided it is proper. By Proposition 2.1c, it is proper iff $\mathcal{F} \subset \mathcal{B}$. On the other hand if $k\mathcal{P}_+ \subset \mathcal{F} \subset \mathcal{F}_1$ and $\mathcal{F}_1$ is full then $\mathcal{F}$ is proper and by (2.22)

$$k\mathcal{B} \cdot \mathcal{F} \subset k\mathcal{B} \cdot \mathcal{F}_1 = \mathcal{F}_1 \subset \mathcal{B}$$

Thus $k\mathcal{B} \cdot \mathcal{F}$ is the smallest full family containing $\mathcal{F}$. If $\mathcal{F}$ is a filter, then $k\mathcal{B} \cdot \mathcal{F}$ is a filter when it is proper by Proposition 2.1d. ■

Of special interest are filters that are maximal with respect to inclusion. Such a filter is called an *ultrafilter*. By Zorn's Lemma every filter is contained in some ultrafilter.

PROPOSITION 2.3. *For a filter $\mathcal{F}$ the following conditions are equivalent.*

(1) $\mathcal{F}$ is an ultrafilter.

(2) $\mathcal{F} = k\mathcal{F}$.

(3) $\mathcal{F}$ is a filterdual.

(4) For all $F \subset S$, either $F \in \mathcal{F}$ or $S \backslash F \in \mathcal{F}$.

With respect to a filterdual family $\mathcal{B}$, an ultrafilter $\mathcal{F}$ is full iff $\mathcal{F} \subset \mathcal{B}$, or equivalently iff $k\mathcal{B} \subset \mathcal{F}$.

PROOF. (2) $\Rightarrow$ (3). This is obvious.

(3) $\Rightarrow$ (4). $S = F \cup (S \backslash F)$ is in $\mathcal{F}$, so by the Ramsey Property (2.15) of the filterdual, $\mathcal{F}$, either $F \in \mathcal{F}$ or $S \backslash F \in \mathcal{F}$.

(4) $\Rightarrow$ (1). If $F \notin \mathcal{F}$, then $S \backslash F \in \mathcal{F}$, so no filter including $\mathcal{F}$ can also contain F. Thus $\mathcal{F}$ is a maximal filter.

(1) $\Rightarrow$ (2). For $F \in k\mathcal{F}$, $[F] \cdot \mathcal{F}$ is a proper family, where $[F]$ is the filter generated by F [cf. (2.3)]. By Proposition 2.1d and (2.8), $[F] \cdot \mathcal{F}$ is a filter including $\mathcal{F}$ and containing F. Then by maximality, $F \in \mathcal{F}$. Thus $k\mathcal{F} \subset \mathcal{F}$, and equality follows from (2.10).

By Proposition 2.2b a filter $\mathcal{F}$ is full iff $k\mathcal{B} \subset \mathcal{F}$ and so iff $k\mathcal{F} \subset \mathcal{B}$. For an ultrafilter, the latter is equivalent to $\mathcal{F} \subset \mathcal{B}$. ■

Notice that $\mathcal{F} = k\mathcal{F}$ does not imply $\mathcal{F}$ is a filter. For example, if S has three points and $\mathcal{F}$ consists of S and the three doubletons, then $\mathcal{F} = k\mathcal{F}$.

We now consider family constructions associated with a monoid action on S.

First recall that we regard a subset G of $S_1 \times S_2$ as a *relation* from S_1 to S_2 with $G^{-1} = \{(y,x) : (x,y) \in G\}$ the *inverse relation* from S_2 to S_1. For $A \subset S_1$, $G(A) = \{y : (x,y) \in G \text{ for some } x \in A\}$ is the *image* of A under G. For $A \subset S_1$ and $B \subset S_2$, note the equivalence:

$$\begin{aligned} G(A) \cap B \neq \emptyset &\Leftrightarrow A \cap G^{-1}(B) \neq \emptyset \\ &\Leftrightarrow A \times B \cap G \neq \emptyset \end{aligned} \tag{2.23}$$

since all of these say there exists $(x,y) \in G$ with $x \in A$ and $y \in B$.

Since $G(A) \subset B$ iff $G(A) \cap S_2 \backslash B = \emptyset$, we have from (2.23):

$$\begin{aligned} G(A) \subset B &\Leftrightarrow G^{-1}(S_2 \backslash B) \subset S_1 \backslash A \\ &\Leftrightarrow A \times (S_2 \backslash B) \cap G = \emptyset \end{aligned} \tag{2.24}$$

For a family $\mathcal{F}_1$ for S_1, define $G\mathcal{F}_1$ to be the family generated by the G images of the elements of $\mathcal{F}_1$:

$$G\mathcal{F}_1 = \{B \in \mathcal{P}_2 : G(A) \subset B \text{ for some } A \in \mathcal{F}_1\} \tag{2.25}$$

For a family $\mathcal{F}_2$ for S_2 define the pullback family:

$$G^*\mathcal{F}_2 = \{A \in \mathcal{P}_1 : G(A) \in \mathcal{F}_2\} \tag{2.26}$$

For a map $g : S_1 \to S_2$, $g^{-1}(S_2 \backslash B) = S_1 \backslash g^{-1}(B)$, so from (2.24) we also have in the map case:

$$g(A) \subset B \Leftrightarrow A \subset g^{-1}(B) \tag{2.27}$$

Then for a map g, $B \in g\mathcal{F}$ iff $g^{-1}(B) \supset A$ for some $A \in \mathcal{F}$, so iff $g^{-1}(B) \in \mathcal{F}$. Thus for a map g:

$$g\mathcal{F}_1 = (g^{-1})^*\mathcal{F}_1 = \{B \in \mathcal{P}_2 : g^{-1}(B) \in \mathcal{F}_1\} \tag{2.28}$$

PROPOSITION 2.4. *Let $G : S_1 \to S_2$ be a relation. Let $\mathcal{F}_1$, $\tilde{\mathcal{F}}_1$ etc. denote families for S_1 and $\mathcal{F}_2$, $\tilde{\mathcal{F}}_2$ denote families for S_2.*

a. The family $G^*\mathcal{F}_2$ is proper iff $\mathcal{F}_2$ is proper and $G(S_1) \in \mathcal{F}_2$. The family $G\mathcal{F}_1$ is proper iff $\mathcal{F}_1$ is proper and $G(A) \neq \emptyset$ for all $A \in \mathcal{F}_1$. In particular, if $G^{-1}(S_2) = S_1$, i.e., $G(x) \neq \emptyset$ for all $x \in S_1$, then $G\mathcal{F}_1$ is proper when $\mathcal{F}_1$ is.

b. The following conditions on $\mathcal{F}_1$, $\mathcal{F}_2$ and G are equivalent:

(1) $A \in \mathcal{F}_1 \Rightarrow G(A) \in \mathcal{F}_2$

(2) $A \in \mathcal{F}_1$ *and* $B \in k\mathcal{F}_2 \Rightarrow (A \times B) \cap G \neq \emptyset$

(3) $G\mathcal{F}_1 \subset \mathcal{F}_2$

(4) $\mathcal{F}_1 \subset G^*\mathcal{F}_2$

(5) $G^{-1}(k\mathcal{F}_2) \subset k\mathcal{F}_1$.

If $G = g$ is a map, then these conditions are equivalent as well to:

(6) $g^{-1}(B) \in \mathcal{F}_1 \Rightarrow B \in \mathcal{F}_2$

c. If $G = g$ is a map, then the following three conditions are equivalent:

(1) $B \in \mathcal{F}_2 \Rightarrow g^{-1}(B) \in \mathcal{F}_1$

(2) $\mathcal{F}_2 \subset g\mathcal{F}_1$

(3) $g^{-1}\mathcal{F}_2 \subset \mathcal{F}_1$

Also the following two conditions are equivalent:

(4) $g(A) \in \mathcal{F}_2 \Rightarrow A \in \mathcal{F}_1$

(5) $g^*\mathcal{F}_2 \subset \mathcal{F}_1$

If g is injective then (1) implies (4). On the other hand if $\mathcal{F}_2 = \mathcal{F}_2 \cdot [g(S_1)]$ then (4) implies (1).

d. The following identities are true:

$$\begin{aligned} G(\cup_\alpha \mathcal{F}_{1\alpha}) &= \cup_\alpha G(\mathcal{F}_{1\alpha}) \\ G^*(\cap_\alpha \mathcal{F}_{2\alpha}) &= \cap_\alpha G^*(\mathcal{F}_{2\alpha}) \\ G^*(k\mathcal{F}_2) &= k(G^{-1}\mathcal{F}_2) \end{aligned} \tag{2.29}$$

The following inclusions hold:

$$\begin{aligned} \mathcal{F}_1 &\subset G^* G \mathcal{F}_1 \\ GG^* \mathcal{F}_2 &\subset \mathcal{F}_2 \\ (G\mathcal{F}_1) \cdot (G\tilde{\mathcal{F}}_1) &\subset G(\mathcal{F}_1 \cdot \tilde{\mathcal{F}}_1) \end{aligned} \tag{2.30}$$

If $\mathcal{F}_1$ is a filter, then $G\mathcal{F}_1$ is a filter if it is proper. If $\mathcal{F}_2$ is a filterdual, then $G^* \mathcal{F}_2$ is a filterdual if it is proper.

e. If $G = g$ is a map, then the following identities are true:

$$\begin{aligned} g(\cap_\alpha \mathcal{F}_{1\alpha}) &= \cap_\alpha g(\mathcal{F}_{1\alpha}) \\ g(k\mathcal{F}_1) &= k(g\mathcal{F}_1) \\ g^{-1}(\mathcal{F}_2) \cdot g^{-1}(\tilde{\mathcal{F}}_2) &= g^{-1}(\mathcal{F}_2 \cdot \tilde{\mathcal{F}}_2) \end{aligned} \tag{2.31}$$

The following inclusions hold, and if in addition g is injective, then equality is true in each case:

$$\begin{aligned} g(\mathcal{F}_1) \cdot g(\tilde{\mathcal{F}}_1) &\subset g(\mathcal{F}_1 \cdot \tilde{\mathcal{F}}_1) \\ g^{-1} g \mathcal{F}_1 \subset \mathcal{F}_1 &\subset g^* g \mathcal{F}_1 \\ g^{-1} \mathcal{F}_2 &\subset g^*(\mathcal{F}_2 \cdot [g(S_1)]) \end{aligned} \tag{2.32}$$

The following holds, and if in addition g is surjective, then equality is true:

$$\mathcal{F}_2 \subset \mathcal{F}_2 \cdot [g(S_1)] = gg^{-1}\mathcal{F}_2 = gg^*(\mathcal{F}_2 \cdot [g(S_1)]) \tag{2.33}$$

Proof. (a) $\emptyset \in G^* \mathcal{F}_2$ iff $\emptyset \in \mathcal{F}_2$, and $S_2 \in G\mathcal{F}_1$ iff $S_1 \in \mathcal{F}_1$; details are left to the reader.

(b) Using (2.23) the equivalence of (1)–(4) are clear. (4) $\Leftrightarrow$ (5) follows from the third equation in (2.29) proved later. In the map case (3) $\Leftrightarrow$ (6) follows from (2.28).

(c) (1) $\Leftrightarrow$ (3) by (1) $\Leftrightarrow$ (3) of (b) applied to $G = g^{-1}$. (1) $\Leftrightarrow$ (2) by (2.28). (4) $\Leftrightarrow$ (5) by definition of g^*. For $A \subset S_1$ observe that:

$$g^{-1}(g(A)) \supset A \qquad \text{(with = if } g \text{ is injective)} \tag{2.34}$$

So (1) implies (4) in the injective case. For $B \subset S_2$:

$$g(g^{-1}(B)) = B \cap g(S_1) \subset B \tag{2.35}$$

So the converse (4) implies (1) is true when $B \in \mathcal{F}_2$ implies $B \cap g(S_1) \in \mathcal{F}_2$.

(d) By (2.23), $G(A)$ meets every B in $\mathcal{F}_2$ iff A meets $G^{-1}(B)$ for every B in $\mathcal{F}_2$. This yields the third equation in (2.29). The other two are simple exercises.

The first two inclusions of (2.30) follow from (3) $\Leftrightarrow$ (4) of (b) applied to $G\mathcal{F}_1 \subset G\mathcal{F}_1$ and $G^*\mathcal{F}_2 \subset G^*\mathcal{F}_2$. If $A \in \mathcal{F}_1$, and $\tilde{A} \in \tilde{\mathcal{F}}_1$ then $G(A) \cap G(\tilde{A})$ contains $G(A \cap \tilde{A})$, so it is in $G(\mathcal{F}_1 \cdot \tilde{\mathcal{F}}_1)$. Then $\mathcal{F}_1 \cdot \mathcal{F}_1 \subset \mathcal{F}_1$ implies $(G\mathcal{F}_1) \cdot (G\mathcal{F}_1) \subset (G\mathcal{F}_1)$. The filterdual result then follows from (2.29).

(e) By (2.28) the family operator g is $(g^{-1})^*$. The first two equations then follow from (2.29). The third equation follows from $g^{-1}(B \cap \tilde{B}) = g^{-1}(B) \cap g^{-1}(\tilde{B})$. Similarly $g(A \cap \tilde{A}) \subset g(A) \cap g(\tilde{A})$ with equality in the injective case yields the first inclusion in (2.32). For the second inclusion apply (3) $\Leftrightarrow$ (4) of (b) and (2) $\Leftrightarrow$ (3) of (c) to $g\mathcal{F}_1 \subset g\mathcal{F}_1$. If $A \in g^*g\mathcal{F}_1$, then $g(A) \supset g(\tilde{A})$ for some $\tilde{A} \in \mathcal{F}_1$. By (2.34), $A \supset g^{-1}g(\tilde{A})$ if g is injective, so $A \in g^{-1}g\mathcal{F}_1$.

Clearly, (2.35) implies the third inclusion of (2.32) and $\mathcal{F}_2 \cdot [g(S_1)] \subset gg^{-1}\mathcal{F}_2$. Applying g to (2.32) we have $gg^{-1}\mathcal{F}_2 \subset gg^*(\mathcal{F}_2 \cdot [g(S_1)])$, which is included in $\mathcal{F}_2 \cdot [g(S_1)]$ by (2.30). This proves (2.33).

If $A \in g^*(\mathcal{F}_2 \cdot [g(S_1)])$, then $g(A) = B \cap g(S_1)$ for some $B \in \mathcal{F}_2$. By (2.35), $g(A) = g(g^{-1}(B))$. So if g is injective, $A = g^{-1}(B) \in g^{-1}\mathcal{F}_2$. ■

REMARK. Assume $\mathcal{B}_2$ is a filterdual family for S_2 and $\mathcal{F}_2$ is full with respect to $\mathcal{B}_2$, i.e., $\mathcal{F}_2 = k\mathcal{B}_2 \cdot \mathcal{F}_2$. If $g(S_1) \in k\mathcal{B}_2$, then $\mathcal{F}_2 = \mathcal{F}_2 \cdot [g(S_1)]$. In particular, if $\mathcal{B}_1$ is a filterdual family for S_1 and $g(k\mathcal{B}_1) \subset k\mathcal{B}_2$, then $g(S_1) \in k\mathcal{B}_2$. ■

Let $\pi_\alpha : S_1 \times S_2 \to S_\alpha$ be the projection map and $\mathcal{F}_\alpha$ a family for S_α ($\alpha = 1,2$). Define $\mathcal{F}_1 \times \mathcal{F}_2$ to be the family for $S_1 \times S_2$ generated by $\{A \times B : A \in \mathcal{F}_1$ and $B \in \mathcal{F}_2\}$.

$$\begin{aligned} \mathcal{F}_1 \times \mathcal{F}_2 &= (\pi_1^{-1}\mathcal{F}_1) \cdot (\pi_2^{-1}\mathcal{F}_2) \\ &= \{C : S_1 \times S_2 \supset C \supset A \times B \text{ for some } A \in \mathcal{F}_1 \text{ and } B \in \mathcal{F}_2\} \end{aligned} \tag{2.36}$$

Now we assume that T is a monoid that acts on S. Recall that all monoids are now assumed abelian. From the action $T \times S \to S$, we obtain the associated maps $g^t : S \to S$ and let $g^t(F)$ and $g^{-t}(F)$ denote the image and preimage of the subset F of S [so that $g^{-t}(F)$ is notation for $(g^t)^{-1}(F)$].

A family $\mathcal{F}$ for S is called $+$ *invariant* if

$$\begin{aligned} &t \in T \text{ and } F \in \mathcal{F} \Rightarrow g^t(F) \in \mathcal{F} \\ &\text{i.e., } g^t\mathcal{F} \subset \mathcal{F} \text{ for all } t \in T \end{aligned} \tag{2.37}$$

A family $\mathcal{F}$ for S is called $-$ *invariant* if

$$t \in T \text{ and } F \in \mathcal{F} \Rightarrow g^{-t}(F) \in F$$
$$\text{i.e., } g^{-t}\mathcal{F} \subset \mathcal{F} \text{ for all } t \in T \tag{2.38}$$

A family $\mathcal{F}$ for S is called an *invariant* family if it is both $+$ and $-$ invariant. So by Proposition 2.4, $\mathcal{F}$ is invariant when for all $t \in T$:

$$F \in \mathcal{F} \Leftrightarrow g^{-t}(F) \in \mathcal{F}$$
$$\text{i.e., } g^{t}\mathcal{F} = \mathcal{F} \tag{2.39}$$

A family $\mathcal{F}$ is called *thick* if for every finite subset $\{t_1, \ldots t_l\}$ of T:

$$F \in \mathcal{F} \Rightarrow g^{-t_1}(F) \cap \cdots \cap g^{-t_l}(F) \in \mathcal{F} \tag{2.40}$$

Clearly a thick family is $-$ invariant, and so it is invariant iff it is $+$ invariant.

If T is a group, then for each $t \in T$, there is an inverse element $-t$ so that the map g^{-t} is the inverse function $(g^t)^{-1}$. Thus for a group action, the notions of $+$ invariance, $-$ invariance, and invariance all agree. In particular a thick family is invariant.

Now we define operations on families related to these properties. For $\mathcal{F}$ a family for S, we define

$$\gamma\mathcal{F} = \{F : g^{-t}(F) \supset g^{-t_1}(F_1) \text{ for some } t, t_1 \in T \text{ and } F_1 \in \mathcal{F}\} \tag{2.41}$$
$$\tilde{\gamma}\mathcal{F} = k\gamma k\mathcal{F} \tag{2.42}$$
$$\tau\mathcal{F} = \{F : g^{-t_1}(F) \cap \cdots \cap g^{-t_l}(F) \in \mathcal{F} \text{ for every finite subset } \{t_1, \ldots, t_l\} \text{ of } T\} \tag{2.43}$$

PROPOSITION 2.5. *Let T be a monoid acting on S and $\mathcal{F}$, $\mathcal{F}_1$, $\mathcal{F}_2$ proper families for S.*

a. $\gamma\mathcal{F}$ is the smallest invariant family containing $\mathcal{F}$. That is, $\gamma\mathcal{F} \supset \mathcal{F}$, $\gamma\mathcal{F}$ is invariant and if $\mathcal{F}_1$ is an invariant family with $\mathcal{F}_1 \supset \mathcal{F}$, then $\mathcal{F}_1 \supset \gamma\mathcal{F}$. Thus $\mathcal{F}$ is invariant iff $\mathcal{F} = \gamma\mathcal{F}$. In particular $\gamma\gamma\mathcal{F} = \gamma\mathcal{F}$. The operator γ preserves inclusions; i.e., $\mathcal{F}_1 \subset \mathcal{F}_2$ implies $\gamma\mathcal{F}_1 \subset \gamma\mathcal{F}_2$.

b. $\mathcal{F}$ is $+$ invariant iff $k\mathcal{F}$ is $-$ invariant, so $\mathcal{F}$ is invariant iff $k\mathcal{F}$ is invariant. For any family $\mathcal{F}$:

$$\gamma k\gamma\mathcal{F} = k\gamma\mathcal{F} \subset k\mathcal{F} \subset \gamma k\mathcal{F} = k\tilde{\gamma}\mathcal{F} = \tilde{\gamma}k\tilde{\gamma}\mathcal{F} \tag{2.44}$$

c. $\tilde{\gamma}\mathcal{F}$ is the largest invariant family contained in $\mathcal{F}$. That is, $\tilde{\gamma}\mathcal{F} \subset \mathcal{F}$, $\tilde{\gamma}\mathcal{F}$ is invariant, and if $\mathcal{F}_1$ is an invariant family with $\mathcal{F}_1 \subset \mathcal{F}$, then $\mathcal{F}_1 \subset \tilde{\gamma}\mathcal{F}$. Thus $\mathcal{F}$ is invariant iff $\tilde{\gamma}\mathcal{F} = \mathcal{F}$. In particular $\gamma\tilde{\gamma}\mathcal{F} = \tilde{\gamma}\tilde{\gamma}\mathcal{F} = \tilde{\gamma}\mathcal{F}$ and $\tilde{\gamma}\gamma\mathcal{F} = \gamma\mathcal{F}$. The operator $\tilde{\gamma}$ preserves inclusions.

d. $\tau\mathcal{F}$ is a proper family, and it is the largest thick family contained in $\mathcal{F}$. That is, $\tau\mathcal{F} \subset \mathcal{F}$, $\tau\mathcal{F}$ is thick, and if $\mathcal{F}_1$ is a thick family with $\mathcal{F}_1 \subset \mathcal{F}$, then $\mathcal{F}_1 \subset \tau\mathcal{F}$. Thus $\mathcal{F}$ is thick iff $\tau\mathcal{F} = \mathcal{F}$. In particular $\tau\tau\mathcal{F} = \tau\mathcal{F}$. The operator τ preserves inclusions.

e. If $\mathcal{F}_1$ and $\mathcal{F}_2$ are thick, then $\mathcal{F}_1 \cdot \mathcal{F}_2$ is. In general for any $\mathcal{F}_1$, $\mathcal{F}_2$:

$$(\tau\mathcal{F}_1)\cdot(\tau\mathcal{F}_2) = \tau((\tau\mathcal{F}_1)\cdot(\tau\mathcal{F}_2)) \subset \tau(\mathcal{F}_1 \cdot \mathcal{F}_2) \tag{2.45}$$

f. Let $\{\mathcal{F}_\alpha\}$ be a collection of families. If each $\mathcal{F}_\alpha$ is $+$ invariant, $-$ invariant, or thick, then $\cap_\alpha \mathcal{F}_\alpha$ and $\cup_\alpha \mathcal{F}_\alpha$ satisfy the corresponding property. In general:

$$\tau \cap_\alpha \mathcal{F}_\alpha = \cap_\alpha \tau \mathcal{F}_\alpha \qquad \tilde{\gamma} \cap_\alpha \mathcal{F}_\alpha = \cap_\alpha \tilde{\gamma} \mathcal{F}_\alpha \tag{2.46}$$

$$\gamma \cup_\alpha \mathcal{F}_\alpha = \cup_\alpha \gamma \mathcal{F}_\alpha \tag{2.47}$$

g. If $\mathcal{F}$ is $+$ invariant, then $\tau\mathcal{F}$ is invariant. If $\mathcal{F}$ is thick, then $\gamma\mathcal{F}$ is thick. In general for any $\mathcal{F}$:

$$\gamma\tau\mathcal{F} = \tau\gamma\tau\mathcal{F} \subset \tau\gamma\mathcal{F} = \gamma\tau\gamma\mathcal{F} \tag{2.48}$$

PROOF. (a) It is clear that $\gamma\mathcal{F}$ is a family containing $\mathcal{F}$ and inclusions are preserved. Also if $\mathcal{F}$ is invariant, then $\gamma\mathcal{F} = \mathcal{F}$ by heredity and (2.39). Then if $\mathcal{F}_1$ is invariant and it contains $\mathcal{F}$, $\mathcal{F}_1 = \gamma\mathcal{F}_1 \supset \gamma\mathcal{F}$. Once we show $\gamma\mathcal{F}$ is itself invariant, then $\gamma\gamma\mathcal{F} = \gamma\mathcal{F}$ follows.

Suppose $t \in T$ and $F \in \gamma\mathcal{F}$. Then: $g^{-t_2}(F) \supset g^{-t_1}(F_1)$ for some $t_1, t_2 \in T$ and $F_1 \in \mathcal{F}$. Because T is abelian:

$$g^{-t_2}(g^{-t}(F)) = g^{-t}(g^{-t_2}(F)) \supset g^{-t}(g^{-t_1}(F_1)) = g^{-(t_1+t)}(F_1)$$

Thus $g^{-t}(F) \in \gamma\mathcal{F}$.

On the other hand, if $g^{-t}(F) \in \gamma\mathcal{F}$, then for some $t_1, t_2 \in T$ and $F_1 \in \mathcal{F}$,

$$g^{-t_2}(g^{-t}(F)) \supset g^{-t_1}(F_1)$$

But

$$g^{-t_2}(g^{-t}(F)) = g^{-(t+t_2)}(F)$$

So $F \in \gamma\mathcal{F}$. Thus $\gamma\mathcal{F}$ is translation invariant by (2.39).

(b) $\mathcal{F}$ is $+$ invariant iff $k\mathcal{F}$ is $-$ invariant by Proposition 2.4b applied to $G = g^t$ for all $t \in T$. Applying this result to $\mathcal{F}$ and $k\mathcal{F}$, we see that $k\mathcal{F}$ is invariant iff $\mathcal{F}$ is. Since k reverses inclusions (2.44) follows [cf. (2.42)].

(c) This follows from (a) and (b) because k reverses inclusions.

(d) It is clear that $\tau\mathcal{F}$ is a family contained in $\mathcal{F}$ and inclusions are preserved. Clearly $\mathcal{F}$ is thick iff $\tau\mathcal{F} = \mathcal{F}$. Thus if $\mathcal{F}_1$ is thick and contained in $\mathcal{F}$, $\mathcal{F}_1 =$

$\tau\mathcal{F}_1 \subset \tau\mathcal{F}$. Since $\mathcal{F}$ is proper, it follows that $\emptyset \notin \tau\mathcal{F}$; $g^{-t}(S) = S$ implies $S \in \tau\mathcal{F}$, so $\tau\mathcal{F}$ is proper. Once we show that $\tau\mathcal{F}$ is itself thick, then $\tau\tau\mathcal{F} = \tau\mathcal{F}$ follows.

Given $F \in \tau\mathcal{F}$ and $\{t_1, \dots, t_l\} \subset T$, we must show that:

$$\tilde{F} = g^{-t_1}(F) \cap \cdots \cap g^{-t_l}(F) \in \tau\mathcal{F}$$

i.e., given $\{s_1, \dots, s_k\} \subset T$, we must show that:

$$g^{-s_1}(\tilde{F}) \cap \cdots \cap g^{-s_k}(\tilde{F}) \in \mathcal{F}$$

But this latter set is

$$\cap_{i=1,j=1}^{l \quad k} g^{-(t_i+s_j)}(F)$$

which is in $\mathcal{F}$ because $F \in \tau\mathcal{F}$.

(e) If $F_1 \in \mathcal{F}_1, F_2 \in \mathcal{F}_2$, and $\{t_1, \dots, t_l\} \subset T$, then

$$g^{-t_1}(F_1 \cap F_2) \cap \cdots \cap g^{-t_l}(F_1 \cap F_2) =$$
$$[g^{-t_1}(F_1) \cap \cdots \cap g^{-t_l}(F_1)] \cap [g^{-t_1}(F_2) \cap \cdots \cap g^{-t_2}(F_2)]$$

which is in $\mathcal{F}_1 \cdot \mathcal{F}_2$ if $\mathcal{F}_1$ and $\mathcal{F}_2$ are thick. Thus $(\tau\mathcal{F}_1) \cdot (\tau\mathcal{F}_2)$ is thick in any case, and it is contained in $\mathcal{F}_1 \cdot \mathcal{F}_2$. Then (2.45) follows from (d).

(f) Each property (2.37), (2.38), and (2.40) holds for $\cap_\alpha \mathcal{F}_\alpha$ and $\cup_\alpha \mathcal{F}_\alpha$ if it holds for all the $\mathcal{F}_\alpha$s. Since τ preserves inclusions:

$$\tau \cap_\alpha \mathcal{F}_\alpha \subset \cap_\alpha \tau\mathcal{F}_\alpha \subset \cap_\alpha \mathcal{F}_\alpha$$

But $\cap_\alpha \tau\mathcal{F}_\alpha$ is thick, so it is included in the largest thick subfamily of $\cap_\alpha \mathcal{F}_\alpha$, namely, $\tau \cap_\alpha \mathcal{F}_\alpha$. A similar argument shows that $\tilde{\gamma} \cap_\alpha \mathcal{F}_\alpha = \cap_\alpha \tilde{\gamma}\mathcal{F}_\alpha$, and a dual argument shows that $\gamma \cup_\alpha \mathcal{F}_\alpha = \cup_\alpha \gamma\mathcal{F}_\alpha$.

(g) $\tau\mathcal{F}$ is thick and hence $-$ invariant in any case. Now assume $\mathcal{F}$ is $+$ invariant. It suffices by Proposition 2.4b to show that $g^{-t}(F) \in \tau\mathcal{F}$ for $t \in T$ implies $F \in \tau\mathcal{F}$. Given $\{t_1, \dots, t_l\}$ we must show that $g^{-t_1}(F) \cap \cdots \cap g^{-t_l}(F) \in \mathcal{F}$. Because T is abelian:

$$g^{-t}[g^{-t_1}(F) \cap \cdots \cap g^{-t_l}(F)] = g^{-t_1}(g^{-t}(F)) \cap \cdots \cap g^{-t_l}(g^{-t}(F))$$

The latter set is in $\mathcal{F}$ because $g^{-t}(F) \in \tau\mathcal{F}$. By $+$ invariance of $\mathcal{F}$, we have $g^{-t_1}(F) \cap \cdots \cap g^{-t_l}(F) \in \mathcal{F}$ as required.

If instead $\mathcal{F}$ is thick, we show that $\gamma\mathcal{F}$ is thick. For $F \in \gamma\mathcal{F}$ and $\{t_1, \dots, t_l\}$, we show that:

$$g^{-t_1}(F) \cap \cdots \cap g^{-t_l}(F) \in \gamma\mathcal{F}$$

By definition of $\gamma\mathcal{F}$, there exist $s, s_1 \in T$ and $F_1 \in \mathcal{F}$ so that $g^{-s}(F) \supset g^{-s_1}(F_1)$. Since T is abelian:

$$g^{-s}(g^{-t_1}(F) \cap \cdots \cap g^{-t_l}(F)) = g^{-t_1}(g^{-s}(F)) \cap \cdots \cap g^{-t_l}(g^{-s}(F))$$
$$\supset g^{-(s_1+t_1)}(F_1) \cap \cdots \cap g^{-(s_1+t_l)}(F_1)$$

which is in the thick family $\mathcal{F}$. Hence $g^{-t_1}(F) \cap \cdots \cap g^{-t_l}(F) \in \gamma\mathcal{F}$.

In general we see that $\gamma\tau\mathcal{F}$ is thick and $\tau\gamma\mathcal{F}$ is invariant. Therefore (2.48) follows. ■

COROLLARY 2.1. *Let T be a monoid acting on S. The largest proper invariant family is the filterdual:*

$$\mathcal{B}_S = \tilde{\gamma}\mathcal{P}_+ = k\gamma k\mathcal{P}_+$$
$$= \{F : g^{-t}(F) \neq \emptyset \text{ for all } t \in T\} \tag{2.49}$$

The smallest proper invariant family is its dual, the filter:

$$k\mathcal{B}_S = \gamma k\mathcal{P}_+ = \{F : g^{-t}(F) = S \text{ for some } t \in T\}$$
$$= \{F : F \supset g^t(S) \text{ for some } t \in T\} \tag{2.50}$$

If $\mathcal{F}$ is any proper invariant family, then $\mathcal{F}$ is full with respect to $\mathcal{B}_S$. If $\mathcal{F}$ is any family full with respect to $\mathcal{B}_S$, then $\tau\mathcal{F}$ is full.

A proper family $\mathcal{F}$ is contained in $\mathcal{B}_S$ iff $\gamma\mathcal{F}$ is proper, in which case $\gamma\mathcal{F} = \gamma(\mathcal{F} \cdot k\mathcal{B}_S)$. A proper family $\mathcal{F}$ contains $k\mathcal{B}_S$ iff $\tilde{\gamma}\mathcal{F}$ is proper.

PROOF. Since $k\mathcal{P}_+ = \{S\}$, $F \in \gamma k\mathcal{P}_+$ iff for some $t, t_1 \in T$ $g^{-t}(F) \supset g^{-t_1}(S) = S$, or equivalently $F \supset g^t(S)$. If $F_1 \supset g^{t_1}(S)$ and $F_2 \supset g^{t_2}(S)$ then

$$\begin{array}{ccc} g^{t_1}(S) & \supset & g^{t_1}(g^{t_2}(S)) \\ & & \| \\ g^{t_2}(S) & \supset & g^{t_2}(g^{t_1}(S)) \end{array}$$

implies $F_1 \cap F_2 \supset g^{t_1+t_2}(S)$. Thus $\gamma k\mathcal{P}_+$ is a filter. A $-$ invariant filter is clearly thick, so $\gamma k\mathcal{P}_+ = k\mathcal{B}_S$ is thick.

F is in the dual $k\gamma k\mathcal{P}_+ = \tilde{\gamma}\mathcal{P}_+$ iff $S \backslash F \notin \gamma k\mathcal{P}_+$, i.e., if for all $t \in T$ $g^{-t}(S\backslash F) \neq S$, which is to say $g^{-t}(F) \neq \emptyset$.

Observe next that for $F_1, F_2 \in \mathcal{P}$:

$$F_1 \equiv F_2 \pmod{k\mathcal{B}_S} \Leftrightarrow F_1 \cap g^t(S) = F_2 \cap g^t(S) \quad \text{for some } t \in T$$
$$\Leftrightarrow g^{-t}(F_1) = g^{-t}(F_2) \quad \text{for some } t \in T \tag{2.51}$$

The first equivalence is clear from (2.50) and for the second note that $g^{-t}(F \cap g^t(S)) = g^{-t}(F)$ and $g^t(g^{-t}(F)) = F \cap g^t(S)$.

In particular if $F_1 \equiv F_2, F_2 \in \mathcal{F}$ and $\mathcal{F}$ is invariant, then $F_1 \in \mathcal{F}$ by (2.51) and invariance. Thus $\mathcal{F}$ is full by Proposition 2.2a.

If $\mathcal{F}$ is full, then by (2.45)

$$k\mathcal{B}_S \cdot \tau\mathcal{F} = \tau k\mathcal{B}_S \cdot \tau\mathcal{F} \subset \tau(k\mathcal{B}_S \cdot \mathcal{F}) = \tau\mathcal{F}$$

because $k\mathcal{B}_S$ is thick. Thus $\tau\mathcal{F}$ is full by Proposition 2.2.

If $\mathcal{F}$ is any proper family, then $k\mathcal{P}_+ \subset \mathcal{F} \subset \mathcal{P}_+$. If $\mathcal{F}$ is invariant, $\gamma k\mathcal{P}_+ \subset \mathcal{F} \subset \tilde{\gamma}\mathcal{P}_+$ by (a) and (c) of Proposition 2.5. For any $\mathcal{F}, \gamma\mathcal{F}$ proper implies $\mathcal{F} \subset \gamma\mathcal{F} \subset \mathcal{B}_S$. On the other hand if $\mathcal{F} \subset \mathcal{B}_S$, then $\gamma\mathcal{F} \subset \gamma\mathcal{B}_S = \mathcal{B}_S$, so $\gamma\mathcal{F}$ is proper. Since $\gamma\mathcal{F}$ is then full, it contains $\mathcal{F} \cdot k\mathcal{B}_S$, so it equals $\gamma(\mathcal{F} \cdot k\mathcal{B}_S)$. Similarly $k\mathcal{B}_S \subset \mathcal{F}$ iff $\tilde{\gamma}\mathcal{F}$ is proper. ∎

It is useful for applications to know when a filter is contained in an invariant filter.

PROPOSITION 2.6. *Assume that $\mathcal{F}$ is a filter and $\mathcal{F}_1$ is a proper, + invariant family. The following conditions are equivalent:*

(1) $\mathcal{F} \subset \tau\mathcal{F}_1$

(2) $(\gamma\mathcal{F}) \cdot (\gamma\mathcal{F}) \subset \tau\mathcal{F}_1$

(3) $\underbrace{(\gamma\mathcal{F}) \cdot (\gamma\mathcal{F}) \ldots \cdot (\gamma\mathcal{F})}_{n \text{ factors}} \subset \mathcal{F}_1$ *for* $n = 1, 2, \ldots$

(4) There exists an invariant filter $\tilde{\mathcal{F}}$ such that $\mathcal{F} \subset \tilde{\mathcal{F}} \subset \mathcal{F}_1$

For any family $\mathcal{F}$, we denote by $\gamma_\infty\mathcal{F}$ the smallest invariant family containing $\mathcal{F}$ and satisfying condition (2.1). It is a filter when it is proper.

PROOF. We show (1) $\Leftrightarrow$ (2), (3) and (1) $\Rightarrow$ (4) $\Rightarrow$ (3).

(1) $\Rightarrow$ (2) and (3). It suffices to prove

$$\mathcal{F} \subset \tau\mathcal{F}_1 \Rightarrow (\gamma\mathcal{F})^n \subset \tau\mathcal{F}_1 \quad n = 1, 2, \cdots \tag{2.52}$$

where $(\gamma\mathcal{F})^n$ is the n-fold product $(\gamma\mathcal{F}) \cdot (\gamma\mathcal{F}) \cdots$.

Let $F_1, \ldots, F_n \in \gamma\mathcal{F}$. There exist $t_i, \tilde{t}_i \in T$ and $\tilde{F}_i \in \mathcal{F}$ so that $g^{-t_i}(F_i) \supset g^{-\tilde{t}_i}(\tilde{F}_i)$ for $i = 1, 2, \ldots, n$. Let $t = t_1 + \ldots + t_n$ and $\tilde{F} = \tilde{F}_1 \cap \ldots \cap \tilde{F}_n$. Note that $\tilde{F} \in \mathcal{F}$ because $\mathcal{F}$ is a filter.

Let $s_i = t_1 + \ldots + t_{i-1} + t_{i+1} + \ldots + t_n$ so that $t = t_i + s_i$. Let $\tilde{s}_i = \tilde{t}_i + s_i$. Then:

$$g^{-t}(F_i) = g^{-s_i} g^{-t_i}(F_i) \supset g^{-\tilde{s}_i}(\tilde{F}_i) \supset g^{-\tilde{s}_i}(\tilde{F})$$

Consequently:

$$g^{-t}(F_1 \cap \ldots \cap F_n) \supset g^{-\tilde{s}_1}(\tilde{F}) \cap \ldots \cap g^{-\tilde{s}_n}(\tilde{F})$$

Since $\tilde{F} \in \mathcal{F} \subset \tau\mathcal{F}_1$ and $\tau\mathcal{F}_1$ is thick, the right-hand element is in $\tau\mathcal{F}_1$. Therefore $g^{-t}(F_1 \cap \cdots \cap F_n) \in \tau\mathcal{F}_1$ by heredity. But $\mathcal{F}_1$ is $+$ invariant, so by Proposition 2.5g, $\tau\mathcal{F}_1$ is invariant. Hence $F_1 \cap \cdots \cap F_n \in \tau\mathcal{F}_1$ as required.

(2) $\Rightarrow$ (1). $\mathcal{F} \subset \gamma\mathcal{F} \subset (\gamma\mathcal{F})\cdot(\gamma\mathcal{F}) \subset \tau\mathcal{F}_1$

(3) $\Rightarrow$ (1). If $F \in \mathcal{F}$ and $\{t_1,\ldots,t_n\} \subset T$, then $g^{-t_i}(F) \in \gamma\mathcal{F}$ for $i = 1,\ldots,n$; therefore:

$$g^{-t_1}(F)\cap\ldots\cap g^{-t_n}(F) \in (\gamma\mathcal{F})^n \subset \mathcal{F}_1$$

by (3). So $F \in \tau\mathcal{F}_1$ by (2.43).

(4) $\Rightarrow$ (3). $\gamma\mathcal{F} \subset \gamma\tilde{\mathcal{F}} = \tilde{\mathcal{F}}$ because $\tilde{\mathcal{F}}$ is invariant. So $(\gamma\mathcal{F})^n \subset (\tilde{\mathcal{F}})^n = \tilde{\mathcal{F}} \subset \mathcal{F}_1$ because $\tilde{\mathcal{F}}$ is a filter.

(1) $\Rightarrow$ (4). With $\tilde{\mathcal{F}}_0 = \mathcal{F}$ we inductively define a filter $\tilde{\mathcal{F}}_k$ so that $\gamma\tilde{\mathcal{F}}_{k-1} \subset \tilde{\mathcal{F}}_k \subset \tau\mathcal{F}_1$. Let

$$\tilde{\mathcal{F}}_{k+1} = \bigcup_{n=1}^{\infty}(\gamma\tilde{\mathcal{F}}_k)^n \tag{2.53}$$

By induction hypothesis, $\tilde{\mathcal{F}}_k$ is a filter contained in $\tau\mathcal{F}_1$, therefore (2.52) implies that $\tilde{\mathcal{F}}_{k+1} \subset \tau\mathcal{F}_1$. Clearly $\gamma\tilde{\mathcal{F}}_k \subset \tilde{\mathcal{F}}_{k+1}$. Because $(\gamma\tilde{\mathcal{F}}_k)^n \cdot (\gamma\tilde{\mathcal{F}}_k)^m = (\gamma\tilde{\mathcal{F}}_k)^{n+m}$, it is clear that $\tilde{\mathcal{F}}_{k+1}$ is a filter. Therefore $\{\tilde{\mathcal{F}}_k\}$ is an increasing sequence of filters containing $\mathcal{F}$ and contained in $\tau\mathcal{F}_1$. Let $\tilde{\mathcal{F}} = \cup_k\tilde{\mathcal{F}}_k$. Because the union is increasing, $\tilde{\mathcal{F}}$ is a filter. Clearly $\mathcal{F} \subset \tilde{\mathcal{F}} \subset \tau\mathcal{F}_1$. If $F \in \gamma\tilde{\mathcal{F}}$, then $F \in \gamma\tilde{\mathcal{F}}_k$ for some k, so $F \in \tilde{\mathcal{F}}_{k+1} \subset \tilde{\mathcal{F}}$. Thus $\tilde{\mathcal{F}}$ is invariant. ■

By induction any invariant filter containing $\mathcal{F}$ contains $\tilde{\mathcal{F}}_k$ and so contains the union. For any family $\mathcal{F}$, denote by $\gamma_\infty\mathcal{F}$ the invariant family $\cup_k\tilde{\mathcal{F}}_k$ constructed inductively by (2.53). The family $\gamma_\infty\mathcal{F}$ is proper iff $\mathcal{F}$ is contained in some invariant filter, in which case it is the smallest invariant filter containing $\mathcal{F}$.

PROPOSITION 2.7. *Assume $\mathcal{F}$ is a filter.*

a. $\mathcal{F}$ is $-$ invariant iff $\mathcal{F}$ is thick. $\gamma\mathcal{F}$ is a filter iff $\gamma\mathcal{F}$ is thick, i.e., iff $\tau\gamma\mathcal{F} = \gamma\mathcal{F}$.

b. $\tau\mathcal{F}$ and $\tau k\tau k\tau\mathcal{F}$ are filters with $\tau\mathcal{F} \subset \tau k\tau k\tau\mathcal{F}$. Also:

$$\tau k\tau k\tau k\tau\mathcal{F} = \tau k\tau\mathcal{F} \tag{2.54}$$

PROOF. (a) Recall that thick implies $-$ invariance in any case. Conversely $F \in \mathcal{F}$ and $\{t_1,\ldots,t_l\} \subset T$ yield $g^{-t_i}(F) \in \mathcal{F}, i = 1,\ldots l$ when $\mathcal{F}$ is $-$ invariant. If $\mathcal{F}$ is a filter, then the intersection is also in $\mathcal{F}$. Therefore $\mathcal{F}$ is thick. Thus if $\gamma\mathcal{F}$ is a filter, it is thick because it is invariant. On the other hand, if $\gamma\mathcal{F}$ is thick then (1) of Proposition 2.6 holds with $\mathcal{F}_1 = \gamma\mathcal{F}$. By (2) of Proposition 2.6 $(\gamma\mathcal{F})\cdot(\gamma\mathcal{F}) \subset \mathcal{F}_1 = \gamma\mathcal{F}$, so $\gamma\mathcal{F}$ is a filter.

(b) Given $\mathcal{F} \cdot \mathcal{F} = \mathcal{F}$, (2.45) implies

$$(\tau\mathcal{F}) \cdot (\tau\mathcal{F}) \subset \tau(\mathcal{F} \cdot \mathcal{F}) = \tau\mathcal{F} \tag{2.55}$$

Therefore $\tau\mathcal{F}$ is a filter because it is proper.

We now alternately apply the inclusions from (2.9) and (2.45) beginning with (2.55):

$$\begin{aligned}
&(\tau\mathcal{F}) \cdot (\tau\mathcal{F}) \subset \tau\mathcal{F} \Rightarrow \\
&(\tau\mathcal{F}) \cdot (k\tau\mathcal{F}) \subset k\tau\mathcal{F} \Rightarrow \\
&(\tau\mathcal{F}) \cdot (\tau k\tau\mathcal{F}) \subset \tau k\tau\mathcal{F} \Rightarrow \\
&(k\tau k\tau\mathcal{F}) \cdot (\tau k\tau\mathcal{F}) \subset k\tau\mathcal{F} \Rightarrow \\
&(\tau k\tau k\tau\mathcal{F}) \cdot (\tau k\tau\mathcal{F}) \subset \tau k\tau\mathcal{F} \Rightarrow \\
&(\tau k\tau k\tau\mathcal{F}) \cdot (k\tau k\tau\mathcal{F}) \subset k\tau k\tau\mathcal{F} \Rightarrow \\
&(\tau k\tau k\tau\mathcal{F}) \cdot (\tau k\tau k\tau\mathcal{F}) \subset \tau k\tau k\tau\mathcal{F}
\end{aligned} \tag{2.56}$$

The last inclusion says that $\tau k\tau k\tau\mathcal{F}$ is a filter, since it is proper.

Now $\tau k\mathcal{F} \subset k\mathcal{F}$, so $k\tau k\mathcal{F} \supset \mathcal{F}$. Hence $\tau k\tau k\mathcal{F} \supset \tau\mathcal{F}$. Applying this to $\tau\mathcal{F}$ yield $\tau k\tau k(\tau\mathcal{F}) \supset \tau(\tau\mathcal{F}) = \tau\mathcal{F}$. Now τk reverses inclusions, so that $\tau k(\tau k\tau k\tau\mathcal{F}) \subset \tau k(\tau\mathcal{F})$. On the other hand, we just showed that

$$\tau k\tau k\tau(\tau k\tau\mathcal{F}) \supset \tau(\tau k\tau\mathcal{F}) = \tau k\tau\mathcal{F}$$

Taking these two inclusions together yields (2.54) ■

COROLLARY 2.2. *Assume $\mathcal{F}$ is a filter. $\mathcal{F}$ is contained in some invariant filter, i.e., $\gamma_\infty\mathcal{F}$ is proper iff $\mathcal{F} \subset \tau\mathcal{B}_S$. A filterdual contains some invariant filterdual iff it contains $k\tau\mathcal{B}_S$.*

PROOF. If $\mathcal{F} \subset \tilde{\mathcal{F}}$ and $\tilde{\mathcal{F}}$ is invariant and proper, then by Corollary 2.1 $\tilde{\mathcal{F}} \subset \mathcal{B}_S$. If $\tilde{\mathcal{F}}$ is also a filter, then by Proposition 2.7a, $\tilde{\mathcal{F}}$ is thick. Then $\mathcal{F} \subset \tilde{\mathcal{F}} = \tau\tilde{\mathcal{F}} \subset \tau\mathcal{B}_S$. On the other hand, $\mathcal{B}_S$ is invariant. Then $\mathcal{F} \subset \tau\mathcal{B}_S$ implies $\mathcal{F}$ is contained in some invariant filter by Proposition 2.6. The filterdual result follows from duality and Proposition 2.5b. ■

We can sharpen some of these results when T acts injectively on S; i.e., each g^t is an injective map.

PROPOSITION 2.8. *Assume that the monoid T acts injectively on S.*

If $\mathcal{F}_1$ and $\mathcal{F}_2$ are $+$ invariant (or $-$ invariant) families, then $\mathcal{F}_1 \cdot \mathcal{F}_2$ is $+$ invariant (resp. $-$ invariant). In general for families $\mathcal{F}_1$ and $\mathcal{F}_2$:

$$\begin{aligned}
(\gamma\mathcal{F}_1) \cdot (\gamma\mathcal{F}_2) &= \gamma((\gamma\mathcal{F}_1) \cdot (\gamma\mathcal{F}_2)) \supset \gamma(\mathcal{F}_1 \cdot \mathcal{F}_2) \\
(\tilde{\gamma}\mathcal{F}_1) \cdot (\tilde{\gamma}\mathcal{F}_2) &= \tilde{\gamma}((\tilde{\gamma}\mathcal{F}_1) \cdot (\tilde{\gamma}\mathcal{F}_2)) \subset \tilde{\gamma}(\mathcal{F}_1 \cdot \mathcal{F}_2)
\end{aligned} \tag{2.57}$$

PROOF. For $t \in T$, (2.31) and (2.32) imply that $g^{-t}(\mathcal{F}_1 \cdot \mathcal{F}_2) = (g^{-t}\mathcal{F}_1) \cdot (g^{-t}\mathcal{F}_2)$ and because g^t is injective that $g^t(\mathcal{F}_1 \cdot \mathcal{F}_2) = (g^t\mathcal{F}_1) \cdot (g^t\mathcal{F}_2)$. The $+$ and $-$ invariance results follow: $g^{\pm t}\mathcal{F}_\alpha \subset \mathcal{F}_\alpha$ for $\alpha = 1,2$ and $t \in T$ imply $g^{\pm t}(\mathcal{F}_1 \cdot \mathcal{F}_2) \subset \mathcal{F}_1 \cdot \mathcal{F}_2$ for all $t \in T$. The equations in (2.57) are now clear. The inclusions follow by monotonicity. ■

PROPOSITION 2.9. *Assume T acts injectively on S and $\mathcal{F}$ is a filter that is invariant and consequently thick. $\tau k \tau k \mathcal{F}$ is an invariant filter containing $\mathcal{F}$ and satisfying:*

$$\tau k \tau k \mathcal{F} \cdot \tau k \mathcal{F} = \tau k \mathcal{F} \tag{2.58}$$

$$\tau k \tau k \mathcal{F} = k(\tau k \mathcal{F} \cdot k \tau k \mathcal{F}) \tag{2.59}$$

PROOF. By Proposition 2.7a $\mathcal{F}$ is thick, so $\tau\mathcal{F} = \mathcal{F}$. Then $\tau k \tau k \mathcal{F} = \tau k \tau k \tau \mathcal{F}$ is a filter containing $\tau\mathcal{F} = \mathcal{F}$ by Proposition 2.7b. (2.58) follows from (2.56). By Proposition 2.5b,g the operators k and τ preserve invariance. Thus $\tau k \tau k \mathcal{F}$ is invariant because $\mathcal{F}$ is.

Proposition 2.1e implies that $\tilde{\mathcal{F}} = k(\tau k \mathcal{F} \cdot k \tau k \mathcal{F})$ is a filter and that (2.58) implies

$$\tau k \tau k \mathcal{F} \subset \tilde{\mathcal{F}} \subset k \tau k \mathcal{F}$$

By Proposition 2.8 $\tilde{\mathcal{F}}$ is an invariant filter, so it is thick. Hence $\tilde{\mathcal{F}} \subset \tau k \tau k \mathcal{F}$, proving (2.59). ■

PROPOSITION 2.10. *Assume T acts injectively on S.*

If $\mathcal{F}$ is a filter containing $k\mathcal{B}_S$, then $\tilde{\gamma}\mathcal{F}$ is an invariant filter contained in $\mathcal{F}$.

If $\mathcal{F}$ is a filterdual contained in $\mathcal{B}_S$, then $\gamma\mathcal{F}$ is an invariant filterdual containing $\mathcal{F}$.

PROOF. By Corollary 2.1, $\tilde{\gamma}\mathcal{F}$ is proper, i.e., nonempty, iff $\mathcal{F} \supset k\mathcal{B}_S$ in which case

$$(\tilde{\gamma}\mathcal{F}) \cdot (\tilde{\gamma}\mathcal{F}) \subset \tilde{\gamma}\mathcal{F}$$

follows from $\mathcal{F} \cdot \mathcal{F} \subset \mathcal{F}$ and (2.57). Thus $\tilde{\gamma}\mathcal{F}$ is a filter. The second part follows by applying the first to $k\mathcal{F}$. ■

When the action is not injective this result need not hold. Consider the map on the countable set diagramed below:

$$\begin{array}{ll} \ldots x_{-3} \longrightarrow x_{-2} \longrightarrow x_{-1} & \\ \qquad\qquad\qquad\qquad\qquad \searrow & \\ & x_0 \longrightarrow x_1 \longrightarrow x_2 \longrightarrow \cdots \\ \qquad\qquad\qquad\qquad\qquad \nearrow & \\ \ldots \tilde{x}_{-3} \longrightarrow \tilde{x}_{-2} \longrightarrow \tilde{x}_{-1} & \end{array} \tag{2.60}$$

Let $\mathcal{F}$ be the ultrafilter $[x_0]$, all subsets containing the point x_0. Then $F \in \gamma\mathcal{F}$ iff $x_n \in F$ for some $n \geq 0$ or $x_n, \tilde{x}_n \in F$ for some $n < 0$. Therefore $\{x_{-1}\} \cup \{\tilde{x}_{-1}\} \in \gamma\mathcal{F}$, but neither $\{x_{-1}\}$ nor $\{\tilde{x}_{-1}\} \in \gamma\mathcal{F}$. Since the Ramsey Property (2.15) fails, $\gamma\mathcal{F}$ is not a filterdual. On the other hand, $F \in \tilde{\gamma}\mathcal{F}$ iff $x_n \in F$ for all $n \geq 0$ and $\{x_n, \tilde{x}_n\} \cap F \neq \emptyset$ for all $n < 0$. Thus $\tilde{\gamma}\mathcal{F}$ is not a filter.

In the injective case, congruence of subsets mod $k\mathcal{B}_S$ (see (2.51)) becomes an especially useful tool. For $F \in \mathcal{P}$ let $[F]_{k\mathcal{B}_S}$ denote the congruence class of F (mod $k\mathcal{B}_S$), i.e.,

$$\begin{aligned} [F]_{k\mathcal{B}_S} &= \{F_1 : F \cap g^t(S) = F_1 \cap g^t(S) \text{ for some } t \in T\} \\ &= \{F_1 : g^{-t}(F) = g^{-t}(F_1) \text{ for some } t \in T\} \end{aligned} \tag{2.61}$$

Now assume that the monoid T satisfies cancellation, so it can be injected canonically into the group of quotients G_T consisting of equivalence classes $[t/s]$ in $T \times T$.

PROPOSITION 2.11. *Assume the monoid T satisfies cancellation and acts injectively on S. Then T acts invertibly on the congruence classes mod $k\mathcal{B}_S$, so the action on congruence classes can be extended to a G_T action well-defined by:*

$$g^{[t/s]}[F]_{k\mathcal{B}_S} = [g^t(g^{-s}(F))]_{k\mathcal{B}_S} = [g^{-s}(g^t(F))]_{k\mathcal{B}_S} \tag{2.62}$$

For any family $\mathcal{F}$ for S:

$$\gamma\mathcal{F} = \{F : g^{[t/s]}[F]_{k\mathcal{B}_S} \cap \mathcal{F} \neq \emptyset \text{ for some } t, s \in T\} \tag{2.63}$$

$$\tilde{\gamma}\mathcal{F} = \{F : g^{[t/s]}[F]_{k\mathcal{B}_S} \subset \mathcal{F} \text{ for all } t, s \in T\} \tag{2.64}$$

$$\gamma_\infty\mathcal{F} = \bigcup_{n=1}^{\infty} (\gamma\mathcal{F})^n \tag{2.65}$$

PROOF. Observe first that (2.34) and (2.35) imply for all $t \in T$ and $F \in \mathcal{P}$:

$$F = g^{-t}(g^t(F)) \supset g^t(g^{-t}(F)) = F \cap g^t(S) \equiv F \ (\text{mod } k\mathcal{B}_S) \tag{2.66}$$

Note next that if $F_1 \cap g^{s_1}(S) = F_2 \cap g^{s_2}(S)$, then $g^t(F_1) \cap g^{t+s_1}(S) = g^t(F_2) \cap g^{t+s_2}(S)$ since, g^t is injective. If $g^{-s_1}(F_1) = g^{-s_2}(F_2)$, then

$$g^{-s_1}(g^{-t}(F_1)) = g^{-t}(g^{-s_1}(F_1)) = g^{-t}(g^{-s_2}(F_2)) = g^{-s_2}(g^{-t}(F_2))$$

Thus $(t, [F]_{k\mathcal{B}_S}) \mapsto [g^t(F)]_{k\mathcal{B}_S}$, and $[g^{-t}(F)]_{k\mathcal{B}_S}$ are well-defined actions of T on the class of mod $k\mathcal{B}_S$ congruence classes. By (2.66) these are inverse actions, so they extend to the group G_T of quotients by:

$$([t/s], [F]_{k\mathcal{B}_S}) \mapsto [g^t(g^{-s}(F))]_{k\mathcal{B}_S} = [g^{-s}(g^t(F))]_{k\mathcal{B}_S}$$

Observe that the right-hand side of (2.63) is clearly an invariant family that contains $\mathcal{F}$. Denoting it temporarily by $\gamma^*\mathcal{F}$, we see that $\gamma\mathcal{F} \subset \gamma^*\mathcal{F}$ because $\gamma\mathcal{F}$ is the smallest invariant family containing $\mathcal{F}$. But if $F \in \gamma^*\mathcal{F}$, then for some $F_1 \in \mathcal{F}$, and $t,s \in T$, $F \equiv g^s(g^{-t}(F_1))$. Then $g^s(g^{-t}(F_1)) \in \gamma\mathcal{F}$ by invariance. By Corollary 2.1 $\gamma\mathcal{F}$ is full. Therefore $F \in \gamma\mathcal{F}$; hence $\gamma^*\mathcal{F} \subset \gamma\mathcal{F}$.

Similarly if $\tilde{\gamma}^*\mathcal{F}$ denotes the right-hand side of (2.64), it is an invariant family contained in $\mathcal{F}$, so $\tilde{\gamma}^*\mathcal{F} \subset \tilde{\gamma}\mathcal{F}$. But if $F \in \tilde{\gamma}\mathcal{F}$, then $g^t(g^{-s}(F)) \in \tilde{\gamma}\mathcal{F}$ by invariance. By Corollary 2.1 $\tilde{\gamma}\mathcal{F}$ is full. Therefore $[g^t(g^{-s}(F))]_{k\mathcal{B}_S} \subset \tilde{\gamma}\mathcal{F} \subset \mathcal{F}$ for all $t,s \in T$; hence $F \in \tilde{\gamma}^*\mathcal{F}$.

Finally let $\gamma_\infty^*\mathcal{F}$ denote the right-hand side of (2.65). Clearly $\mathcal{F} \subset \gamma\mathcal{F} \subset \gamma_\infty^*\mathcal{F} \subset \gamma_\infty\mathcal{F}$. By Proposition 2.8, $\gamma_\infty^*\mathcal{F}$ is invariant. If $\gamma_\infty^*\mathcal{F}$ is $\mathcal{P}$, then so is $\gamma_\infty\mathcal{F}$. Otherwise $\gamma_\infty^*\mathcal{F}$ is an invariant filter containing $\mathcal{F}$, so it contains $\gamma_\infty\mathcal{F}$. Thus the inductive construction of (2.53) stabilizes immediately in the injective case. ■

PROPOSITION 2.12. *Assume T satisfies cancellation and acts injectively on S.*

a. If $\mathcal{F}$ is an invariant family, then for any family $\mathcal{F}_1$:

$$(\gamma\mathcal{F}_1)\cdot\mathcal{F} = \gamma(\mathcal{F}_1\cdot\mathcal{F})$$
$$(\tilde{\gamma}\mathcal{F}_1)\cdot\mathcal{F} = \tilde{\gamma}(\mathcal{F}_1\cdot\mathcal{F}) \tag{2.67}$$

In particular for the full extension of $\mathcal{F}_1$:

$$\gamma\mathcal{F}_1 = \gamma(\mathcal{F}_1\cdot k\mathcal{B}_S)$$
$$\tilde{\gamma}\mathcal{F}_1 = \tilde{\gamma}(\mathcal{F}_1\cdot k\mathcal{B}_S) \tag{2.68}$$

b. On the class of full families, T acts reversibly, so the action can be extended to a G_T action well-defined by:

$$g^{[t/s]}\mathcal{F} = g^t g^{-s}\mathcal{F} = g^{-s}g^t\mathcal{F} \tag{2.69}$$

The action preserves arbitrary unions and intersections, and for a full family $\mathcal{F}$ and $t,s \in T$:

$$g^{[t/s]}(k\mathcal{F}) = kg^{[t/s]}\mathcal{F} \tag{2.70}$$

For full families $\mathcal{F}_1$ and $\mathcal{F}_2$ and $t,s \in T$:

$$g^{[t/s]}(\mathcal{F}_1\cdot\mathcal{F}_2) = (g^{[t/s]}\mathcal{F}_1)\cdot(g^{[t/s]}\mathcal{F}_2) \tag{2.71}$$

For any full family $\mathcal{F}$:

$$\gamma\mathcal{F} = \bigcup_{t,s} g^{[t/s]}\mathcal{F} \tag{2.72}$$

$$\tilde{\gamma}\mathcal{F} = \bigcap_{t,s} g^{[t/s]}\mathcal{F} \tag{2.73}$$

c. If $\mathcal{F}$ is a full, thick family, then $\tilde{\gamma}\mathcal{F}$ is thick. In general for any full family $\mathcal{F}$, we have:

$$g^{[t/s]}\tau\mathcal{F} = \tau g^{[t/s]}\mathcal{F} \qquad t,s \in T \tag{2.74}$$

and

$$\tilde{\gamma}\tau\mathcal{F} = \tau\tilde{\gamma}\mathcal{F} \tag{2.75}$$

PROOF. (a) By (2.57), $(\gamma\mathcal{F}_1)\cdot\mathcal{F} = (\gamma\mathcal{F}_1)\cdot(\gamma\mathcal{F}) \supset \gamma(\mathcal{F}_1\cdot\mathcal{F})$. On the other hand, if $F \in \gamma\mathcal{F}_1$, then by (2.63) there exist $t_1, t_2 \in T$ and $F_1 \in \mathcal{F}_1$ such that $g^{-t_1}(F_1) \equiv g^{-t_2}(F)$. Now if $\tilde{F} \in \mathcal{F}$, then $F \cap \tilde{F}$ satisfies

$$\begin{aligned} g^{-t_2}(F\cap\tilde{F}) &= g^{-t_2}(F)\cap g^{-t_2}(\tilde{F}) \equiv g^{-t_1}(F_1)\cap g^{-t_1}(g^{t_1}g^{-t_2}(\tilde{F})) \\ &= g^{-t_1}(F_1\cap g^{t_1}g^{-t_2}(\tilde{F})) \end{aligned}$$

Because $\mathcal{F}$ is invariant, $g^{t_1}g^{-t_2}(\tilde{F}) \in \mathcal{F}$, so $F\cap\tilde{F} \in \gamma(\mathcal{F}_1\cdot\mathcal{F})$ by (2.63). The equation for $\tilde{\gamma}$ is proved similarly using (2.64). In particular with $\mathcal{F} = k\mathcal{B}_S$ (2.67) implies (2.68), because by Corollary 2.1 an invariant family is full. Thus $(\gamma\mathcal{F}_1)\cdot k\mathcal{B}_S = \gamma\mathcal{F}_1$ and similarly for $\tilde{\gamma}\mathcal{F}_1$.

(b) By Proposition 2.2a, a family is full when it is saturated by the congruence mod $k\mathcal{B}_S$ equivalence relation. Thus (2.69) follows from (2.62). In particular for a full family $\mathcal{F}$ we have:

$$g^t g^{-t}\mathcal{F} = g^{-t}g^t\mathcal{F} = \mathcal{F} \qquad t \in T \tag{2.76}$$

Observe now that (2.71) follows from (2.31) and (2.32) (each g^t is injective). Applied with $\mathcal{F}_1 = \mathcal{F}$ and $\mathcal{F}_2 = k\mathcal{F}$, it follows that:

$$g^{[t/s]}(k\mathcal{F}) \subset kg^{[t/s]}(\mathcal{F})$$

Replacing $\mathcal{F}$ by $g^{[s/t]}(\mathcal{F})$ and applying $g^{[s/t]}$ to the inclusion, we obtain the reverse (with s and t interchanged). Thus (2.70) follows.

Equation (2.72) follows from (2.63) and (2.73) from (2.64) once we note that for a full family $\mathcal{F}$, the congruence class $g^{[t/s]}[F]_{k\mathcal{B}_S}$ meets $\mathcal{F}$ iff it is contained in $\mathcal{F}$.

(c) A full family $\mathcal{F}$ is thick iff $t_1,\dots,t_l \in T$ and $F \in \mathcal{F}$ imply

$$\cap_{i=1}^{l} g^{[0/t_i]}[F]_{k\mathcal{B}_S} \in \mathcal{F}$$

(2.74) follows because the G_T action preserves intersection. (2.75) follows from (2.73) and (2.46). From (2.75) we see that $\tau\mathcal{F} = \mathcal{F}$ implies $\tau\tilde{\gamma}\mathcal{F} = \tilde{\gamma}\mathcal{F}$ when $\mathcal{F}$ is full. ■

Suppose now that S is a uniform space. We say that a family $\mathcal{F}$ is *open* when:

$$F \in \mathcal{F} \Rightarrow F \supset V(F_1) \text{ for some } F_1 \in \mathcal{F} \text{ and } V \in \mathcal{U}_S \tag{2.77}$$

For any family $\mathcal{F}$, the *interior* is

$$u\mathcal{F} = \{F \in \mathcal{P} : F \supset V(F_1) \text{ for some } F_1 \in \mathcal{F} \text{ and } V \in \mathcal{U}_S\} \tag{2.78}$$

That is, $u\mathcal{F}$ is generated by $\{V(F_1) : F_1 \in \mathcal{F} \text{ and } V \in \mathcal{U}_S\}$. The *closure* of $\mathcal{F}$ is

$$\tilde{u}\mathcal{F} = \{F \in \mathcal{P} : V(F) \in \mathcal{F} \text{ for all } V \in \mathcal{U}_S\} \tag{2.79}$$

We call a family $\mathcal{F}$ *closed* if $\tilde{u}\mathcal{F} = \mathcal{F}$, i.e., if $V(F) \in \mathcal{F}$ for all $V \in \mathcal{U}_S$ implies $F \in \mathcal{F}$.

Notice that if S is *discrete* so that the diagonal $1_S \in \mathcal{U}_S$ then $1_S(F) = F$ implies that every family is open and closed and $u\mathcal{F} = \mathcal{F} = \tilde{u}\mathcal{F}$ for every family $\mathcal{F}$.

PROPOSITION 2.13. *Assume S is a uniform space. Let $\mathcal{F}$, $\mathcal{F}_1$, $\mathcal{F}_2$, etc., be proper families for S.*

a. The family $u\mathcal{F}$ is a proper family and it is the largest open family contained in $\mathcal{F}$. That is, $u\mathcal{F} \subset \mathcal{F}$, $u\mathcal{F}$ is open, and if $\mathcal{F}_1$ is an open family with $\mathcal{F}_1 \subset \mathcal{F}$, then $\mathcal{F}_1 \subset u\mathcal{F}$. Thus $\mathcal{F}$ is open iff $u\mathcal{F} = \mathcal{F}$. In particular $uu\mathcal{F} = u\mathcal{F}$.

b. The following identities hold:

$$\tilde{u}k\mathcal{F} = ku\mathcal{F} = \{F : V(F) \in k\mathcal{F} \text{ for all } V \in \mathcal{U}_S\} \tag{2.80}$$

$$k\tilde{u}k\mathcal{F} = u\mathcal{F} \subset \mathcal{F} \subset \tilde{u}\mathcal{F} = kuk\mathcal{F} \tag{2.81}$$

$$u\tilde{u}\mathcal{F} = u\mathcal{F} = uu\mathcal{F} \quad \text{and} \quad \tilde{u}u\mathcal{F} = \tilde{u}\mathcal{F} = \tilde{u}\tilde{u}\mathcal{F} \tag{2.82}$$

c. The operators u and $\tilde{u}$ preserve inclusions and:

$$u\mathcal{F}_1 \subset \mathcal{F}_2 \Leftrightarrow \mathcal{F}_1 \subset \tilde{u}\mathcal{F}_2 \tag{2.83}$$

$$(u\mathcal{F}_1) \cdot (u\mathcal{F}_2) = u(\mathcal{F}_1 \cdot (u\mathcal{F}_2)) \tag{2.84}$$

$$u(\mathcal{F}_1 \cap \mathcal{F}_2) = (u\mathcal{F}_1) \cap (u\mathcal{F}_2) \tag{2.85}$$

$$\tilde{u}(\mathcal{F}_1 \cup \mathcal{F}_2) = (\tilde{u}\mathcal{F}_1) \cup (\tilde{u}\mathcal{F}_2) \tag{2.86}$$

If $\mathcal{F}$ is a filter, then $u\mathcal{F}$ is a filter. If $\mathcal{F}$ is a filterdual, then $\tilde{u}\mathcal{F}$ is a filterdual.

d. For any collection of families $\{\mathcal{F}_\alpha\}$:

$$u(\cup_\alpha \mathcal{F}_\alpha) = \cup_\alpha u\mathcal{F}_\alpha \qquad \tilde{u}(\cap_\alpha \mathcal{F}_\alpha) = \cap_\alpha \tilde{u}\mathcal{F}_\alpha \tag{2.87}$$

$$u(\cap_\alpha u\mathcal{F}_\alpha) = u(\cap_\alpha \mathcal{F}_\alpha) \qquad \tilde{u}(\cup_\alpha \tilde{u}\mathcal{F}_\alpha) = \tilde{u}(\cup_\alpha \mathcal{F}_\alpha) \tag{2.88}$$

e. If $\tilde{\mathcal{F}}$ is a proper family for a uniform space $\tilde{S}$, then with the product uniformity on $S \times \tilde{S}$:

$$u(\mathcal{F} \times \tilde{\mathcal{F}}) = (u\mathcal{F}) \times (u\tilde{\mathcal{F}}) \tag{2.89}$$

f. Assume $g : S \to \tilde{S}$ is a uniformly continuous map:

$$\tilde{u}g\tilde{u}\mathcal{F} = \tilde{u}g\mathcal{F} = \tilde{u}gu\mathcal{F}$$
$$ugu\mathcal{F} = ug\mathcal{F} = ug\tilde{u}\mathcal{F} \tag{2.90}$$
$$[g(S)] \cdot ug\mathcal{F} \subset gu\mathcal{F} \tag{2.91}$$

with equality in (2.91) if g is also a uniformly open map onto its range $g(S)$.
If $\tilde{\mathcal{F}}$ is an open family for $\tilde{S}$, then $g^{-1}\tilde{\mathcal{F}}$ is open. For any family $\tilde{\mathcal{F}}$ for $\tilde{S}$:

$$g^{-1}u\tilde{\mathcal{F}} = ug^{-1}u\tilde{\mathcal{F}} \subset ug^{-1}\tilde{\mathcal{F}} \tag{2.92}$$

with equality in (2.92) if g is a uniformly open surjection to $\tilde{S}$.

PROOF. (a) $\emptyset \notin \tilde{u}\mathcal{F}$, $S \in u\mathcal{F}$, and $u\mathcal{F} \subset \mathcal{F} \subset \tilde{u}\mathcal{F}$ imply that $u\mathcal{F}$ and $\tilde{u}\mathcal{F}$ are proper. It is obvious that $\mathcal{F}_1 \subset \mathcal{F}$ and $\mathcal{F}_1$ open imply $\mathcal{F}_1 \subset u\mathcal{F}$. If $F \in u\mathcal{F}$, then $F \supset V(F_1)$ with $F_1 \in \mathcal{F}$, $V \in \mathcal{U}_S$. Choose $W \in \mathcal{U}_S$ with $W \circ W \subset V$ and let $F_2 = W(F_1)$. $F_2 \in u\mathcal{F}$ and $F \supset W(F_2)$. Thus $u\mathcal{F}$ is open. Since u clearly preserves inclusions, the rest of (a) is clear.

(b) Recall that the symmetric elements $V \in \mathcal{U}_S$ with $V = V^{-1}$ generate $\mathcal{U}_S$. By (2.23) applied to $G = V = V^{-1}$, $V(F) \cap F_1 \neq \emptyset$ iff $F \cap V(F_1) \neq \emptyset$. Hence $F \in ku\mathcal{F}$ iff $S \backslash F \notin u\mathcal{F}$ iff for all $F_1 \in \mathcal{F}$ and symmetric $V \in \mathcal{U}_S$, $V(F_1) \cap F \neq \emptyset$ iff for all $F_1 \in \mathcal{F}$ and symmetric $V \in \mathcal{U}_S$, $F_1 \cap V(F) \neq \emptyset$ iff for all $V \in \mathcal{U}_S$, $V(F) \in k\mathcal{F}$. This proves (2.80). Consequently:

$$k\tilde{u}k\mathcal{F} = kku\mathcal{F} = u\mathcal{F} \qquad \tilde{u}kk\mathcal{F} = kuk\mathcal{F}$$

proving (2.81).

Observe next that $u\mathcal{F} \subset u\tilde{u}\mathcal{F}$. If $F \in u\tilde{u}\mathcal{F}$, then there exists $F_1 \in \tilde{u}\mathcal{F}$ and $V \in \mathcal{U}_S$ such that $F \supset V(F_1)$. Again choose $W \in \mathcal{U}_S$ with $W^2 \subset V$. Since $F \supset W(F_2)$ with $F_2 = W(F_1) \in \mathcal{F}$, $F \in u\mathcal{F}$. This proves the first equation of (2.82). The rest follow by duality; i.e., apply k fore and aft then use (2.81).

(c) Preserving inclusions is clear. If $u\mathcal{F}_1 \subset \mathcal{F}_2$ then $\mathcal{F}_1 \subset \tilde{u}\mathcal{F}_1 = \tilde{u}u\mathcal{F}_1 \subset \tilde{u}\mathcal{F}_2$. Similarly $\mathcal{F}_1 \subset \tilde{u}\mathcal{F}_2$ implies

$$u\mathcal{F}_1 \subset u\tilde{u}\mathcal{F}_2 = u\mathcal{F}_2 \subset \mathcal{F}_2$$

proving (2.83).

Now if $F_1, F_2 \in \mathcal{P}$, and V_1, V_2, W are symmetric in $\mathcal{U}_S$ with $V_1 \circ V_2 \subset W$, we have

$$V_1(F_1) \cap V_2(F_2) \subset V_1(F_1 \cap W(F_2)) \tag{2.93}$$

For if $x \in V_1(F_1) \cap V_2(F_2)$, then there exists $x_1 \in F_1$ with $x \in V_1(x_1)$. Notice that $x_1 \in V_1 \circ V_2(F_2) \subset W(F_2)$, so $x_1 \in F_1 \cap W(F_2)$. Hence x is in the right-hand side expression. This inclusion shows that $u(\mathcal{F}_1 \cdot (u\mathcal{F}_2)) \subset (u\mathcal{F}_1) \cdot (u\mathcal{F}_2)$. Reversing the argument, we have

$$V_1(F_1 \cap V_2(F_2)) \subset V_1(F_1) \cap W(F_2) \tag{2.94}$$

which yields the reverse inclusion in (2.84). In particular if $\mathcal{F}$ is a filter, then

$$(u\mathcal{F}) \cdot (u\mathcal{F}) = u(\mathcal{F} \cdot (u\mathcal{F})) \subset u(\mathcal{F} \cdot \mathcal{F}) = u\mathcal{F}$$

Thus $u\mathcal{F}$ is a filter. The $\tilde{u}$ result for filterduals follows from duality, i.e., (2.80).

By monotonicity $\tilde{u}(\mathcal{F}_1 \cup \mathcal{F}_2) \supset (\tilde{u}\mathcal{F}_1) \cup (\tilde{u}\mathcal{F}_2)$. If $F \notin (\tilde{u}\mathcal{F}_1) \cup (\tilde{u}\mathcal{F}_2)$ then there exist $V_1, V_2 \in \mathcal{U}_S$ such that $V_\alpha(F) \notin \mathcal{F}_\alpha$ for $\alpha = 1,2$. Let $V = V_1 \cap V_2$. Then $V(F) \notin \mathcal{F}_\alpha$ for $\alpha = 1,2$ by heredity, i.e., $V(F) \notin \mathcal{F}_1 \cup \mathcal{F}_2$, so $F \notin \tilde{u}(\mathcal{F}_1 \cup \mathcal{F}_2)$. This proves (2.86). (2.85) then follows from duality and (2.7).

(d) It is an easy exercise to prove that the operator u commutes with union. Clearly by monotonicity $u(\cap_\alpha u\mathcal{F}_\alpha) \subset u(\cap_\alpha \mathcal{F}_\alpha)$. On the other hand, for each α, $\mathcal{F}_\alpha \supset \cap_\alpha \mathcal{F}_\alpha$, so $u\mathcal{F}_\alpha \supset u(\cap_\alpha \mathcal{F}_\alpha)$. We intersect over α to obtain $u(\cap u\mathcal{F}_\alpha) \supset uu(\cap_\alpha \mathcal{F}_\alpha) = u(\cap_\alpha \mathcal{F}_\alpha)$. The remaining results of (2.87) and (2.88) follow from duality.

(e) (2.89) is clear from the definition of the product uniformity.

(f) For every symmetric $\tilde{V} \in \mathcal{U}_{\tilde{S}}$, uniform continuity implies there exists a symmetric $V \in \mathcal{U}_S$ such that $g \circ V \subset \tilde{V} \circ g$, so $V \circ g^{-1} \subset g^{-1} \circ \tilde{V}$. Therefore if $\tilde{F} \in ug\mathcal{F}$, then $\tilde{V}(g(F)) \subset \tilde{F}$ for some $F \in \mathcal{F}$ and $\tilde{V} \in \mathcal{U}_{\tilde{S}}$. Then for some $V \in \mathcal{U}_S$, $\tilde{F}$ contains $g(V(F)) = g(V(F)) \cap g(S)$. This proves inclusion (2.91). Similarly we show $g^{-1}u\tilde{\mathcal{F}} \subset ug^{-1}\tilde{\mathcal{F}}$. It follows that $ug\mathcal{F} = uug\mathcal{F} \subset ugu\mathcal{F} \subset ug\mathcal{F}$, so $ug\mathcal{F} = ugu\mathcal{F}$. Applied to $\tilde{u}\mathcal{F}$, this says $ug\tilde{u}\mathcal{F} = ugu\tilde{u}\mathcal{F} = ugu\mathcal{F}$ by (2.82). Similarly

$$g^{-1}u\tilde{\mathcal{F}} = g^{-1}uu\tilde{\mathcal{F}} \subset ug^{-1}u\tilde{\mathcal{F}} \subset g^{-1}u\tilde{\mathcal{F}}$$

proving (2.92). The remaining identities follow from duality [cf. (2.31)].

If g is uniformly open onto $g(S)$, then for every $V \in \mathcal{U}_S$ (symmetric) there exists $\tilde{V} \in \mathcal{U}_S$ (symmetric) such that:

$$(\tilde{V} \circ g) \cap (S \times g(S)) \subset g \circ V \tag{2.95}$$

From this we obtain equality in (2.91). When in addition $g(S) = \tilde{S}$, we can invert in (2.95) to obtain $g^{-1} \circ \tilde{V} \subset V \circ g^{-1}$ and obtain equality in (2.92). ■

From (2.85) and (2.87) we see that the collection of open families is closed under finite intersections and arbitrary unions. Furthermore $\tilde{u}$ acts like a closure operator: It is monotone and preserves arbitrary intersections and finite unions.

For any subset A of S, $u[A]$ is the filter of uniform neighborhoods of A. In particular for $x \in S$, $u[x]$ is the filter of neighborhoods of x. Notice that $[x] = k[x]$ is the ultrafilter generated by x; hence:

$$u[x] = \{F \in \mathcal{P}_+ : x \in \operatorname{Int} F\}$$
$$ku[x] = \tilde{u}[x] = \{F \in \mathcal{P}_+ : x \in \overline{F}\} \tag{2.96}$$

By (2.83), $\{x\} \in \tilde{u}\mathcal{F}$ iff $u[x] \subset \mathcal{F}$. For a filter $\mathcal{F}$ for S, we say that $\mathcal{F}$ *converges* to x if $u[x] \subset \mathcal{F}$; in which case x is unique by the Hausdorff property. We say that x is a *limit point* of a filter $\mathcal{F}$ if $u[x] \subset k\mathcal{F}$, i.e., $\mathcal{F} \subset ku[x]$. By (2.96) we have for any family $\mathcal{F}$:

$$\mathcal{F} \subset ku[x] \Leftrightarrow x \in \bigcap_{F \in \mathcal{F}} \overline{F} \tag{2.97}$$

So the set of limit points of a filter $\mathcal{F}$ is the intersection $\cap_{F \in \mathcal{F}} \overline{F}$. Notice that if S is compact this intersection is nonempty for a filter.

PROPOSITION 2.14. *Assume $g : S \to \tilde{S}$ is a closed, uniformly continuous map. If $\tilde{A}$ is a compact subset of $\tilde{S}$, then:*

$$g^{-1}u[\tilde{A}] = u[g^{-1}(\tilde{A})] \tag{2.98}$$

PROOF. If $F \in u[g^{-1}(\tilde{A})]$ then for some open $V \in \mathcal{U}_S$, $F \supset V(g^{-1}(\tilde{A}))$. Because g is a closed map, the image $g(S \backslash V(g^{-1}(\tilde{A})))$ is a closed set disjoint from $\tilde{A}$ and so, by compactness, disjoint from $\tilde{V}(\tilde{A})$ for some $\tilde{V} \in \mathcal{U}_{\tilde{S}}$. Hence, $F \supset g^{-1}(\tilde{V}(\tilde{A}))$ proving that $u[g^{-1}(\tilde{A})] \subset g^{-1}u[\tilde{A}]$. The reverse inclusion is (2.92). ∎

PROPOSITION 2.15. *Assume S is a uniform space and T acts on S by uniformly continuous maps; i.e., each $g^t : S \to S$ is uniformly continuous. Let $\mathcal{F}$ be a family for S.*

If $\mathcal{F}$ is $+$ invariant, then $\tilde{u}\mathcal{F}$ is $+$ invariant. If $\mathcal{F}$ is $-$ invariant, then $u\mathcal{F}$ is $-$ invariant. If $\mathcal{F}$ is thick, then $u\mathcal{F}$ is thick. In general for any family $\mathcal{F}$:

$$u\tau u\mathcal{F} = u\tau\mathcal{F} = \tau u\tau\mathcal{F} \tag{2.99}$$

PROOF. If $g^{-t}\mathcal{F} \subset \mathcal{F}$, then by (2.92) $g^{-t}u\mathcal{F} \subset u\mathcal{F}$. Similarly if $g^t\mathcal{F} \subset \mathcal{F}$, then by (2.90) $g^t\tilde{u}\mathcal{F} \subset \tilde{u}\mathcal{F}$.

Now assume $F \in u\tau\mathcal{F}$. Then there exists $F_1 \in \tau\mathcal{F}$ and V symmetric in $\mathcal{U}_S$, so that $F \supset V^2(F_1)$. Given $\{t_1, \dots, t_k\}$, choose $W \in \mathcal{U}_S$ so that $g^{t_i} \circ W \subset V \circ g^{t_i}$ for $i = 1, \dots, k$.

$$\cap_i g^{-t_i}(V(F_1)) \supset \cap_i W(g^{-t_i}(F_1)) \supset W(\cap_i g^{-t_i}(F_1))$$

The intersection is in the thick family $\tau\mathcal{F}$, so its W neighborhood is in $u\tau\mathcal{F}$; hence:

$$V(F_1) \in \tau u\tau\mathcal{F} \qquad F \supset V^2(F_1) \in u\tau u\tau\mathcal{F}$$

Hence $u\tau\mathcal{F} \subset u\tau u\tau\mathcal{F}$. On the other hand,

$$u\tau u\tau\mathcal{F} \subset u\tau u\mathcal{F} \subset u\tau\mathcal{F}$$

and

$$u\tau u\tau\mathcal{F} \subset \tau u\tau\mathcal{F} \subset u\tau\mathcal{F}$$

So all the inclusions are equations yielding (2.99).

In particular if $\mathcal{F}$ is thick, then $u\mathcal{F} = u\tau\mathcal{F} = \tau u\tau\mathcal{F}$ is thick. ∎

In subsequent chapters we apply all of these family concepts primarily to the case where $S = T$ itself and the action $T \times T \to T$ is addition, so that $g^t : T \to T$ is translation by t, i.e., $g^t(s) = s + t$. Then T satisfies cancellation precisely when T acts injectively on itself. We define the *tail* subsets:

$$T_t = g^t(T) = \{s + t : s \in T\}$$

The filter $k\mathcal{B}_T$ is the family generated by $\{T_t : t \in T\}$ and $\mathcal{B}_T = \{S : S \cap T_t \neq \emptyset$ for all $t \in T\}$. We call a family $\mathcal{F}$ of subsets of T $+$ or $-$ *translation invariant* if it is $+$ or $-$ invariant with respect to this translation action.

Recall that a uniform monoid T satisfies cancellation and it has an invariant uniformity $\mathcal{U}_T$ uniquely defined by the topology on T and the Interior Condition (cf. Corollary 1.1).

PROPOSITION 2.16. *Assume T is a uniform monoid. Let $\mathcal{F}$ be a family of subsets of T.*

a. The filter $k\mathcal{B}_T$ is open. If $\mathcal{F}$ is full (with respect to $\mathcal{B}_T$), then $u\mathcal{F}$ and $\tilde{u}\mathcal{F}$ are full. For any family $\mathcal{F}$:

$$u(\mathcal{F} \cdot k\mathcal{B}_T) = (u\mathcal{F}) \cdot k\mathcal{B}_T \tag{2.100}$$

b. Assume that $\mathcal{F}$ is a full family. The G_T action commutes with the operators $u, \tilde{u}$, i.e., for $t, s \in T$:

$$\begin{aligned} g^{[t/s]}\tilde{u}\mathcal{F} &= \tilde{u}g^{[t/s]}\mathcal{F} \\ g^{[t/s]}u\mathcal{F} &= ug^{[t/s]}\mathcal{F} \end{aligned} \tag{2.101}$$

$$\begin{aligned} \tilde{\gamma}\tilde{u}\mathcal{F} &= \tilde{u}\tilde{\gamma}\mathcal{F} \\ u\tilde{\gamma}u\mathcal{F} &= u\tilde{\gamma}\mathcal{F} \\ \gamma u\mathcal{F} &= u\gamma\mathcal{F} \end{aligned} \tag{2.102}$$

If $\mathcal{F}$ is translation invariant, then $u\mathcal{F}$ and $\tilde{u}\mathcal{F}$ are translation invariant. If $\mathcal{F}$ is open, then $\gamma\mathcal{F}$ is open. If $\mathcal{F}$ is closed, then $\tilde{\gamma}\mathcal{F}$ is closed.

PROOF. (a) $k\mathcal{B}_T$ is open by the Interior Condition; then (2.100) follows from (2.84). If $\mathcal{F}$ is full, then $\mathcal{F} \cdot k\mathcal{B}_T = \mathcal{F}$, so $(u\mathcal{F}) \cdot k\mathcal{B}_T = u\mathcal{F}$, i.e., $u\mathcal{F}$ is full. Then $\tilde{u}\mathcal{F} = kuk\mathcal{F}$ is full by Proposition 2.2c.

By Proposition 2.11, g^t and g^{-t} act on mod $k\mathcal{B}_T$ congruence classes, so by Proposition 2.2a, $g^{\pm t}\mathcal{F}$ is full when $\mathcal{F}$ is.

(b) Because T is a uniform monoid, $g^t : T \to T_t$ is a uniform isomorphism; i.e., it is uniformly open as well as uniformly continuous. Then for a full family $\mathcal{F}$ we prove:

$$g^t u\mathcal{F} = [T_t] \cdot ug^t\mathcal{F} = ug^t\mathcal{F}$$

The first equation follows from (2.91) and the second holds because by Proposition 2.12 $g^t\mathcal{F}$ is full, so by (a) $ug^t\mathcal{F}$ is full. The second equation of (2.101) then follows by invertibility of the action. The third equation of (2.102) follows from (2.72) and (2.87). After using duality the remaining (middle) equation of (2.102) follows from:

$$\begin{aligned} u\tilde{\gamma}u\mathcal{F} &= u\tilde{u}\tilde{\gamma}u\mathcal{F} = u\tilde{\gamma}\tilde{u}u\mathcal{F} \\ &= u\tilde{\gamma}\tilde{u}\mathcal{F} = u\tilde{u}\tilde{\gamma}\mathcal{F} = u\tilde{\gamma}\mathcal{F} \end{aligned}$$

The final assertions follow directly from (2.102). ■

For a uniform monoid, there is also a nice topological picture of thickness.

PROPOSITION 2.17. *Let T be a uniform monoid and $\mathcal{F}$ an open, thick family of subsets of T. Let K be a totally bounded subset of T (e.g., a compact subset) and let $F \in \mathcal{F}$; then:*

$$\{t : g^t(K) \subset F\} = \cap\{g^{-s}(F) : s \in K\} \in \mathcal{F}$$

PROOF. Clearly t is in either set iff $s+t \in F$ for all $s \in K$. We must prove this set is in $\mathcal{F}$.

Because $\mathcal{U}_T$ is invariant and $\mathcal{F}$ is open, there exist $F_1 \in \mathcal{F}$ and V a symmetric, invariant element of $\mathcal{U}_T$ such that $V(F_1) \subset F$. Because K is totally bounded, there is a finite subset $\{s_1, \dots, s_n\}$ of K such that $\{V(s_i) : i = 1, \dots, n\}$ covers K. Let $F_2 = \cap_{i=1}^n g^{-s_i}(F_1)$, which is in $\mathcal{F}$ because $\mathcal{F}$ is thick. It suffices to show that $F_2 \subset \cap\{g^{-s}(F) : s \in K\}$. With $t \in F_2$ and $s \in K$, there exists s_i such that $s \in V(s_i)$, so, by invariance of V, $s+t \in V(s_i+t)$. But $t \in F_2$ implies $s_i+t \in F_1$, so $s+t \in V(F_1) \subset F$. ■

Now we apply these results to the two uniform monoids of greatest interest: the nonnegative integers $\mathbf{Z}_+$ under addition with the zero/one metric and the nonnegative reals $\mathbf{R}_+$ under addition with the usual metric. Since $\mathbf{Z}_+$ is discrete, we have $\mathcal{F} = u\mathcal{F}$ for every family for $\mathbf{Z}_+$.

PROPOSITION 2.18. *With $T = \mathbf{Z}_+$ or $\mathbf{R}_+$ let $\mathcal{F}$ be a translation invariant proper family for T.*

a. For a subset F of T the following are equivalent:

(1) $F \in u\tau\mathcal{F}$

(2) For every $n \in \mathbf{Z}_+$, $\{t : [t, t+n] \subset F\} \in \mathcal{F}$

(3) For every $n \in \mathbf{Z}_+$, $\{t \geq n : [t-n, t] \subset F\} \in \mathcal{F}$

(4) For every $n \in \mathbf{Z}_+$. $\{t \geq n : [t-n, t+n] \subset F\} \in \mathcal{F}$

b. For a subset F of T the following are equivalent:

(1) $F \in ku\tau\mathcal{F} = \tilde{u}k\tau\mathcal{F}$

(2) For some $n \in \mathbf{Z}_+$, $F + [0,n] = \{t_1 + t_2 : t_1 \in F$ and $t_2 \in T$ with $t_2 \leq n\} \in k\mathcal{F}$

(3) For some $n \in \mathbf{Z}_+$, $\{t : [t, t+n] \cap F \neq \emptyset\} \in k\mathcal{F}$

(4) For some $n \in \mathbf{Z}_+$, $(F + [-n, n]) \cap T = V_n(F) \in k\mathcal{F}$.

PROOF. (a) (1) $\Rightarrow$ (2). This follows from Proposition 2.17 with $K = [0,n]$, since $g^t([0,n)] = [t, t+n]$ in T.

(2) $\Rightarrow$ (4). Since $\mathcal{F}$ is translation invariant we have $g^n\{t : [t, t+2n] \subset F\} \in \mathcal{F}$; but this is the set described by (4).

(4) $\Rightarrow$ (3). This is true by heredity.

(3) $\Rightarrow$ (1). Given $\{t_1, \ldots, t_k\} \subset T$, choose $n \in \mathbf{Z}_+$ so that $n \geq \max\{t_i\}$. Since $\mathcal{F}$ is translation invariant

$$g^{-(n+1)}\{t \geq n+2 : [t-n-2, t] \subset F\} \in \mathcal{F}$$

This set is contained in $\{t \geq 1 : [t-1, t+n+1] \subset F\}$, which is contained in the V_1 neighborhood of $\{s : s + t_i \in F$ for $i = 1, \ldots, k\}$, i.e., $g^{-t_1}(F) \cap \ldots \cap g^{-t_k}(F)$. Thus with $F_1 = \{t \in T : V_1(t) \subset F\}, F_1 \in \tau\mathcal{F}, F \in u\tau\mathcal{F}$.

(b) Notice first that:

$$\begin{aligned} F + [0,n] &= \{t \in T : [t-n, t] \cap F \neq \emptyset\} \\ (F + [-n, n]) \cap T &= \{t \in T : [t-n, t+n] \cap F \neq \emptyset\} \end{aligned} \tag{2.103}$$

Now we apply the results of (a). $F \in ku\tau\mathcal{F}$ iff $T \backslash F \notin u\tau\mathcal{F}$, so by (3) in (a) iff for some $n \in \mathbf{Z}_+$ $\{t \geq n : [t-n, t] \subset S \backslash F\} \notin \mathcal{F}$ iff for some $n \in \mathbf{Z}_+$ $[0,n] \cup \{t \geq n : [t-n, t] \cap F \neq \emptyset\} \in k\mathcal{F}$. Since $k\mathcal{F}$ is full this shows (1) $\Leftrightarrow$ (3) in (b) by (2.103). Similarly (1) $\Leftrightarrow$ (4) in (b) by (1) $\Leftrightarrow$ (4) in (a) and (2.103). An even more direct but analogous duality argument shows (1) $\Leftrightarrow$ (3) in (b) from (1) $\Leftrightarrow$ (2) in (a). ∎

For any uniform monoid, we call $u\tau\mathcal{B}_T$ the family of *thick sets* in T. Then $F \subset T$ is thick for $T = \mathbf{Z}_+$ or $\mathbf{R}_+$ when it contains arbitrarily long runs, i.e., for every $n \in \mathbf{Z}_+$, $[t, t+n] \subset F$ for infinitely many t. Observe that by (2.99) we have

$$u\tau\mathcal{B}_T = u\tau u\mathcal{B}_T = \tau u\tau\mathcal{B}_T \tag{2.104}$$

For $T = \mathbf{Z}_+$ or $\mathbf{R}_+$, $ku\tau\mathcal{B}_T = \tilde{u}k\tau\mathcal{B}_T$ is the family of *syndetic sets* in T. By Proposition 2.18b, $F \subset T$ is syndetic when for some compact set K and some $n \in \mathbf{Z}_+$, $F + K \supset T_n$, or equivalently for some $n \in \mathbf{Z}_+$, $F + [0,n] \supset T_n$. By increasing n if necessary in Proposition 2.18b, we see that F is syndetic when for some $n \in \mathbf{Z}_+$, $[t, t+n] \cap F \neq \emptyset$ for all $t \in T$; i.e., every interval of length n meets F. For the interior of the family of syndetic sets (2.81) and (2.82) imply

$$uku\tau\mathcal{B}_T = uk\tau\mathcal{B}_T \tag{2.105}$$

For any uniform monoid T, we call the family $u\tau k\tau\mathcal{B}_T$ the family of *replete sets* in T. Repeatedly applying (2.82) and (2.99) we see that:

$$u\tau k\tau = u\tau uk\tau = u\tau u\tilde{u}k\tau = u\tau uku\tau = u\tau ku\tau$$

then:

$$u\tau k\tau\mathcal{B}_T = u\tau(ku\tau\mathcal{B}_T) \tag{2.106}$$

So applying Proposition 2.18a to the family $\mathcal{F}$ of syndetic sets, we see for $T = \mathbf{Z}_+$ or $\mathbf{R}_+$ $F \subset T$ is replete iff for every $n \in \mathbf{Z}_+$ $\{t : [t, t+n] \subset F\}$ is syndetic. Applying Proposition 2.9 to the filter $k\mathcal{B}_T$, we see that $\tau k\tau\mathcal{B}_T$ is a filter, so by Proposition 2.13c the replete sets form an open filter. Furthermore we have

$$(u\tau k\tau\mathcal{B}_T) \cdot u\tau\mathcal{B}_T = u\tau\mathcal{B}_T \tag{2.107}$$

and

$$(u\tau k\tau\mathcal{B}_T) \cdot uk\tau\mathcal{B}_T = uk\tau\mathcal{B}_T \tag{2.108}$$

Recall that $u(\mathcal{F}_1 \cdot (u\mathcal{F})) \subset u(\mathcal{F}_1 \cdot \mathcal{F}_2)$.

Let $\mathbf{Z}^*$ denote the set of positive integers; for $T = \mathbf{R}_+$ or $\mathbf{Z}_+$ let $\mathcal{A}_T$ be the family generated by $\{t\mathbf{Z}^* : t \in T \text{ with } t > 0\}$. Thus $S \in \mathcal{A}_T$ iff S contains an infinite subsemigroup of T.

We can regard $\mathbf{Z}^*$ as a discrete monoid under multiplication. Then $\mathcal{B}_{\mathbf{Z}^*}$ is $k\mathcal{A}_{\mathbf{Z}}$ (strictly speaking, $S \in k\mathcal{A}_{\mathbf{Z}}$ iff $S \cap \mathbf{Z}^* = S \backslash 0 \in \mathcal{B}_{\mathbf{Z}^*}$). Thus $kA_{\mathbf{Z}}$ is the largest $\mathbf{Z}^*$ invariant family. The larger family $\mathcal{B}_{\mathbf{Z}}$ is $\mathbf{Z}^* +$ invariant, but not $\mathbf{Z}^* -$ invariant.

3

Recurrence

Recall that a relation $G : X_1 \to X_2$ is a subset of $X_1 \times X_2$. For $A \times B \subset X_1 \times X_2$:

$$\begin{aligned} A \cap G^{-1}(B) \neq \emptyset &\Leftrightarrow G(A) \cap B \neq \emptyset \\ &\Leftrightarrow (A \times B) \cap G \neq \emptyset \end{aligned} \tag{3.1}$$

We say that subsets $A_1, A_2 \subset X_1$ *meet* when $A_1 \cap A_2 \neq \emptyset$. Thus, $G(A)$ meets B iff A meets $G^{-1}(B)$; furthermore:

$$\begin{aligned} G(A) \subset B &\Leftrightarrow G^{-1}(X_2 \backslash B) \subset X_1 \backslash A \\ &\Leftrightarrow A \subset G^{-1}(B) \qquad \text{when } G \text{ is a map} \end{aligned} \tag{3.2}$$

Observe that $G(A) \subset B$ iff $G(A)$ does not meet $X_2 \backslash B$.

Now let $\varphi : T \times X \to X$ be an action of a monoid T on X. Associated with $A, B \subset X$ we define the *meeting time set* $N^{\varphi}(A,B)$ and the *inclusion time set* $J^{\varphi}(A,B)$, subsets of T, by:

$$\begin{aligned} N^{\varphi}(A,B) &= \{t : f^t(A) \cap B \neq \emptyset\} = \{t : A \cap f^{-t}(B) \neq \emptyset\} \\ J^{\varphi}(A,B) &= \{t : f^t(A) \subset B\} = \{t : A \subset f^{-t}(B)\} = T \backslash N^{\varphi}(A, X \backslash B) \end{aligned} \tag{3.3}$$

We usually omit the superscript φ when the action is understood.

Clearly if $A \neq \emptyset$, then $J(A,B) \subset N(A,B)$ and if $A_1 \times B_1 \subset A \times B$, then $N(A_1,B_1) \subset N(A,B)$ and $J(A,B_1) \subset J(A_1,B)$. Furthermore if φ is reversible so that the reverse action $\overline{\varphi}$ is defined [cf. (1.17)] then

$$\begin{aligned} N^{\varphi}(A,B) &= N^{\overline{\varphi}}(B,A) \\ J^{\varphi}(A,B) &= J^{\overline{\varphi}}(X \backslash B, X \backslash A) \end{aligned} \tag{3.4}$$

Recall that $g^t : T \to T$ denotes translation by t on T and $T_t = g^t(T)$ for all $t \in T$.

PROPOSITION 3.1. *Let $\varphi : T \times X \to X$ be an action and $A, B \subset X$. For all $t \in T$:*

$$N(A, f^{-t}(B)) = g^{-t}(N(A,B)) = N(f^t(A), B)$$
$$J(A, f^{-t}(B)) = g^{-t}(J(A,B)) = J(f^t(A), B) \tag{3.5}$$

Let $\varphi_1 : T \times X_1 \to X_1$ be an action, $h : \varphi \to \varphi_1$ an action map, and $A \times B_1 \subset X \times X_1$:

$$N^{\varphi}(A, h^{-1}(B_1)) = N^{\varphi_1}(h(A), B_1)$$
$$J^{\varphi}(A, h^{-1}(B_1)) = J^{\varphi_1}(h(A), B_1) \tag{3.6}$$

If φ_π is the product of a family of T actions $\{\varphi_\alpha : T \times X_\alpha \to X_\alpha\}$ and A_α, $B_\alpha \subset X_\alpha$ with $A_\alpha \neq \emptyset$ for all α, then:

$$N^{\varphi_\pi}(\prod_\alpha A_\alpha, \prod_\alpha B_\alpha) = \cap_\alpha N^{\varphi_\alpha}(A_\alpha, B_\alpha)$$
$$J^{\varphi_\pi}(\prod_\alpha A_\alpha, \prod_\alpha B_\alpha) = \cap_\alpha J^{\varphi_\alpha}(A_\alpha, B_\alpha) \tag{3.7}$$

PROOF. $f^s(A)$ meets $f^{-t}(B)$ iff $f^{t+s}(A)$ meets B and $f^s(f^t(A))$ meets B iff $f^{s+t}(A)$ meets B, proving (3.5) for N.

Since $h \circ f^s = f_1^s \circ h$, A meets $f^{-s}h^{-1}(B_1)$ iff A meets $h^{-1}f_1^{-s}(B_1)$ and iff $h(A)$ meets $f_1^{-s}(B_1)$ by (3.1), proving (3.6) for N.

The results for J are similar, and (3.7) is obvious. ■

The map $\varphi : T \times X \to X$ is a subset of $T \times X \times X$, so it can also be regarded as a relation from T to $X \times X$ associating to $t \in T$ the points in $f^t \subset X \times X$. For $F \subset T$, we denote the image under this relation by:

$$f^F = \cup\{f^t : t \in F\} \subset X \times X \tag{3.8}$$

We leave to the reader the easy proof of the equivalence of different notations in Lemma 3.1.

LEMMA 3.1. *Let $\varphi : T \times X \to X$ be an action and $F \times A \times B \subset T \times X \times X$.*

a. The following are equivalent:

(1) $(F \times A \times B) \cap \varphi \neq \emptyset$

(2) $(A \times B) \cap f^F \neq \emptyset$

(3) $f^F(A) \cap B \neq \emptyset$

(4) $A \cap (f^F)^{-1}(B) \neq \emptyset$

(5) $F \cap N^{\varphi}(A,B) \neq \emptyset$

b. The following are equivalent:

(1) $\varphi(F \times A) \subset B$

(2) $F \times A \subset \varphi^{-1}(B)$

(3) $f^F(A) \subset B$

(4) $A \subset \cap\{f^{-t}(B) : t \in F\}$

(5) $F \subset J^{\varphi}(A,B)$

Assume now that $\mathcal{A}$ and $\tilde{\mathcal{A}}$ are families for X. We define $\mathcal{N}^{\varphi}(\mathcal{A},\tilde{\mathcal{A}})$ to be the family for T generated by $\{N^{\varphi}(A,B) : A \in \mathcal{A} \text{ and } \tilde{A} \in \tilde{\mathcal{A}}\}$ and $\mathcal{N}^{\varphi}(\mathcal{A})$ to be the family for T generated by $\{N^{\varphi}(A,A) : A \in \mathcal{A}\}$. It suffices to let A vary over a set of generators for $\mathcal{A}$ and $\tilde{A}$ vary over a set of generators for $\tilde{\mathcal{A}}$.

PROPOSITION 3.2. *Let* $\varphi : T \times X \to X$ *be an action and* $\mathcal{A}$, $\tilde{\mathcal{A}}$, *etc., be families for* X.

a. If $\mathcal{A} \subset \mathcal{A}_1$ *and* $\tilde{\mathcal{A}} \subset \tilde{\mathcal{A}}_1$ *then* $\mathcal{N}(\mathcal{A},\tilde{\mathcal{A}}) \subset \mathcal{N}(\mathcal{A}_1,\tilde{\mathcal{A}}_1)$ *and* $\mathcal{N}(\mathcal{A}) \subset \mathcal{N}(\mathcal{A}_1)$. *For collections of families* $\{\mathcal{A}_\alpha\}$ *and* $\{\tilde{\mathcal{A}}_\alpha\}$:

$$\begin{aligned} \mathcal{N}(\cup_\alpha \mathcal{A}_\alpha, \cup_\beta \tilde{\mathcal{A}}_\beta) &= \cup_{\alpha,\beta} \mathcal{N}(\mathcal{A}_\alpha, \tilde{\mathcal{A}}_\beta) \\ \mathcal{N}(\cup_\alpha \mathcal{A}_\alpha) &= \cup_\alpha \mathcal{N}(\mathcal{A}_\alpha) \end{aligned} \tag{3.9}$$

For all $t \in T$:

$$\mathcal{N}(\mathcal{A}, f^{-t}\tilde{\mathcal{A}}) = g^{-t}\mathcal{N}(\mathcal{A},\tilde{\mathcal{A}}) = \mathcal{N}(f^t\mathcal{A},\tilde{\mathcal{A}}) \tag{3.10}$$

b. $\mathcal{N}(\mathcal{A}) \subset \mathcal{N}(\mathcal{A},\mathcal{A})$. *If* $\mathcal{A}$ *is a filter for* X, *then* $\mathcal{N}(\mathcal{A}) = \mathcal{N}(\mathcal{A},\mathcal{A})$ *is a filter for* T. *If in addition* $\tilde{\mathcal{A}}$ *is a filter for* X, *then* $\mathcal{N}(\mathcal{A},\tilde{\mathcal{A}})$ *is a filter for* T.

c. Let $\varphi_1 : T \times X_1 \to X_1$ *be an action and* $\mathcal{A}_1$, $\tilde{\mathcal{A}}_1$ *be families for* X_1.

$$\begin{aligned} \mathcal{N}(\mathcal{A} \times \mathcal{A}_1, \tilde{\mathcal{A}} \times \tilde{\mathcal{A}}_1) &= \mathcal{N}(\mathcal{A},\tilde{\mathcal{A}}) \cdot \mathcal{N}(\mathcal{A}_1,\tilde{\mathcal{A}}_1) \\ \mathcal{N}(\mathcal{A} \times \mathcal{A}_1) &= \mathcal{N}(\mathcal{A}) \cdot \mathcal{N}(\mathcal{A}_1) \end{aligned} \tag{3.11}$$

If $h : \varphi \to \varphi_1$ *is an action map, then:*

$$\mathcal{N}^{\varphi}(\mathcal{A}, h^{-1}\tilde{\mathcal{A}}_1) = \mathcal{N}^{\varphi_1}(h\mathcal{A}, \tilde{\mathcal{A}}_1) \tag{3.12}$$

$$\mathcal{N}^{\varphi_1}(h\mathcal{A}, h\tilde{\mathcal{A}}) \subset \mathcal{N}^{\varphi}(\mathcal{A},\tilde{\mathcal{A}}) \tag{3.13}$$

$$\mathcal{N}^{\varphi_1}(h\mathcal{A}) \subset \mathcal{N}^{\varphi}(\mathcal{A}) \tag{3.14}$$

d. Assume that φ *is a uniform action. If* $\tilde{\mathcal{A}}$ *is an open family for* X, *then* $\mathcal{N}(\mathcal{A},\tilde{\mathcal{A}})$ *and* $\mathcal{N}(\tilde{\mathcal{A}})$ *are open families for* T.

PROOF. (a) This part is obvious with (3.10) following from (3.5). The inclusion in (b) is clear. The filter results follow from:

$$\begin{aligned} &N(A \cap A_1, \tilde{A} \cap \tilde{A}_1) \subset N(A, \tilde{A}) \cap N(A_1, \tilde{A}_1) \\ &N(A \cap \tilde{A}, A \cap \tilde{A}) \subset N(A, \tilde{A}) \end{aligned} \tag{3.15}$$

(3.11) and (3.12) follow from (3.7) and (3.6). Because $h^{-1}h(\tilde{A}) \supset \tilde{A}$, (3.13) and (3.14) also follow from (3.6).

Now assume that φ is a uniform action and $\tilde{A} \in \tilde{\mathcal{A}}$, an open family for X. There exists V symmetric in $\mathcal{U}_X$ and $\tilde{A}_1 \in \tilde{\mathcal{A}}$ such that $\tilde{A} \supset V(\tilde{A}_1)$. Because the action is uniform, there exists $W \in \mathcal{U}_T$ such that $(t_1, t) \in W$ implies $(f^{t_1}(x), f^t(x)) \in V$ for all $x \in X$. It follows that $W(N(A, \tilde{A}_1)) \subset N(A, \tilde{A})$ for any $A \subset X$. Thus $\mathcal{N}(\mathcal{A}, \tilde{\mathcal{A}})$ is open for any family $\mathcal{A}$. A fortiori $W(N(\tilde{A}_1, \tilde{A}_1)) \subset N(\tilde{A}, \tilde{A})$, so $\mathcal{N}(\tilde{\mathcal{A}})$ is open as well. ■

When $\mathcal{A} = [A]$ we write $\mathcal{N}(A, \tilde{\mathcal{A}})$ for $\mathcal{N}([A], \tilde{\mathcal{A}})$, omitting brackets. Similarly, we omit brackets when $\tilde{\mathcal{A}} = [B]$. For example $\mathcal{N}(A, B)$ is $\mathcal{N}([A], [B]) = [N(A, B)]$. Our most important use of this notation occurs when φ is a uniform action and $\mathcal{A}$ or $\tilde{\mathcal{A}}$ is of the form $[A]$, the filter generated by the nonempty set A, or $u[A]$, the filter of uniform neighborhoods of A. Thus for $A, B \subset X$:

$$\begin{aligned} \mathcal{N}^{\varphi}(A, u[B]) &= \{F : F \supset N(A, V(B)) \text{ for some } V \in \mathcal{U}_X\} \\ \mathcal{N}^{\varphi}(u[A], u[B]) &= \{F : F \supset N(V(A), V(B)) \text{ for some } V \in \mathcal{U}_X\} \end{aligned} \tag{3.16}$$

By Proposition 3.2b:

$$\mathcal{N}^{\varphi}(u[A]) = \mathcal{N}^{\varphi}(u[A], u[A]) \tag{3.17}$$

As usual we omit the superscript when the action is understood. Notice that for $V_1, V_2 \in \mathcal{U}_X$ $N(V_1(A), V_2(B))$ contains $N(V(A), V(B))$ with $V = V_1 \cap V_2$. From this it easily follows that:

$$\mathcal{N}(u[A], u[B]) = \cup\{\mathcal{N}(V(A), u[B]) : V \in \mathcal{U}_X\} \tag{3.18}$$

We call $A, B \subset X$ *separated* if for some $V \in \mathcal{U}_X$, $V(A) \cap V(B) = \emptyset$. If there exists $W \in \mathcal{U}_X$ such that $W(A) \cap B = \emptyset$ then with $V \in \mathcal{U}_X$ satisfying $V = V^{-1}$ and $V \circ V \subset W$, $V(A) \cap V(B) = \emptyset$ and so A and B are separated. Clearly, if A and B are separated they have disjoint closures and the converse is true as well if either A or B has compact closure. If $f^T(A)$ and B are separated then $\mathcal{N}(A, u[B]) = \mathcal{P}(T)$. Otherwise, $\mathcal{N}(A, u[B])$ and $\mathcal{N}(u[A], u[B])$ are proper families.

PROPOSITION 3.3. *Let $\varphi : T \times X \to X$ be a uniform action. Assume $A, B \subset X$.*

a. Each $\mathcal{N}(A,u[B])$ and $\mathcal{N}(u[A],u[B])$ is an open family and a filter when proper. For all $t \in T$:

$$\mathcal{N}(f^t(A),u[B]) = g^{-t}\mathcal{N}(A,u[B]) \tag{3.19}$$

$$\mathcal{N}(u[f^t(A)],u[B]) \subset g^{-t}\mathcal{N}(u[A],u[B]) \subset \mathcal{N}(u[A],u[f^{-t}(B)]) \tag{3.20}$$

b. Let $\varphi_1 : T \times X_1 \to X_1$ be a uniform action. Assume $A_1, B_1 \subset X_1$.

$$\mathcal{N}^{\varphi\times\varphi_1}(A \times A_1, u[B \times B_1]) = \mathcal{N}^{\varphi}(A,u[B]) \cdot \mathcal{N}^{\varphi_1}(A_1,u[B_1])$$

$$\mathcal{N}^{\varphi\times\varphi_1}(u[A \times A_1], u[B \times B_1]) = \mathcal{N}^{\varphi}(u[A],u[B]) \cdot \mathcal{N}^{\varphi_1}(u[A_1],u[B_1]) \tag{3.21}$$

Assume $h : \varphi \to \varphi_1$ is a uniformly continuous action map and $h \times h(A \times B) \subset A_1 \times B_1$.

$$\mathcal{N}^{\varphi_1}(A_1,u[B_1]) \subset \mathcal{N}^{\varphi}(A,u[B])$$

$$\mathcal{N}^{\varphi_1}(u[A_1],u[B_1]) \subset \mathcal{N}^{\varphi}(u[A],u[B]) \tag{3.22}$$

If, in addition either h is a closed map and B_1 is compact, or $h : X \to h(X)$ is uniformly open and $B_1 \subset h(X)$, then:

$$\mathcal{N}^{\varphi_1}(h(A),u[B_1]) = \mathcal{N}^{\varphi}(A,u[h^{-1}(B_1)]) \tag{3.23}$$

If, in addition either h is surjective as well as closed and A_1 as well as B_1 is compact, or $h : X \to X_1$ is a uniformly open surjective map, then:

$$\mathcal{N}^{\varphi_1}(u[A_1],u[B_1]) = \mathcal{N}^{\varphi}(u[h^{-1}(A_1)],u[h^{-1}(B_1)]) \tag{3.24}$$

PROOF. The families are open by Proposition 3.2d and they are filters when proper by Proposition 3.2b. The product result (3.21) follows from (3.11) and (2.89). (3.19) and (3.20) follow from (3.10) and:

$$u[h(A)] \subset hu[A] \qquad h^{-1}u[B_1] \subset u[h^{-1}(B_1)]$$

(2.91) and (2.92), applied with $h = f^t$. Similarly we obtain (3.22) from (3.13). Under the hypotheses of (3.23) $h^{-1}u[B_1] = u[h^{-1}(B_1)]$ by (2.92) or (2.98). Under the alternative hypotheses of (3.24) we also have

$$u[A_1] = hh^{-1}u[A_1] = hu[h^{-1}(A_1)]$$

The equations follow from (3.12). ∎

We define various notions of invariance with respect to an action φ for a subset B of X.

$$\begin{aligned} B \text{ is } + \text{ invariant} &\Leftrightarrow f^t(B) \subset B \\ B \text{ is } - \text{ invariant} &\Leftrightarrow f^{-t}(B) \subset B \\ B \text{ is } \pm \text{ invariant} &\Leftrightarrow f^{-t}(B) = B \\ B \text{ is invariant} &\Leftrightarrow f^t(B) = B \end{aligned} \tag{3.25}$$

for all $t \in T$. By (3.2) B is + invariant iff $B \subset f^{-t}(B)$ for all t, so B is $\pm$ invariant iff it is both + and − invariant. Invariance implies + invariance; for a surjective action, $\pm$ invariance implies invariance. Notice that X is always $\pm$ invariant, but it is invariant only when the action is surjective.

PROPOSITION 3.4. *For a uniform action $\varphi : T \times X \to X$ assume that $A, B \subset X$ with B + invariant. The families $\mathcal{N}(A, u[B])$ and $\mathcal{N}(u[A], u[B])$ are thick and translation − invariant families of subsets of T. If B is invariant and $\pm$ invariant and either the action is closed and B is compact, or the action is uniformly open, then $\mathcal{N}(A, u[B]) \cdot k\mathcal{B}_T$ is a translation invariant family.*

PROOF. Because $f^t(B) \subset B$ we have from (3.19) and (3.22) applied to $f^t : \varphi \to \varphi$ that:

$$\begin{aligned} g^{-t}(\mathcal{N}(A, u[B])) \subset g^{-t}(\mathcal{N}(A, u[f^t(B)])) &= \mathcal{N}(f^t(A), u[f^t(B)]) \\ \subset \mathcal{N}(A, u[f^{-t} f^t(B)]) &\subset \mathcal{N}(A, u[B]) \end{aligned} \tag{3.26}$$

Hence the family $\mathcal{N}(A, u[B])$ is − invariant. Because it is closed under intersection, the family is also thick. The result for $\mathcal{N}(u[A], u[B])$ then follows from (3.18).

In (3.26) the first inclusion is an equality when B is invariant; the next inclusion is an equality when f^t is a closed map and B is compact or when f^t is uniformly open by (3.23). Finally $f^{-t} f^t(B) = B$ when B is $\pm$ invariant. Thus applying (2.31) we obtain

$$\begin{aligned} g^{-t}(\mathcal{N}(A, u[B])) &= \mathcal{N}(A, u[B]) \\ g^{-t}(\mathcal{N}(A, u[B]) \cdot k\mathcal{B}_T) &= \mathcal{N}(A, u[B]) \cdot k\mathcal{B}_T \end{aligned}$$

It follows from Proposition 2.12b that the full family $\mathcal{N}(A, u[B]) \cdot k\mathcal{B}_T$ is invariant. ∎

REMARK. If B is a nonempty invariant subset, then $\mathcal{N}(B, u[B])$ is $k\mathcal{P}_+ = \{T\}$, which is not invariant when T is not a group. ∎

Now for any proper family $\mathcal{F}$ of subsets of T and any nonempty subset A of X, we define

$$\omega_{\mathcal{F}}\varphi[A] = \bigcap_{F \in k\mathcal{F}} \{\overline{f^F(A)}\} \tag{3.27}$$

For example if $F \in \mathcal{P}_+$ and $[F] = \{F_1 \in \mathcal{P}_+ : F_1 \supset F\}$ is the filter generated by F [cf. (2.3)], then with $\mathcal{F} = k[F]$ we have $\omega_{\mathcal{F}}\varphi[A] = \overline{f^F(A)}$.

The most important applications occur in the compact case with $\mathcal{F}$ a filterdual.

PROPOSITION 3.5. *Let $\varphi : T \times X \to X$ be a uniform action with X compact and let $\mathcal{F}$ be a filterdual. If U is a neighborhood of $\omega_{\mathcal{F}}\varphi[A]$, then there exists $F \in k\mathcal{F}$ such that $\overline{f^F(A)} \subset U$. In particular, $\omega_{\mathcal{F}}\varphi[A]$ is nonempty.*

PROOF. By compactness $\omega_{\mathcal{F}}\varphi[A] \subset \operatorname{Int} U$ implies there exists $\{F_1, \dots, F_k\} \subset k\mathcal{F}$ such that:

$$\cap_{i=1}^k \overline{f^{F_i}(A)} \subset \operatorname{Int} U$$

As $k\mathcal{F}$ is a filter,

$$F = \cap_{i=1}^k F_i \in k\mathcal{F} \qquad \overline{f^F(A)} \subset \operatorname{Int} U$$

Applied to $U = \emptyset$ this shows that $\omega_{\mathcal{F}}\varphi[A]$ is nonempty. ■

PROPOSITION 3.6. *Let $\varphi : T \times X \to X$ be a uniform action, A, A_1, A_2 nonempty subsets of X and $\mathcal{F}, \mathcal{F}_1, \mathcal{F}_2$ proper families for T.*

a. For $y \in X$ the following are equivalent:

(1) $y \in \omega_{\mathcal{F}}\varphi[A]$

(2) For all $F \in k\mathcal{F}$ and U a neighborhood of y, $A \cap (f^F)^{-1}(U) \neq \emptyset$

(3) For all U a neighborhood of y, $N^{\varphi}(A, U) \in \mathcal{F}$

(4) $\mathcal{N}(A, u[y]) \subset \mathcal{F}$

b. $\omega_{\mathcal{F}_1}\varphi[A] \cap \omega_{\mathcal{F}_2}\varphi[A] = \omega_{\mathcal{F}_1 \cap \mathcal{F}_2}\varphi[A]$.

In particular $\mathcal{F}_1 \subset \mathcal{F}_2$ implies $\omega_{\mathcal{F}_1}\varphi[A] \subset \omega_{\mathcal{F}_2}\varphi[A]$. Furthermore $\omega_{u\mathcal{F}}\varphi[A] = \omega_{\mathcal{F}}\varphi[A] = \omega_{\tilde{u}\mathcal{F}}\varphi[A]$. If $A_1 \subset A_2$, then $\omega_{\mathcal{F}}\varphi[A_1] \subset \omega_{\mathcal{F}}\varphi[A_2]$, with equality if A_1 is dense in A_2. In general if $\mathcal{F}$ is a filterdual, then $\omega_{\mathcal{F}}\varphi[A_1] \cup \omega_{\mathcal{F}}\varphi[A_2] = \omega_{\mathcal{F}}\varphi[A_1 \cup A_2]$.

c. If $\varphi_1 : T \times X_1 \to X_1$ is a uniform action and $h : \varphi \to \varphi_1$ is a continuous action map, then:

$$h(\omega_{\mathcal{F}}\varphi[A]) \subset \omega_{\mathcal{F}}\varphi_1[h(A)] \tag{3.28}$$

with equality if X is compact and $\mathcal{F}$ is a filterdual. On the other hand, for any family $\mathcal{F}$ if h is an embedding, then:

$$\omega_{\mathcal{F}}\varphi[A] = h^{-1}(\omega_{\mathcal{F}}\varphi_1[h(A)]) \tag{3.29}$$

In particular, for $t \in T$:

$$f^t(\omega_{\mathcal{F}}\varphi[A]) \subset \omega_{\mathcal{F}}\varphi[f^t(A)] \tag{3.30}$$

with equality if either the action is uniformly reversible or X is compact and $\mathcal{F}$ is filterdual.

d. If $\mathcal{F}$ is a translation + invariant family then

$$\omega_{\mathcal{F}}\varphi[f^t(A)] \subset \omega_{\mathcal{F}}\varphi[A] \tag{3.31}$$

with equality if $\mathcal{F}$ is translation invariant.

e. If $\varphi_1 \times \varphi_2$ is the product of two uniform actions, then:

$$\omega_{\mathcal{F}_1}\varphi_1[A_1] \times \omega_{\mathcal{F}_2}\varphi_2[A_2] \subset \omega_{\mathcal{F}_1 \cdot \mathcal{F}_2}(\varphi_1 \times \varphi_2)[A_1 \times A_2] \tag{3.32}$$

If φ_π is the product of the actions $\{\varphi_\alpha : T \times X_\alpha \to X_\alpha\}$ and A_α is a nonempty subset of X_α for each α, then:

$$\omega_{\mathcal{F}}\varphi_\pi[\Pi_\alpha A_\alpha] \subset \Pi_\alpha \omega_{\mathcal{F}}\varphi_\alpha[A_\alpha] \tag{3.33}$$

with equality when $\mathcal{F}$ is a filter.

PROOF. (a) The equivalence of (1), (2), and (3) follows from Lemma 3.1a. Note that for U open, U meets $f^F(A)$ iff it meets the closure $\overline{f^F(A)}$. The equivalence of (3) with (4) is obvious.

(b) $\mathcal{N}(A, u[y])$ is contained in $\mathcal{F}_1$ and in $\mathcal{F}_2$ iff it is contained in $\mathcal{F}_1 \cap \mathcal{F}_2$. By Proposition 3.3a, $\mathcal{N}(A, u[y])$ is open. Then $\mathcal{N}(A, u[y]) \subset \mathcal{F}$ implies

$$\mathcal{N}(A, u[y]) = u\mathcal{N}(A, u[y]) \subset u\mathcal{F}$$

Hence $\omega_{\mathcal{F}} = \omega_{u\mathcal{F}}$. In particular $\omega_{\tilde{u}\mathcal{F}} = \omega_{u\tilde{u}\mathcal{F}}$. But by (2.82), $u\tilde{u}\mathcal{F} = u\mathcal{F}$.

Monotonicity in A is obvious. If $F \in k\mathcal{F}$ and U is open containing $y \in \omega_{\mathcal{F}}\varphi[A_2]$ then $(f^F)^{-1}(U)$ is an open set meeting A_2 and therefore meeting A_1 if A_1 is dense in A_2. Thus equality holds for A_1 dense in A_2. By monotonicity,

$$\omega_{\mathcal{F}}\varphi[A_1] \cup \omega_{\mathcal{F}}\varphi[A_2] \subset \omega_{\mathcal{F}}\varphi[A_1 \cup A_2]$$

If y is not in $\omega_{\mathcal{F}}\varphi[A_1] \cup \omega_{\mathcal{F}}\varphi[A_2]$, then there exist neighborhoods U_1 and U_2 of y and subsets F_1 and F_2 in $k\mathcal{F}$ so that $A_\alpha \cap (f^{F_\alpha})^{-1}(U_\alpha) = \emptyset$ for $\alpha = 1,2$. Let $U = U_1 \cap U_2$ and $F = F_1 \cap F_2$.

$$(A_1 \cup A_2) \cap (f^F)^{-1}(U) = \emptyset$$

and if $\mathcal{F}$ is a filterdual, then $F \in k\mathcal{F}$. Hence $y \notin \omega_{\mathcal{F}}\varphi[A_1 \cup A_2]$.

(c) If $y \in \omega_{\mathcal{F}}\varphi[A]$ and U_1 is a neighborhood of $h(y)$ in X_1 then $N^{\varphi_1}(h(A), U_1) = N^{\varphi}(A, h^{-1}(U_1)) \in \mathcal{F}$ since $h^{-1}(U_1)$ is a neighborhood of y. Thus $h(y) \in \omega_{\mathcal{F}}\varphi_1[h(A)]$. If h is an embedding, i.e., a homeomorphism of X onto $h(X)$, then for every neighborhood U of y there exists U_1 a neighborhood of $h(y)$ such that $h^{-1}(U_1) \subset U$. So conversely $h(y) \in \omega_{\mathcal{F}}\varphi_1[h(A)]$ implies $y \in \omega_{\mathcal{F}}\varphi[A]$.

Now if $z \in \omega_{\mathcal{F}}\varphi_1[h(A)]$, then for all $F \in k\mathcal{F}$ $\overline{f_1^F(h(A))}$ contains z. If X is compact, then $h \circ f^t = f_1^t \circ h$ implies

$$\overline{f_1^F(h(A))} = \overline{h(f^F(A))} = h(\overline{f^F(A)})$$

Thus $\overline{f^F(A)}$ meets the compact set $h^{-1}(z)$. In particular $h^{-1}(z)$ is nonempty. If $k\mathcal{F}$ is a filter, then $\{h^{-1}(z) \cap \overline{f^F(A)} : F \in k\mathcal{F}\}$ is a filterbase of compacta, so the intersection $h^{-1}(z) \cap \omega_{\mathcal{F}}\varphi[A]$ is nonempty, proving the reverse inclusion of (3.28). Inclusion (3.30) follows from (3.28) and (3.29) applied to $h = f^t$.

(d) By (3.19), $\mathcal{N}(f^t(A), u[y]) = g^{-t}\mathcal{N}(A, u[y])$. If $\mathcal{N}(f^t(A), u[y]) \subset \mathcal{F}$ and $\mathcal{F}$ is translation $+$ invariant, then $\mathcal{N}(A, u[y]) \subset \mathcal{F}$, with the reverse implication if $\mathcal{F}$ is translation $-$ invariant.

(e) The inclusion (3.33) follows from (3.28) applied to the projection maps from the product to the factors. By (3.21), $\mathcal{N}^{\varphi_\alpha}(A_\alpha, u[y_\alpha]) \subset \mathcal{F}_\alpha$ $(\alpha = 1, 2)$ implies $\mathcal{N}^{\varphi_1 \times \varphi_2}(A_1 \times A_2, u[(y_1, y_2)]) \subset \mathcal{F}_1 \cdot \mathcal{F}_2$. This proves (3.32). When $\mathcal{F} = \mathcal{F} \cdot \mathcal{F}$, i.e., $\mathcal{F}$ is a filter, a similar argument yields equality in (3.33). Recall that if $U = \Pi_\alpha U_\alpha$ is a basic neighborhood of a point y in the product space, then $U_\alpha = X_\alpha$, and so $N^{\varphi_\alpha}(A_\alpha, U_\alpha) = T$, for all but finitely many values of α. ∎

Using (3.27) with A the singleton x, we define the $\mathcal{F}$ limit relation $\omega_{\mathcal{F}}\varphi \subset X \times X$ by:

$$\omega_{\mathcal{F}}\varphi(x) = \bigcap_{F \in k\mathcal{F}} \overline{f^F(x)} \subset X \tag{3.34}$$

By Proposition 3.6a, $y \in \omega_{\mathcal{F}}\varphi(x)$ when for every neighborhood U of y the hitting time set $N(x, U) = \{t : f^t(x) \in U\}$ is in the family $\mathcal{F}$. For the special case $\mathcal{F} = \mathcal{B}_T$ we omit the subscript in (3.27) and (3.34). This definition of $\omega\varphi$ agrees with the usual one [see Akin (1993)] when $T = \mathbf{Z}_+$ or $\mathbf{R}_+$.

Proposition 3.7. *Assume that $\mathcal{F}$ is a translation $+$ invariant family. For all $t \in T$ and $x \in X$:*

$$f^t(\omega_{\mathcal{F}}\varphi(x)) \subset \omega_{\mathcal{F}}\varphi(f^t(x)) \subset \omega_{\mathcal{F}}\varphi(x) \tag{3.35}$$

with equality for the right inclusion if $\mathcal{F}$ is translation invariant and equality for the left either if the action is uniformly reversible or if X is compact and $\mathcal{F}$ is a filterdual family. Thus in general each $\omega_{\mathcal{F}}\varphi(x)$ is a closed $+$ invariant subset of X.

PROOF. Apply (3.30) and (3.31) with $A = x$. ■

There is an odd dichotomy between $\omega_{\mathcal{F}}\varphi(x)$ and $\omega_{k\mathcal{F}}\varphi(x)$.

PROPOSITION 3.8. *Let $\varphi : T \times X \to X$ be a uniform action, $\mathcal{F}$ a proper family for T, and $x \in X$.*

Either $\omega_{\mathcal{F}}\varphi(x)$ and $\omega_{k\mathcal{F}}\varphi(x)$ are a common singleton set or at least one of the two sets $\omega_{\mathcal{F}}\varphi(x)$, $\omega_{k\mathcal{F}}\varphi(x)$ is empty. Thus if $\omega_{\mathcal{F}}\varphi(x)$ contains more than one point, then $\omega_{k\mathcal{F}}\varphi(x)$ is empty.

Suppose that X is compact and $\mathcal{F}$ is a filterdual. The set $\omega_{\mathcal{F}}\varphi(x)$ is nonempty; if it is a singleton set, then it equals $\omega_{k\mathcal{F}}\varphi(x)$. Otherwise $\omega_{k\mathcal{F}}\varphi(x)$ is empty.

PROOF. If $y \in \omega_{\mathcal{F}}\varphi(x)$ and U is a neighborhood of y, then $F_U = N(x,U) \in \mathcal{F}$. If U is closed

$$\omega_{k\mathcal{F}}\varphi(x) = \cap_{F \in \mathcal{F}} \overline{f^F(x)} \subset \overline{f^{F_U}(x)} \subset U$$

Intersecting over the closed neighborhoods of y, we see that $\omega_{k\mathcal{F}}\varphi(x) \subset \{y\}$. In particular if $y_1 \neq y$ is another element of $\omega_{\mathcal{F}}\varphi(x)$, then

$$\omega_{k\mathcal{F}}\varphi(x) \subset \{y\} \cap \{y_1\} = \emptyset$$

If neither $\omega_{\mathcal{F}}\varphi(x)$ nor $\omega_{k\mathcal{F}}\varphi(x)$ is empty, then they are this common singleton $\{y\}$.

If X is compact and $\mathcal{F}$ is a filterdual, then $\omega_{\mathcal{F}}\varphi(x)$ is nonempty by Proposition 3.5. If it contains more than one point, then $\omega_{k\mathcal{F}}\varphi(x)$ is empty. Assume that $\omega_{\mathcal{F}}\varphi(x) = \{y\}$. We prove that $y \in \omega_{k\mathcal{F}}\varphi(x)$, so $\omega_{\mathcal{F}}\varphi(x), \omega_{k\mathcal{F}}\varphi(x)$ must be this common singleton.

Let U be a neighborhood of y. By Proposition 3.5, $U \supset \omega_{\mathcal{F}}\varphi(x)$ implies $\overline{f^F(x)} \subset U$ for some $F \in k\mathcal{F}$. Hence $F \subset N(x,U)$, so $N(x,U) \in k\mathcal{F}$. Thus $y \in \omega_{k\mathcal{F}}\varphi(x)$. ■

The relation $\omega_{\mathcal{F}}\varphi$ is usually not a closed subset of $X \times X$. To obtain a closed relation, we define:

$$\Omega_{\mathcal{F}}\varphi = \bigcap_{F \in k\mathcal{F}} \overline{f^F} \subset X \times X \tag{3.36}$$

Let $1 : T \times X \to X$ denote the *trivial action* obtained by projecting on the second coordinate $1(t,x) = x$, so that the time t map is the identity 1_X for all $t \in T$. Recall that the map 1_X is just the diagonal subset of $X \times X$. If we take the product action $1 \times \varphi$ on $X \times X$, then the image of the subset 1_X under the time t map $1_X \times f^t$ is the subset $f^t \subset X \times X$. Taking the union over all t in $F \subset T$ we obtain

$$(1_X \times f)^F(1_X) = f^F \subset X \times X \tag{3.37}$$

So comparing (3.36) with (3.27) we see that:

$$\Omega_{\mathcal{F}}\varphi = \omega_{\mathcal{F}}(1 \times \varphi)[1_X] \tag{3.38}$$

PROPOSITION 3.9. *Let $\varphi : T \times X \to X$ be a uniform action and $\mathcal{F}, \mathcal{F}_1, \mathcal{F}_2$ proper families for T.*

a. For $(x,y) \in X \times X$ the following are equivalent:

(1) $y \in \Omega_{\mathcal{F}}\varphi(x)$

(2) For all $F \in k\mathcal{F}$ and W a neighborhood of x:

$$y \in \overline{f^F(W)}$$

(3) For all $F \in k\mathcal{F}$ and U a neighborhood of y:

$$x \in \overline{(f^F)^{-1}(U)}$$

(4) For all neighborhoods $W \times U$ of (x,y) in $X \times X$:

$$N^{\varphi}(W,U) \in \mathcal{F}$$

(5) $\mathcal{N}^{\varphi}(u[x], u[y]) \subset \mathcal{F}$

b. $\Omega_{\mathcal{F}_1}\varphi \cap \Omega_{\mathcal{F}_2}\varphi = \Omega_{\mathcal{F}_1 \cap \mathcal{F}_2}\varphi$. *In particular if $\mathcal{F}_1 \subset \mathcal{F}_2$, then $\Omega_{\mathcal{F}_1}\varphi \subset \Omega_{\mathcal{F}_2}\varphi$. Furthermore $\Omega_{u\mathcal{F}}\varphi = \Omega_{\mathcal{F}}\varphi = \Omega_{\tilde{u}\mathcal{F}}\varphi$.*

c. If $\varphi_1 : T \times X_1 \to X_1$ is a uniform action and $h : \varphi \to \varphi_1$ is a continuous action map, then:

$$\Omega_{\mathcal{F}}\varphi \subset (h \times h)^{-1}(\Omega_{\mathcal{F}}\varphi_1) \tag{3.39}$$

with equality if h is a dense embedding.

If $x \in X$, then:

$$h(\Omega_{\mathcal{F}}\varphi(x)) \subset \Omega_{\mathcal{F}}\varphi_1(h(x)) \tag{3.40}$$

with equality either if h is a homeomorphism or if X is compact, $\mathcal{F}$ is a filterdual, and h is open at x.

In particular with $t \in T$ and $x \in X$:

$$f^t(\Omega_{\mathcal{F}}\varphi(x)) \subset \Omega_{\mathcal{F}}(f^t(x)) \tag{3.41}$$

with equality if φ is reversible.

d. If $\mathcal{F}$ is translation + invariant, then each $\Omega_{\mathcal{F}}\varphi(x)$ is a closed φ + invariant subset of X; if $\mathcal{F}$ is translation invariant, then with $t \in T$ and $x \in X$:

$$f^t(\Omega_{\mathcal{F}}\varphi(x)) \subset \Omega_{\mathcal{F}}\varphi(x) \subset \Omega_{\mathcal{F}}\varphi(f^t(x)) \tag{3.42}$$

with equality if the action is reversible. Equality holds on the left if X is compact and $\mathcal{F}$ is a translation invariant filterdual.

e. If the action φ is uniformly reversible and $\overline{\varphi}$ is the reverse action, then:

$$\Omega_{\mathcal{F}}\overline{\varphi} = (\Omega_{\mathcal{F}}\varphi)^{-1} \tag{3.43}$$

i.e., $y \in \Omega_{\mathcal{F}}\overline{\varphi}(x)$ iff $x \in \Omega_{\mathcal{F}}\varphi(y)$.

f. If $\varphi_1 \times \varphi_2$ is the product of two uniform actions, then:

$$\Omega_{\mathcal{F}_1}\varphi_1 \times \Omega_{\mathcal{F}_2}\varphi_2 \subset \Omega_{\mathcal{F}_1 \cdot \mathcal{F}_2}(\varphi_1 \times \varphi_2) \tag{3.44}$$

If φ_π is the product of the uniform actions $\{\varphi_\alpha : T \times X_\alpha \to X_\alpha\}$ and $x = \{x_\alpha\} \in \Pi_\alpha X_\alpha$, then:

$$\Omega_{\mathcal{F}}\varphi_\pi(x) \subset \Pi_\alpha \Omega_{\mathcal{F}}\varphi_\alpha(x_\alpha) \tag{3.45}$$

with equality when $\mathcal{F}$ is a filter.

PROOF. (a) $(x,y) \in \Omega_{\mathcal{F}}\varphi$ iff every basic neighborhood $W \times U$ of (x,y) meets f^F for every $F \in k\mathcal{F}$. So the equivalence of (1), (2), and (3) is clear. The equivalence with (4) then follows from Lemma 3.1a. (4) $\Leftrightarrow$ (5) is obvious from the definition of $\mathcal{N}^{\varphi}(u[x], u[y])$.

From the equivalence of (1) and (2) we see that:

$$\Omega_{\mathcal{F}}\varphi(x) = \cap\{\omega_{\mathcal{F}}\varphi[W] : W \text{ a neighborhood of } x\} \tag{3.46}$$

(b) Apply Proposition 3.6b to $1 \times \varphi$ with $A = 1_X$, then use (3.38).

(c) For (3.39) observe that $h \times h$ maps $1 \times \varphi$ on $X \times X$ to $1 \times \varphi_1$ on $X_1 \times X_1$ and $h \times h(1_X) \subset 1_{X_1}$. Then the inclusion follows from (3.28) with $A = 1_X$ and (3.38).

If h is a dense embedding, then equality in (3.39) follows from (3.29), and:

$$\omega_{\mathcal{F}}(1 \times \varphi_1)[(h \times h)(1_X)] = \omega_{\mathcal{F}}(1 \times \varphi_1)[1_{X_1}]$$

which holds, since $h \times h(1_X)$ is dense in 1_{X_1}.

The inclusion (3.40) for all $x \in X$ is equivalent to (3.39) [see (3.2)]. We prove the reverse inclusion assuming X is compact, $\mathcal{F}$ is a filterdual, and h is open at x.

For U any neighborhood of x, since $h(U)$ is a neighborhood of $h(x)$, (3.46) implies

$$\Omega_{\mathcal{F}}\varphi_1(h(x)) \subset \omega_{\mathcal{F}}\varphi_1[h(U)] = h(\omega_{\mathcal{F}}\varphi[U])$$

This equality holds by (3.28) because X is compact and $\mathcal{F}$ is a filterdual. Shrinking U toward x, the compacta $\omega_{\mathcal{F}}\varphi[U]$ shrink to $\Omega_{\mathcal{F}}\varphi(x)$ by (3.46), so intersecting over U, we have $\Omega_{\mathcal{F}}\varphi_1(h(x)) \subset h(\Omega_{\mathcal{F}}\varphi(x))$ the reverse inclusion (3.40).

If h is a homeomorphism, then the reverse inclusion follows from (3.40) applied to h^{-1}.

As usual (3.41) follows from (3.40), with $h = f^t$.

(d) Given a neighborhood U_1 of $f^t(x)$ let U be a neighborhood of x contained in $f^{-t}(U_1)$, so that $f^t(U) \subset U_1$. Then by (3.30) and (3.31), $\mathcal{F}$ translation + invariant implies

$$f^t(\Omega_{\mathcal{F}}\varphi(x)) \subset f^t(\omega_{\mathcal{F}}\varphi[U]) \subset \omega_{\mathcal{F}}\varphi[f^t(U)] \subset \omega_{\mathcal{F}}\varphi[U]$$

We intersect over neighborhoods U of x to obtain $f^t(\Omega_{\mathcal{F}}\varphi(x)) \subset \Omega_{\mathcal{F}}\varphi(x)$ by (3.46).

When $\mathcal{F}$ is translation invariant, we use equality in (3.31) to obtain

$$\omega_{\mathcal{F}}\varphi[U] = \omega_{\mathcal{F}}\varphi[f^t(U)] \subset \omega_{\mathcal{F}}\varphi[U_1]$$

We intersect first over U as before, then over U_1 to obtain the right inclusion of (3.42). In the reversible case, the extremes are equal by (c).

If $\mathcal{F}$ is a filterdual as well as translation invariant and X is compact, then $f^t(\omega_{\mathcal{F}}\varphi[U]) = \omega_{\mathcal{F}}\varphi[U]$. These sets decrease to $\Omega_{\mathcal{F}}\varphi(x)$ as U decreases toward x. By compactness and continuity of f^t, they shrink to $f^t(\Omega_{\mathcal{F}}\varphi(x))$ as well.

(e) Symmetry between (2) and (3) of (a) yields (3.43).

(f) Apply Proposition 3.5e to (3.38). ■

As before we drop the subscript in the case $\mathcal{F} = \mathcal{B}_T$.

As usual there are special results in the compact case.

PROPOSITION 3.10. *Let $\varphi : T \times X \to X$ be a uniform action and $\mathcal{F}$ a filterdual for T. If A is a compact subset of X, then:*

$$\Omega_{\mathcal{F}}\varphi(A) = \cap\{\omega_{\mathcal{F}}\varphi[U] : U \text{ is a neighborhood of } A\} \tag{3.47}$$

If in addition X is compact, then:

$$\omega_{\mathcal{F}}\varphi[A] \subset \Omega_{\mathcal{F}}\varphi(A) = \cap\{\Omega_{\mathcal{F}}\varphi(U) : U \text{ is a neighborhood of } A\} \tag{3.48}$$

PROOF. By (3.46), $\Omega_{\mathcal{F}}\varphi(A)$ is contained in the intersection on the right-hand side of (3.47). If y is not in $\Omega_{\mathcal{F}}\varphi(A)$, then by (3.46) and compactness, there is an open cover $\{U_1, \dots, U_k\}$ of A and a subset $\{F_1, \dots, F_k\}$ of $k\mathcal{F}$ such that $\overline{f^{F_i}(U_i)}$ does not contain y. With $U = \cup_{i=1}^k U_i$ and $F = \cap_{i=1}^k F_i$, we have $y \notin \overline{f^F(U)}$. Since $k\mathcal{F}$ is a filter, $F \in k\mathcal{F}$, so y is not in the intersection.

If X is compact, then for $F \in k\mathcal{F}$, $\overline{f^F}(A)$ is the projection on the second factor of the compact subset $\overline{f^F} \cap (A \times X) \subset X \times X$. Hence $\overline{f^F(A)} \subset \overline{f^F}(A)$. Now we intersect over $F \in k\mathcal{F}$. Since $\overline{f^F} \cap (A \times X)$ is a filterbase of compacta, intersection commutes with projection. Thus:

$$\omega_{\mathcal{F}}\varphi[A] = \cap_{F\in k\mathcal{F}} \overline{f^F(A)} \subset \cap_{F\in k\mathcal{F}} (\overline{f^F}(A)) = (\cap_{F\in k\mathcal{F}} \overline{f^F})(A) = \Omega_{\mathcal{F}}\varphi(A)$$

Similarly restricting to closed neighborhoods U of A, $\Omega_{\mathcal{F}}\varphi(U)$ is a filterbase of compacta, so $\cap_U \Omega_{\mathcal{F}}\varphi(U) = \Omega_{\mathcal{F}}\varphi(\cap_U U) = \Omega_{\mathcal{F}}\varphi(A)$. ∎

REMARK. Thus if X is compact, A a closed subset of X, $\mathcal{F}$ a filterdual, and V a neighborhood of $\Omega_{\mathcal{F}}\varphi(A)$, it follows from (3.47) that $\omega_{\mathcal{F}}\varphi[U] \subset \operatorname{Int} V$ for some neighborhood U of A. Then by Proposition 3.5, $f^F(U) \subset V$ for some $F \in k\mathcal{F}$. Thus for every neighborhood V of $\Omega_{\mathcal{F}}\varphi(A)$, there exists a neighborhood U of A such that $J^{\varphi}(U,V) \in k\mathcal{F}$. ∎

We now consider the situation dual to that of Proposition 3.6.

PROPOSITION 3.11. *Let $\varphi : T \times X \to X$ be a uniform action, B, B_1, B_2 nonempty subsets of X, and $\mathcal{F}, \mathcal{F}_1, \mathcal{F}_2$ proper families for T.*

a. We say that $x \in X$ $\mathcal{F}$ adheres *to B when the following four equivalent conditions hold:*

(1) For every $V \in \mathcal{U}_X$ and $F \in k\mathcal{F}$, $x \in (f^F)^{-1}(V(B))$

(2) For every $V \in \mathcal{U}_X$, $N(x, V(B)) \in \mathcal{F}$

(3) For every $V \in \mathcal{U}_X$, there exists $F \in \mathcal{F}$ such that $f^F(x) \subset V(B)$

(4) $\mathcal{N}^{\varphi}(x, u[B]) \subset \mathcal{F}$

For x to $\mathcal{F}$ adhere to B, it is sufficient that:

(5) For all $F \in k\mathcal{F}$, $\overline{f^F(x)} \cap B \neq \emptyset$

If B is compact, condition (5) is necessary as well.

b. x $\mathcal{F}$ adheres to B iff x $\mathcal{F}$ adheres to $\overline{B}$. If $B_1 \subset B_2$, then x $\mathcal{F}$ adheres to B_1 implies x $\mathcal{F}$ adheres to B_2. x $\mathcal{F}$ adheres to a point y iff $y \in \omega_{\mathcal{F}}\varphi(x)$. In particular if $\omega_{\mathcal{F}}\varphi(x) \cap B \neq \emptyset$, then x $\mathcal{F}$ adheres to B. If x $k\mathcal{F}$ adheres to B and B is closed, then $\omega_{\mathcal{F}}\varphi(x) \subset B$.

c. x $\mathcal{F}_1$ adheres to B and $\mathcal{F}_2$ adheres to B iff x $\mathcal{F}_1 \cap \mathcal{F}_2$ adheres to B. In particular if $\mathcal{F}_1 \subset \mathcal{F}_2$ and x $\mathcal{F}_1$ adheres to B, then x $\mathcal{F}_2$ adheres to B. Furthermore x $\mathcal{F}$ adheres to B iff x $u\mathcal{F}$ adheres to B.

d. If $\varphi_1 : T \times X_1 \to X_1$ is a uniform action and $h : \varphi \to \varphi_1$ is a uniformly continuous action map, then x $\mathcal{F}$ adheres to B with respect to φ implies $h(x)$ $\mathcal{F}$ adheres to $h(B)$ with respect to φ_1. In particular if $t \in T$ and x $\mathcal{F}$ adheres to B, then $f^t(x)$ $\mathcal{F}$ adheres to $f^t(B)$; if B is $\varphi+$ invariant, then $f^t(x)$ $\mathcal{F}$ adheres to B.

e. If $\mathcal{F}$ is translation + invariant and $t \in T$, then $f^t(x)$ $\mathcal{F}$ adheres to B implies x $\mathcal{F}$ adheres to B, which implies x $\mathcal{F}$ adheres to $f^t(B)$. If $\mathcal{F}$ is translation invariant, then $f^t(x)$ $\mathcal{F}$ adheres to B iff x $\mathcal{F}$ adheres to B.

f. If B is φ + invariant and x $\mathcal{F}$ adheres to B, then x $\tau\mathcal{F}$ adheres to B. If B is invariant as well as $\pm$ invariant, X is compact, and $\mathcal{F}$ is a full family, then x $\tilde{\gamma}\tau\mathcal{F}$ adheres to B (where $\tilde{\gamma}\tau\mathcal{F} = \tau\tilde{\gamma}\mathcal{F}$ is the largest invariant, thick family contained in $\mathcal{F}$, cf. Proposition 2.12).

g. If B is compact, B_1 is closed, x $\mathcal{F}$ adheres to B, and x $\mathcal{F}_1$ adheres to B_1, then x $\mathcal{F} \cdot \mathcal{F}_1$ adheres to $B \cap B_1$. In particular if $\mathcal{F} = \mathcal{F}_1$ is a filter, then x $\mathcal{F}$ adheres to $B \cap B_1$. Assume $\{B_\alpha\}$ is a family of nonempty compacta in X and x $\mathcal{F}$ adheres to B_α for all α. If either $\mathcal{F}$ is a filter or $\{B_\alpha\}$ is a filterbase, then x $\mathcal{F}$ adheres to $B = \cap_\alpha B_\alpha$.

h. Assume $\mathcal{F}$ is a filterdual. If B is compact, then x $\mathcal{F}$ adheres to B iff $\omega_{\mathcal{F}}\varphi(x) \cap B \neq \emptyset$, i.e., iff x $\mathcal{F}$ adheres to some point of B. If in addition X is compact, then x $k\mathcal{F}$ adheres to B iff $\omega_{\mathcal{F}}\varphi(x) \subset B$.

PROOF. (a) The equivalence of (1)–(4) follows from the usual application of Lemma 3.1a. These are also equivalent to the condition that $f^F(x)$ meet $V(B)$ for every $V \in \mathcal{U}_X$ and every $F \in k\mathcal{F}$. So condition (5) is sufficient in any case and necessary when B is compact.

(b) These are easy, e.g., $V(B) = V(\overline{B})$ for all $V \in \mathcal{U}_X$. If x $k\mathcal{F}$ adheres to B, then it cannot also $\mathcal{F}$ adhere to a point y with $y \notin \overline{B}$. Hence $\omega_{\mathcal{F}}\varphi(x) \subset \overline{B}$.

(c) These are clear from (4) of (a) and the observation that $\mathcal{N}(x, u[B])$ is open.

(d) Apply (3.22)

(e) By (3.19):

$$\mathcal{N}(f^t(x), u[B]) = g^{-t}\mathcal{N}(x, u[B])$$

By (3.22):

$$g^{-t}\mathcal{N}(x, u[f^t(B)]) \subset \mathcal{N}(x, u[f^{-t}f^t(B)]) \subset \mathcal{N}(x, u[B])$$

Results follow from condition (4) of (a).

(f) By Proposition 3.4, $\mathcal{N}(x, u[B])$ is thick when B is + invariant, so $\mathcal{N}(x, u[B]) \subset \mathcal{F}$ implies $\mathcal{N}(x, u[B]) \subset \tau\mathcal{F}$. If X is compact and B is $\pm$ invariant as well as invariant, then $\mathcal{N}(x, u[B]) \cdot k\mathcal{B}_T$ is translation invariant, so when $\mathcal{F}$ is full, $\mathcal{N}(x, u[B])$ is contained in the full family $\tilde{\gamma}\tau\mathcal{F}$.

(g) Given $V \in \mathcal{U}_X$, choose $V_1 \in \mathcal{U}_X$ and open so that $V_1 \circ V_1 \subset V$. The set $B \backslash V_1(B \cap B_1)$ is compact and disjoint from the closed set B_1. Then there exists $V_2 \in \mathcal{U}_X$ symmetric, with $V_2 \subset V_1$ and such that

$$(V_2 \circ V_2(B \backslash V_1(B \cap B_1))) \cap B_1 = \emptyset$$

Hence:

$$V_2(B\backslash V_1(B\cap B_1))\cap V_2(B_1)=\emptyset$$

then

$$V_2(B)\cap V_2(B_1)\subset V_2\circ V_1(B\cap B_1)\subset V(B\cap B_1)$$

Hence:

$$N(x,V(B\cap B_1))\supset N(x,V_2(B))\cap N(x,V_2(B_1))$$

The latter is in $\mathcal{F}\cdot\mathcal{F}_1$. In particular if $\mathcal{F}_1=\mathcal{F}$ is a filter, then $\mathcal{F}\cdot\mathcal{F}_1=\mathcal{F}$ implies x $\mathcal{F}$ adheres to $B\cap B_1$.

For the family of compacta $\{B_\alpha\}$ and $V\in\mathcal{U}_X$, there exist indices $\{\alpha_1,\cdots,\alpha_k\}$ and $V_1\in\mathcal{U}_X$ so that:

$$V(B)\supset V_1(B_{\alpha_1}\cap\cdots\cap B_{\alpha_k})$$

If $\mathcal{F}$ is a filter, then by the preceding x $\mathcal{F}$ adheres to $B_{\alpha_1}\cap\cdots\cap B_{\alpha_k}$. Therefore $N(x,V(B))$ is in $\mathcal{F}$, showing that x $\mathcal{F}$ adheres to B. If $\{B_\alpha\}$ is a filterbase, then there exists α^* so that $B_{\alpha^*}\subset B_{\alpha_1}\cap\cdots\cap B_{\alpha_k}$. Hence $V(B)\supset V_1(B_{\alpha^*})$. As x $\mathcal{F}$ adheres to B_{α^*} it follows that $N(x,V(B))\in\mathcal{F}$ as before.

(h) Since B is compact, x $\mathcal{F}$ adheres to B implies $\overline{f^F(x)}\cap B\neq\emptyset$ for all $F\in k\mathcal{F}$. Since $\mathcal{F}$ is a filterdual, this collection is a filterbase of compacta so the intersection $\omega_{\mathcal{F}}\varphi(x)\cap B$ is nonempty. The reverse implication is in (b). If X is compact and U is a neighborhood of B, then by Proposition 3.5 $\omega_{\mathcal{F}}\varphi(x)\subset B$ implies there exists $F\in k\mathcal{F}$ such that $\overline{f^F(x)}\subset U$. Then x $k\mathcal{F}$ adheres to B by (3) of (a). Again the reverse implication was proved in (b). ■

PROPOSITION 3.12. *Let $\varphi: T\times X\to X$ be a uniform action with X compact, B a closed subset of X and $\mathcal{F}$ a full family for T.*

a. x $\mathcal{F}$ adheres to B iff x $\mathcal{F}$ adheres to $B\cap\omega\varphi(x)$.

b. If $\mathcal{F}$ is translation invariant, then the following conditions are equivalent:

(1) x $\tau\mathcal{F}$ adheres to B.

(2) There exists a closed $\varphi+$ invariant subset B_1 such that $B_1\subset B$ and x $\mathcal{F}$ adheres to B_1.

(3) There exists a nonempty, closed φ invariant subset B_1 such that $B_1\subset B\cap\omega\varphi(x)$ and x $\tau\mathcal{F}$ adheres to B_1.

PROOF. (a) By Proposition 3.11h, x $k\mathcal{B}_T$ adheres to $\omega\varphi(x)$. If x $\mathcal{F}$ adheres to B, then by Proposition 3.11g, x $\mathcal{F}\cdot k\mathcal{B}_T$ adheres to $B\cap\omega\varphi(x)$. Since $\mathcal{F}$ is full, $\mathcal{F}\cdot k\mathcal{B}_T=\mathcal{F}$, so x $\mathcal{F}$ adheres to $B\cap\omega\varphi(x)$. The converse is obvious.

(b) (3) $\Rightarrow$ (2) is obvious. If x $\mathcal{F}$ adheres to B_1 and B_1 is $\varphi+$ invariant, then by (f) x $\tau\mathcal{F}$ adheres to B_1. Since $B_1\subset B$, x $\tau\mathcal{F}$ adheres to B. Hence (2) $\Rightarrow$ (1).

(1) $\Rightarrow$ (3). By (a) x $\tau\mathcal{F}$ adheres to $\tilde{B}=B\cap\omega\varphi(x)$. Thus $\mathcal{N}(x,u[\tilde{B}])$ is a filter contained in $\tau\mathcal{F}$. By Proposition 2.6, there exists a translation invariant filterdual $\tilde{\mathcal{F}}$ such that $\mathcal{N}(x,u[\tilde{B}])\subset k\tilde{\mathcal{F}}\subset\tau\mathcal{F}$. Then x $k\tilde{\mathcal{F}}$ adheres to $\tilde{B}$, which implies by Proposition 3.11h, that $B_1=\omega_{\tilde{\mathcal{F}}}\varphi(x)\subset\tilde{B}$, and x $k\tilde{\mathcal{F}}$ adheres to B_1. Hence x $\tau\mathcal{F}$ adheres to B_1. Because $\tilde{\mathcal{F}}$ is a translation invariant filterdual, Proposition 3.6c and d imply that B_1 is φ invariant. ■

PROPOSITION 3.13. *A point x does* not *$\mathcal{F}$ adhere to a closed set B iff there exists a closed set B_1 separated from B such that x does $k\mathcal{F}$ adhere to B_1.*

PROOF. If $V(B)\cap V(B_1)=\emptyset$ for some $V\in\mathcal{U}_X$, then:

$$N(x,V(B))\cap N(x,V(B_1))=\emptyset$$

If $N(x,V(B_1))\in k\mathcal{F}$, then $N(x,V(B))\notin\mathcal{F}$. Conversely suppose $V\in\mathcal{U}_X$, V open, and $N(x,V(B))\notin\mathcal{F}$. Let B_1 be the closed set $X\backslash V(B)$ which is separated from B. $N(x,B_1)=T\backslash N(x,V(B))\in k\mathcal{F}$ by definition of $k\mathcal{F}$, (2.5). A fortiori x $k\mathcal{F}$ adheres to B_1. ■

PROPOSITION 3.14. *Let $\varphi:T\times X\to X$ be a uniform action.*

a. φ is called a minimal action *if it satisfies the following two equivalent conditions:*

(1) For all $x\in X$, $\omega\varphi(x)=X$.

(2) If A is a closed, nonempty, $+$ invariant subset of X then $A=X$.

These conditions imply:

(3) If A is a closed, nonempty, invariant subset of X, then $A=X$.

If X is compact, then condition (3) is sufficient for minimality.

b. If $\varphi_1:T\times X_1\to X_1$ is a uniform action and $h:\varphi\to\varphi_1$ is a surjective, continuous action map, then φ minimal implies φ_1 minimal.

c. If B is any $+$ invariant subset of X with φ minimal, then the subsystem φ_B obtained by restricting φ to B is a minimal system (Note: we do not assume that B is closed). For any uniform action, we call a subset B a minimal subset *if B is a nonempty, closed, $+$ invariant subset such that the subsystem φ_B is minimal. Any compact, nonempty, $+$ invariant subset contains a minimal subset.*

PROOF. (a) (1) $\Rightarrow$ (2). If $x \in A$, then $X = \omega\varphi(x) \subset A$ because A is closed and $+$ invariant.

(2) $\Rightarrow$ (1). For any $x \in X$ and $t \in T$, $f^{T_t}(x) = \{f^s(x) : s = t + s_1 \text{ for some } s_1 \in T\}$ is $+$ invariant as is its closure. Then $\overline{f^{T_t}(X)} = X$ by (2). The collection of sets $T_t = g^t(T)$ generates $k\mathcal{B}_T$. Intersecting as t varies over T, $\omega\varphi(x) = \omega_{\mathcal{B}_T}\varphi(x) = X$.

(2) $\Rightarrow$ (3) is obvious. If A is compact, nonempty and $+$ invariant, then:

$$\omega\varphi[A] = \cap_t \overline{f^{T_t}(A)} = \cap_t f^t(A)$$

is nonempty, invariant (cf. Propositions 3.5 and 3.6c,d) and contained in A. Then if (3) holds, $\omega\varphi[A] = X$ and a fortiori $A = X$, proving (2).

(b) By (3.28), $h(\omega\varphi(x)) \subset \omega\varphi_1(h(x))$. Thus if h is surjective, $\omega\varphi(x) = X$ implies $\omega\varphi_1(h(x)) = X_1$.

(c) By (3.29), with h the inclusion map of B into X, for $x \in B$ we have:

$$\omega\varphi_B(x) = B \cap \omega\varphi(x)$$

Then $\omega\varphi(x) = X$ implies $\omega\varphi_B(x) = B$. The existence of minimal subsets in any compact $+$ invariant set follows from the usual Zorn's Lemma argument. ∎

PROPOSITION 3.15. *Let $\varphi : T \times X \to X$ be a uniform action with X compact, B a compact minimal subset of X, $x \in X$, and $\mathcal{F}$ a translation invariant family for T. The following conditions are equivalent:*

(1) x $\tau k\tau\mathcal{F}$ adheres to B.

(2) x $k\tau\mathcal{F}$ adheres to B.

(3) $B \cap \omega_{k\tau\mathcal{F}}\varphi(x) \neq \emptyset$

(4) $B \subset \omega_{k\tau\mathcal{F}}\varphi(x)$

PROOF. (4) $\Rightarrow$ (3). B is nonempty.

(3) $\Rightarrow$ (2). Apply Proposition 3.11b.

(1) $\Leftrightarrow$ (2). Apply Proposition 3.11f since B is $+$ invariant.

(2) $\Rightarrow$ (4). We prove the contrapositive beginning with $y \in B$ such that $y \notin \omega_{k\tau\mathcal{F}}\varphi(x)$; i.e., x does not $k\tau\mathcal{F}$ adhere to y. Then by Proposition 3.13 there exists B_1 a closed set such that $y \notin B_1$ but x $\tau\mathcal{F}$ adheres to B_1. By Proposition 3.12, there exists B_2 closed, nonempty, invariant with $B_2 \subset B_1$ such that x $\tau\mathcal{F}$ adheres to B_2. The subset $B_2 \cap B$ of B is closed and $+$ invariant but it is not all of B because $y \notin B_2$. Therefore by minimality it is empty; i.e., B_2 and B are disjoint. Since they are compact, they are separated. So by Proposition 3.13, x does not $k\tau\mathcal{F}$ adhere to B; thus not (4) $\Rightarrow$ not (2). ∎

If $\varphi : T \times X \to X$ is a uniform action, then we call x a *minimal point* if $x \in M$ for some minimal subset M of X. By Proposition 3.14 this says $\omega\varphi(x)$ is a minimal subset and $x \in \omega\varphi(x)$. In that case, $\omega\varphi(x) = M$ is the unique minimal set containing x.

If $\mathcal{F}$ is a family for T, then we call x an $\mathcal{F}$ *recurrent point* if $x \in \omega_{\mathcal{F}}\varphi(x)$, and an $\mathcal{F}$ *nonwandering point* if $x \in \Omega_{\mathcal{F}}\varphi(x)$. As usual we omit $\mathcal{F}$ when $\mathcal{F} = \mathcal{B}_T$. In general for any relation F on X, we denote by $|F|$ the set of points x such that $x \in F(x)$ or equivalently $(x,x) \in F$. Then $|F|$ is a closed subset of X when F is a closed relation. Thus $|\omega_{\mathcal{F}}\varphi|$ and $|\Omega_{\mathcal{F}}\varphi|$ are the set of $\mathcal{F}$ recurrent points and the closed set of $\mathcal{F}$ nonwandering points, respectively. We denote by $|\varphi|$ the closed set of *fixed points* for the action:

$$|\varphi| = \cap_{t \in T}|f^t| = \{x : f^t(x) = x \text{ for all } t \in T\} \tag{3.49}$$

THEOREM 3.1. *Let $\varphi : T \times X \to X$ be a uniform action with X compact. Let B be a closed subset of X and x a point of X.*

a. By definition, $\omega_{\mathcal{B}_T}\varphi(x) = \omega\varphi(x)$; $\omega_{k\mathcal{B}_T}\varphi(x) = \emptyset$ unless $\omega\varphi(x)$ is a singleton $\{y\}$. In that case, y is a fixed point to which the filter generated by $\{f^{T_t}(x)\}$ converges and $\omega_{k\mathcal{B}_T}\varphi(x) = \omega\varphi(x) = \{y\}$.

$\omega_{k\tau\mathcal{B}_T}\varphi(x) = \emptyset$ unless $\omega\varphi(x)$ contains a unique minimal subset M, in which case $\omega_{k\tau\mathcal{B}_T}\varphi(x) = M$; $\omega_{\tau\mathcal{B}_T}\varphi(x)$ is the possibly empty set of fixed points in $\omega\varphi(x)$, i.e., $\omega_{\tau\mathcal{B}_T}\varphi(x) = |\varphi| \cap \omega\varphi(x)$.

$\omega_{k\tau k\tau\mathcal{B}_T}\varphi(x)$ is the mincenter of $\omega\varphi(x)$, that is, the closure of the union of all minimal subsets of $\omega\varphi(x)$; $\omega_{\tau k\tau\mathcal{B}_T}\varphi(x) = \emptyset$ unless there is a fixed point y that is the unique minimal subset of $\omega\varphi(x)$. In that case, $\omega_{\tau k\tau\mathcal{B}_T}\varphi(x) = \omega_{k\tau k\tau\mathcal{B}_T}\varphi(x) = \{y\}$.

b. x adheres to B; i.e., x $\mathcal{B}_T$ adheres to B iff $\omega\varphi(x) \cap B \neq \emptyset$; x $k\mathcal{B}_T$ adheres to B iff $\omega\varphi(x) \subset B$.

x $k\tau\mathcal{B}_T$ adheres to B iff $B \cap M \neq \emptyset$ for every minimal subset M of $\omega\varphi(x)$; x $\tau\mathcal{B}_T$ adheres to B iff B contains some minimal subset M of $\omega\varphi(x)$.

x $k\tau k\tau\mathcal{B}_T$ adheres to B iff B meets the mincenter of $\omega\varphi(x)$; x $\tau k\tau\mathcal{B}_T$ adheres to B iff B contains the mincenter of $\omega\varphi(x)$.

c. x is a $\mathcal{B}_T$ recurrent point iff $x \in \omega\varphi(x)$; x is a $k\mathcal{B}_T$ recurrent point iff x is a $\tau\mathcal{B}_T$ recurrent point iff x is a fixed point, i.e., $x \in |\varphi|$.

x is a $k\tau\mathcal{B}_T$ recurrent point iff x is a minimal point; i.e., $x \in \omega\varphi(x)$ and $\omega\varphi(x)$ is a minimal subset. x is a $k\tau k\tau\mathcal{B}_T$ recurrent point iff the minimal points of $\omega\varphi(x)$ are dense in $\omega\varphi(x)$.

PROOF. First recall the results that hold when $\mathcal{F}$ is a filterdual. By Proposition 3.8, $\omega_{\mathcal{F}}\varphi(x)$ is nonempty, while $\omega_{k\mathcal{F}}\varphi(x)$ is empty unless $\omega_{\mathcal{F}}\varphi(x)$ is a singleton, in which case, $\omega_{k\mathcal{F}}\varphi(x) = \omega_{\mathcal{F}}\varphi(x)$. Furthermore by Proposition 3.11h, x $\mathcal{F}$ adheres to B (or x $k\mathcal{F}$ adheres to B) iff $\omega_{\mathcal{F}}\varphi(x) \cap B \neq \emptyset$ (resp. $\omega_{\mathcal{F}}\varphi(x) \subset B$).

$\mathcal{B}_T$ is a filterdual and $\omega\varphi(x)$ is $\omega_{\mathcal{B}_T}\varphi(x)$. Then results for $\mathcal{B}_T$ and $k\mathcal{B}_T$ follow.

$\mathcal{F} = \tau\mathcal{B}_T$. By Proposition 3.12b, x $\tau\mathcal{B}_T$ adheres to B implies that $B \cap \omega\varphi(x)$ contains some closed invariant set and hence some minimal set by compactness. Conversely if M is a minimal subset of $B \cap \omega\varphi(x)$ then x $\mathcal{B}_T$ adheres to M since $M \subset \omega\varphi(x)$. Therefore x $\tau\mathcal{B}_T$ adheres to B by Proposition 3.12b. In particular x $\tau\mathcal{B}_T$ adheres to a point y iff y contains and hence it is a minimal subset of $B \cap \omega\varphi(x)$, i.e., $f^t(y) = y$ for all $t \in T$. The collection of all such points y, $\omega_{\tau\mathcal{B}_T}\varphi(x)$, is thus the set of fixed points in $\omega\varphi(x)$ if any.

$\mathcal{F} = k\tau\mathcal{B}$. x does not $k\tau\mathcal{B}_T$ adhere to B iff x $\tau\mathcal{B}_T$ adheres to some closed nonempty set B_1 disjoint from B. Thus x does not $k\tau B_T$ adhere to B iff there is a minimal subset of $\omega\varphi(x)$ disjoint from B. Contrapositively x $k\tau\mathcal{B}_T$ adheres to B iff B meets every minimal subset of $\omega\varphi(x)$. In particular x $k\tau\mathcal{B}_T$ adheres to y iff y meets every minimal subset of $\omega\varphi(x)$. Since distinct minimal subsets are disjoint, $\omega_{k\tau\mathcal{B}_T}\varphi(x)$ is empty unless $\omega\varphi(x)$ contains a unique minimal subset M, in which case $\omega_{k\tau\mathcal{B}_T}\varphi(x) = M$.

Since $k\mathcal{B}_T$ is a translation invariant filter, $\tau k\tau\mathcal{B}_T = \tau k\tau k(k\mathcal{B}_T)$ is a translation invariant filter by Proposition 2.9, and so $\mathcal{F} = k\tau k\tau\mathcal{B}_T$ is a translation invariant filterdual. x does not $\mathcal{F}$ adhere to B iff x $k\mathcal{F} = \tau k\tau\mathcal{B}_T$ adheres to some closed set B_1 disjoint from B (Proposition 3.13) and so iff x $k\tau\mathcal{B}_T$ adheres to some invariant subset B_2 contained in $B_1 \cap \omega\varphi(x)$ (Proposition 3.12). Thus x does not $\mathcal{F}$ adhere to B iff x $k\tau\mathcal{B}_T$ adheres to some closed invariant subset B_2 of $\omega\varphi(x)$ disjoint from B. As we saw earlier, x $k\tau\mathcal{B}_T$ adheres to B_2 iff B_2 meets every minimal subset M of $\omega\varphi(x)$. Since B_2 is invariant, $B_2 \cap M \neq \emptyset$ implies $B_2 \supset M$. Thus x $k\tau\mathcal{B}_T$ adheres to the closed invariant set B_2 iff B_2 contains the mincenter of $\omega\varphi(x)$. Contrapositively x $\mathcal{F}$ adheres to B iff B meets the mincenter of $\omega\varphi(x)$. In particular x $\mathcal{F}$ adheres to y iff y is in the mincenter of $\omega\varphi(x)$. Therefore $\omega_{\mathcal{F}}\varphi(x)$ is the mincenter of $\omega\varphi(x)$. It follows that x $k\mathcal{F}$ adheres to B iff $\omega_{\mathcal{F}}\varphi(x) \subset B$ and $\omega_{k\mathcal{F}}\varphi(x)$ is empty unless $\omega_{\mathcal{F}}\varphi(x)$ is a singleton. ■

REMARK. Since our actions are uniform, we can replace $\tau\mathcal{B}_T$ and $k\tau\mathcal{B}_T$ by the thick sets $u\tau\mathcal{B}_T$ and their dual, $ku\tau\mathcal{B}_T$ [cf. (2.104)]. Recall that by (2.105) $uku\tau\mathcal{B}_T = uk\tau\mathcal{B}_T$. Similarly we can replace $\tau k\tau\mathcal{B}_T$ by the family of replete sets, $u\tau k\tau\mathcal{B}_T$ [cf. (2.106)]. These replacements leave the notions of adherence and limit set unaffected. ■

Recall that for any collection $\mathcal{A}$ of nonempty subsets of T the family $[\mathcal{A}]$, generated by $\mathcal{A}$, is $\{F \in \mathcal{P} : F \supset F_1 \text{ for some } F_1 \in \mathcal{A}\}$. For example $k\mathcal{B}_T$ is generated by the set of tails $\{T_t : t \in T\}$, where $T_t = g^t(T)$. We say that $\mathcal{A}$ generates $\mathcal{F}$, if $[\mathcal{A}] = \mathcal{F}$ and we call $\mathcal{F}$ *countably generated* if $\mathcal{F}$ is generated by some countable collection $\mathcal{A}$.

PROPOSITION 3.16. *Assume that the uniform monoid T is separable metric; i.e., T has a countable dense set and $\mathcal{U}_T$ is countably generated.*

a. The family $k\mathcal{B}_T$ is a countably generated filter.

b. If $\mathcal{F}$ and $\mathcal{F}_1$ are countably generated families, then $\mathcal{F} \cdot \mathcal{F}_1$, $u\mathcal{F}$, and $u\gamma\mathcal{F} = \gamma u(\mathcal{F} \cdot k\mathcal{B}_T)$ are countably generated families. If $\mathcal{F}$ is contained in some filter, then the smallest filter containing $\mathcal{F}$ is countably generated.

c. Assume $\mathcal{F}$ is a countably generated filter and $\mathcal{F} \subset \tau\mathcal{B}_T$. Recall that $\gamma_\infty\mathcal{F}$ is the smallest translation invariant filter containing $\mathcal{F}$. The translation invariant filter $\gamma_\infty u(\mathcal{F} \cdot k\mathcal{B}_T)$ is countably generated.

PROOF. It is convenient to use an invariant metric d_T defined via the group G_T. Let T_0 be a countable dense subset of T.

(a) Given $t \in T$, there exist $\varepsilon > 0$ and $t_1 \in T$ such that $T_t \supset V_\varepsilon(T_{t_1})$ because T is uniform [cf. (1.6)]. If $d_T(t_1, t_0) < \varepsilon$ and $t_0 \in T_0$, then by invariance of the metric, $T_{t_0} \subset V_\varepsilon(T_{t_1}) \subset T_t$. Hence $\{T_{t_0} : t_0 \in T_0\}$ generates $k\mathcal{B}_T$.

(b) If $\mathcal{F}$ and $\mathcal{F}_1$ are generated by $\mathcal{A}$ and $\mathcal{A}_1$, then $\mathcal{F} \cdot \mathcal{F}_1$ is generated by $\{F \cap F_1 : F \in \mathcal{A} \text{ and } F \in \mathcal{A}_1\}$. In particular $\mathcal{F} \cdot k\mathcal{B}_T$ is generated by $\{F \cap T_{t_0} : F \in \mathcal{A}$ and $t_0 \in T_0\}$. $u\mathcal{F}$ is generated by $\{V_\varepsilon(F) : F \in \mathcal{A}$ and $\varepsilon > 0$ rational $\}$. The smallest family containing $\mathcal{F}$ and closed under intersection is generated by intersections of finite subcollections of $\mathcal{A}$. By (2.62) and (2.101) $\gamma u(\mathcal{F} \cdot k\mathcal{B}_T) = u\gamma(\mathcal{F} \cdot k\mathcal{B}_T)$ is generated by $\{V_\varepsilon(g^{t_1}(g^{-t_2}(F) \cap T_{t_3})) : t_1, t_2, t_3 \in T_0$, and $\varepsilon > 0$ rational and $F \in \mathcal{A}\}$. By Corollary 2.1 of Proposition 2.5, $\gamma\mathcal{F}$ is full and equals $\gamma(\mathcal{F} \cdot k\mathcal{B}_T)$.

(c) By (2.65) $\gamma_\infty u(\mathcal{F} \cdot k\mathcal{B}_T) = \bigcup_n (\gamma u(\mathcal{F} \cdot k\mathcal{B}_T))^n$, so the result follows from (b). ■

Notice that if $\mathcal{A}$ generates $k\mathcal{F}$, the intersections in (3.27), (3.34), and (3.36) need only be taken over the Fs in $\mathcal{A}$.

PROPOSITION 3.17. *Let $\varphi : T \times X \to X$ be a uniform action and $\mathcal{F}$ a proper family for T.*

a. If X is a separable metric space, then there exists $\mathcal{F}_1 \supset \mathcal{F}$ such that $k\mathcal{F}_1$ is countably generated and $\Omega_\mathcal{F}\varphi = \Omega_{\mathcal{F}_1}\varphi$. If $\mathcal{F}$ is a filterdual, then $\mathcal{F}_1$ can be chosen to be a filterdual.

b. Assume X is a Baire space, e.g., a locally compact space or a complete metric space, and $k\mathcal{F}$ is countably generated. If $\mathcal{U}$ is a countable collection of open subsets of X, then there is a residual subset, i.e., a dense G_δ, $R_\mathcal{U}$ such that $x \in R_\mathcal{U}$ and $\Omega_\mathcal{F}\varphi(x) \cap U \neq \emptyset$ implies $N(x, U) \in \mathcal{F}$ for every U in $\mathcal{U}$.

c. Assume X is compact and $k\mathcal{F}$ is a countably generated filter. If $\mathcal{U}$ is a countable collection of open subsets of X, then there is a residual subset $R_\mathcal{U}$ such that $x \in R_\mathcal{U}$ and $\Omega_\mathcal{F}\varphi(x) \cap U \neq \emptyset$ implies $\omega_\mathcal{F}\varphi(x) \cap \overline{U} \neq \emptyset$ for all U in $\mathcal{U}$.

d. Assume X is a completely metrizable space and $k\mathcal{F}$ is a countably generated family. If X_0 is a separable subset of X, then there is a residual subset R_{X_0} of X such that for $x \in R_{X_0}$, $\Omega_\mathcal{F}\varphi(x) \cap X_0 = \omega_\mathcal{F}\varphi(x) \cap X_0$. In particular if X is separable, then there is a residual subset R such that for $x \in R$, $\Omega_\mathcal{F}\varphi(x) = \omega_\mathcal{F}\varphi(x)$.

PROOF. (a) $X \times X \backslash \Omega_{\mathcal{F}} \varphi$ satisfies the second axiom of countability and therefore the Lindelöf property. The collection $\{X \times X \backslash \overline{f^F} : F \in k\mathcal{F}\}$ is an open cover for this set. Let $\mathcal{A} \subset k\mathcal{F}$ index a countable subcover so that $\Omega_{\mathcal{F}} \varphi = \cap\{\overline{f^F} : F \in \mathcal{A}\}$. Let $\mathcal{F}_1$ be the family whose dual $k\mathcal{F}_1$ is $[\mathcal{A}]$. Since $k\mathcal{F}_1 \subset k\mathcal{F}$, $\mathcal{F} \subset \mathcal{F}_1$. If $k\mathcal{F}$ is a filter, we can take finite intersections of elements of $\mathcal{A}$ to obtain the countably generated filter $k\tilde{\mathcal{F}}_1$, containing $k\mathcal{F}_1$ and still contained in the filter $k\mathcal{F}$. Then $\mathcal{F} \subset \tilde{\mathcal{F}}_1 \subset \mathcal{F}_1$. Thus:

$$\Omega_{\mathcal{F}} \varphi \subset \Omega_{\tilde{\mathcal{F}}_1} \varphi \subset \Omega_{\mathcal{F}_1} \varphi \subset \cap\{\overline{f^F} : F \in \mathcal{A}\} = \Omega_{\mathcal{F}} \varphi$$

(b) Suppose $\mathcal{A}$ is countable and $[\mathcal{A}] = k\mathcal{F}$. For U open:

$$(f^F)^{-1}(U) = \cup\{(f^t)^{-1}(U) : t \in F\}$$

is open and therefore has a nowhere dense boundary. $\{(f^F)^{-1}(U) : F \in \mathcal{A}$ and $U \in \mathcal{U}\}$ is a countable collection of open sets. Let $R_{\mathcal{U}}$ be the complement of the union of the boundaries of this collection. Assume $x \in R_{\mathcal{U}}$, $U \in \mathcal{U}$, and $y \in \Omega_{\mathcal{F}} \varphi(x) \cap U$. For all $F_1 \in k\mathcal{F}$, F_1 contains some $F \in \mathcal{A}$. Because $y \in \Omega_{\mathcal{F}} \varphi(x)$ and U is a neighborhood of y, $x \in \overline{(f^F)^{-1}(U)}$. But since $x \in R_{\mathcal{U}}$ it is not in the boundary of $(f^F)^{-1}(U)$, therefore:

$$x \in (f^F)^{-1}(U) \subset (f^{F_1})^{-1}(U)$$

Thus $f^{F_1}(x) \cap U \neq \emptyset$, so $N(x,U) \cap F_1 \neq \emptyset$. Since this is true for all $F_1 \in k\mathcal{F}$, $N(x,U) \in \mathcal{F}$.

(c) Continuing from (b), if $k\mathcal{F}$ is a filter, then $k\mathcal{F} \cdot [N(x,U)]$ is proper, since $N(x,U) \in \mathcal{F}$; therefore it is a filter whose dual we denote $\mathcal{F}_1$. Thus $\mathcal{F}_1$ is a filterdual contained in $\mathcal{F}$. Since X is compact, $\omega_{\mathcal{F}_1} \varphi(x)$ is a nonempty subset of $\omega_{\mathcal{F}} \varphi(x)$ by Proposition 3.5. For $F \in k\mathcal{F}$:

$$F_1 = F \cap N(x,U) \in k\mathcal{F}_1$$

and so $\overline{f^{F_1}(x)} \supset \omega_{\mathcal{F}_1} \varphi(x)$. But $f^{F_1}(x) \subset U$, so $\omega_{\mathcal{F}_1} \varphi(x)$ is a nonempty subset of $\omega_{\mathcal{F}} \varphi(x) \cap \overline{U}$.

(d) If X_0 is a separable subset of a metric space X, then X_0 has a countable base; therefore we can choose a countable family $\mathcal{U}$ of open subsets of X such that $\{U : U \in \mathcal{U}\}$ includes a base for neighborhoods in X of every point of X_0 [e.g., let $\mathcal{U} = \{V_\varepsilon(z) : \varepsilon$ rational and z in a countable dense subset of $X_0\}$]. With $x \in R_{\mathcal{U}}$ and $y \in \Omega_{\mathcal{F}} \varphi(x) \cap X_0$, we have $N(x,U) \in \mathcal{F}$ for every $U \in \mathcal{U}$. Since these include a base for the neighborhoods of y in X, we have $y \in \omega_{\mathcal{F}} \varphi(x)$ by Proposition 3.6a. Because X is completely metrizable, it is a Baire space. ■

4

Transitive and Central Systems

We now consider topologically transitive and central systems.

PROPOSITION 4.1. *Let $\varphi : T \times X \to X$ be a uniform action and $\mathcal{F}, \mathcal{F}_1, \mathcal{F}_2$ be proper families for T.*

a. The following conditions are equivalent. When they hold, we say that φ is $\mathcal{F}$ central.

(1) $1_X \subset \Omega_{\mathcal{F}} \varphi$

(2) For every nonempty open subset U of X, $N^{\varphi}(U,U) \in \mathcal{F}$

(3) For every $x \in X$, $\mathcal{N}^{\varphi}(u[x]) \subset \mathcal{F}$

(4) For every $F \in k\mathcal{F}$, $x \in X$, and U open containing x, $x \in \overline{(f^F)^{-1}(U)}$

(5) For every $F \in k\mathcal{F}$, $x \in X$, and U open containing x, $x \in \overline{(f^F)(U)}$

b. The following conditions are equivalent. When they hold, we say that φ is $\mathcal{F}$ transitive.

(1) $X \times X = \Omega_{\mathcal{F}} \varphi$

(2) For all nonempty open subsets U, V of X, $N^{\varphi}(U,V) \in \mathcal{F}$

(3) For all points $x,y \in X$, $\mathcal{N}(u[x],u[y]) \subset \mathcal{F}$

(4) For every $F \in k\mathcal{F}$ and nonempty open subset U of X, $(f^F)^{-1}(U)$ is open and dense in X.

(5) For every $F \in k\mathcal{F}$ and nonempty open subset U of X, $f^F(U)$ is dense in X.

c. An action φ is $\mathcal{F}_1 \cap \mathcal{F}_2$ central (or $\mathcal{F}_1 \cap \mathcal{F}_2$ transitive) iff φ is both $\mathcal{F}_1$ and $\mathcal{F}_2$ central (resp. $\mathcal{F}_1$ and $\mathcal{F}_2$ transitive). In particular if $\mathcal{F}_1 \subset \mathcal{F}_2$ and φ is $\mathcal{F}_1$ central (or $\mathcal{F}_1$ transitive), then it is $\mathcal{F}_2$ central (resp. $\mathcal{F}_2$ transitive). φ is $\mathcal{F}$ central (or $\mathcal{F}$ transitive) iff it is $u\mathcal{F}$ central (resp. $u\mathcal{F}$ transitive) iff it is $\tilde{u}\mathcal{F}$ central (resp. $\tilde{u}\mathcal{F}$ transitive).

d. Let $\varphi_1 : T \times X_1 \to X_1$ be a uniform action and $h : \varphi \to \varphi_1$ a continuous action map. If h is dense and φ is $\mathcal{F}$ central (or $\mathcal{F}$ transitive), then φ_1 is $\mathcal{F}$ central (resp. $\mathcal{F}$ transitive). If h is a dense embedding and φ_1 is $\mathcal{F}$ central (or $\mathcal{F}$ transitive), then φ is $\mathcal{F}$ central (resp. $\mathcal{F}$ transitive).

e. If φ is uniformly reversible and $\mathcal{F}$ central (or $\mathcal{F}$ transitive), then the reverse action $\overline{\varphi}$ is $\mathcal{F}$ central (resp. $\mathcal{F}$ transitive).

PROOF. Equivalences (a) and (b) are immediate from Proposition 3.9a. Proposition 3.9b and e imply Proposition 4.1c and e above.

Now suppose that $h : \varphi \to \varphi_1$ is a dense map of actions. By (3.39) $(h \times h)(\Omega_{\mathcal{F}}\varphi) \subset \Omega_{\mathcal{F}}\varphi_1$. Since $(h \times h)(1_X) = 1_{h(X)}$ is dense in 1_{X_1},

$$(h \times h)(X \times X) = h(X) \times h(X)$$

is dense in $X_1 \times X_1$, and $\Omega_{\mathcal{F}}\varphi_1$ is closed, it follows that φ $\mathcal{F}$ central (or $\mathcal{F}$ transitive) implies φ_1 is $\mathcal{F}$ central (resp. $\mathcal{F}$ transitive). When h is a dense embedding, $\Omega_{\mathcal{F}}\varphi = (h \times h)^{-1}(\Omega_{\mathcal{F}}\varphi_1)$ which implies the converse result. ■

We call φ central (or transitive) when it is $\mathcal{B}_T$ central (resp. $\mathcal{B}_T$ transitive). By Proposition 4.1c the notions of $\mathcal{F}$ centrality and $\mathcal{F}$ transitivity become weaker as $\mathcal{F}$ becomes larger. Therefore the weakest notion occurs when $\mathcal{F}$ is $\mathcal{P}_+ = \{F \subset T : F \neq \emptyset\}$ with $k\mathcal{P}_+ = \{T\}$. By Proposition 4.1b we see that:

$$\varphi \text{ is } \mathcal{P}_+ \text{ transitive} \Leftrightarrow \overline{f^T} = X \times X \tag{4.1}$$

As we will soon see, $\mathcal{P}_+$ transitivity is very close to transitivity, i.e., $\mathcal{B}_T$ transitivity. On the other hand, any system is $\mathcal{P}_+$ central because the zero element $0 \in N^{\varphi}(U,U)$ whenever U is nonempty. Equivalently $1_X = f^0 \subset \overline{f^T}$.

We define families of subsets:

$$\begin{aligned}
\mathcal{F}_{\varphi} &= [\{N^{\varphi}(U,U) : U \text{ open and nonempty in } X\}] \\
&= \bigcup_{x \in X} \mathcal{N}^{\varphi}(u[x]) \\
\overline{\mathcal{F}}_{\varphi} &= \mathcal{F}_{\varphi} \cdot k\mathcal{B}_T & (4.2) \\
\mathcal{T}_{\varphi} &= [\{N^{\varphi}(U,V) : U,V \text{ open and nonempty in } X\}] \\
&= \bigcup_{x,y \in X} \mathcal{N}^{\varphi}(u[x],u[y]) \\
\overline{\mathcal{T}}_{\varphi} &= \mathcal{T}_{\varphi} \cdot k\mathcal{B}_T & (4.3)
\end{aligned}$$

From Lemma 3.1a we see that $F \in k\mathcal{F}_\varphi$ iff $1_X \subset \overline{f^F}$ and $F \in k\mathcal{T}_\varphi$ iff $X \times X \subset \overline{f^F}$. Thus:

$$\begin{aligned} k\mathcal{F}_\varphi &= \{F : 1_X \subset \overline{f^F}\} \\ k\mathcal{T}_\varphi &= \{F : X \times X = \overline{f^F}\} \end{aligned} \tag{4.4}$$

Clearly φ is $\mathcal{F}$ central (or $\mathcal{F}$ transitive) iff $\mathcal{F}_\varphi \subset \mathcal{F}$ (resp. $\mathcal{T}_\varphi \subset \mathcal{F}$).

Recall that the action φ is called *dense* when $f^t(X)$ is dense in X for all $t \in T$. Of course a surjective action is dense, and the converse is true when X is compact.

PROPOSITION 4.2. *Let $\varphi : T \times X \to X$ be a uniform action.*

a. The following are equivalent:

(1) φ is central (i.e., $\mathcal{B}_T$ central).

(2) $\mathcal{F}_\varphi \subset \mathcal{B}_T$

(3) $\overline{\mathcal{F}}_\varphi \subset \mathcal{P}_+$

When φ is central, then it is a dense action and $\overline{\mathcal{F}}_\varphi$ is a full family.

b. The following are equivalent:

(1) φ is $\mathcal{P}_+$ transitive.

(2) If U is an open, nonempty, and $-$ invariant subset of X, then U is dense.

(3) $\mathcal{T}_\varphi$ is a proper family.

When φ is $\mathcal{P}_+$ transitive, then $\mathcal{F}_\varphi$ is a filter.

c. The following are equivalent:

(1) φ is transitive (i.e., $\mathcal{B}_T$ transitive).

(2) φ is central and $\mathcal{P}_+$ transitive.

(3) φ is dense and $\mathcal{P}_+$ transitive.

(4) $\mathcal{T}_\varphi \subset \mathcal{B}_T$

(5) $\overline{\mathcal{T}}_\varphi \subset \mathcal{P}_+$

When φ is transitive, then $\overline{\mathcal{F}}_\varphi$ is a full filter and:

$$\gamma\mathcal{F}_\varphi = \gamma\overline{\mathcal{F}}_\varphi = \overline{\mathcal{T}}_\varphi \tag{4.5}$$

PROOF. Since $0 \in N(U,U)$ whenever $U \neq \emptyset$, $\mathcal{T}_\varphi \supset \mathcal{F}_\varphi \neq \emptyset$. Recall that $\emptyset \neq \mathcal{F} \subset \mathcal{B}_T$ iff $\mathcal{F} \cdot k\mathcal{B}_T$ is proper and therefore full (cf. Proposition 2.2). Then in (a) (1) $\Leftrightarrow$ (2) $\Leftrightarrow$ (3) and in (b) (1) $\Leftrightarrow$ (3) are clear. (1) $\Leftrightarrow$ (2) in (b) follows from condition (4) of Proposition 4.1b because $k\mathcal{P}_+ = \{T\}$ and a set U is $-$ invariant iff $(f^T)^{-1}(U) = U$. Furthermore for any set U, $(f^T)^{-1}(U)$ is $-$ invariant.

If φ is central, then $k\mathcal{B}_T \subset k\mathcal{F}_\varphi$, so by (4.4), $1_X \subset \overline{f^{T_t}}$ for all $t \in T$. Projecting to the second coordinate, we see that $X = \overline{f^{T_t}(X)}$. But:

$$f^{T_t}(X) = f^t(f^T(X)) \subset f^t(X)$$

therefore $f^t(X)$ is dense in X.

If φ is $\mathcal{P}_+$ transitive and U_1, U_2 are nonempty open sets, then $N(U_1,U_2) \neq \emptyset$, so there exists $s \in T$ such that $U_3 = U_1 \cap f^{-s}(U_2)$ is a nonempty open set. By (3.6) with $h = f^s$:

$$\begin{aligned} N(U_3,U_3) &\subset N(U_1,U_1) \cap N(f^{-s}(U_2), f^{-s}(U_2)) \\ &= N(U_1,U_1) \cap N(f^s f^{-s}(U_2), U_2) \subset N(U_1,U_1) \cap N(U_2,U_2) \end{aligned}$$

By (4.2) we see that $\mathcal{F}_\varphi$ is a filter. If in addition $\mathcal{F}_\varphi \subset \mathcal{B}_T$, i.e., when (2) in (c) holds, then $\overline{\mathcal{F}}_\varphi$ is full filter (cf. Proposition 2.1d).

(c) (1) $\Leftrightarrow$ (4) $\Leftrightarrow$ (5) follows as in (a). (1) $\Rightarrow$ (2) is obvious, and (2) $\Rightarrow$ (3) follows from (a). Now assume φ is $\mathcal{P}_+$ transitive and dense. Observe that for $F \subset T$ and $t \in T$, $f^{g^t(F)}$ is the image of f^F under the map $1 \times f^t$. If f^F is dense in $X \times X$, the image $f^{g^t(F)}$ is dense because $1 \times f^t$ is a dense map. Thus $F \in k\mathcal{T}_\varphi$ implies $g^t(F) \in k\mathcal{T}_\varphi$. Since φ is $\mathcal{P}_+$ transitive, $T \in k\mathcal{T}_\varphi$ and so $T_t = g^t(T) \in k\mathcal{T}_\varphi$. Thus $k\mathcal{B}_T \subset k\mathcal{T}_\varphi$ and $\mathcal{T}_\varphi \subset \mathcal{B}_T$. This shows that (3) $\Rightarrow$ (4). Furthermore we showed that $k\mathcal{T}_\varphi$ is translation $+$ invariant, so by Proposition 2.5b, $\mathcal{T}_\varphi$ is translation $-$ invariant, and so $\overline{\mathcal{T}}_\varphi$ is translation $-$ invariant. We see this directly by applying (3.5) with $s,t \in T$:

$$g^{-t}(N(U,V) \cap T_s) = N(U, f^{-t}(V)) \cap g^{-t}(T_s)$$

Since f^t is a dense map $f^{-t}(V)$ is nonempty when V is.

Similarly we apply (3.5) with $A = f^{-t}(U)$:

$$\begin{aligned} g^t(N(U,V)) \supset g^t(N(f^t f^{-t}(U), V)) = \\ g^t g^{-t}(N(f^{-t}(U), V)) = N(f^{-t}(U), V) \cap T_t \end{aligned}$$

Because translation maps g^t are injective it follows that $\overline{\mathcal{T}}_\varphi$ is translation invariant. Hence $\overline{\mathcal{T}}_\varphi \supset \overline{\mathcal{F}}_\varphi$ implies

$$\overline{\mathcal{T}}_\varphi = \gamma \overline{\mathcal{T}}_\varphi \supset \gamma \overline{\mathcal{F}}_\varphi$$

The family $\gamma\mathcal{F}_\varphi$ is full and equals $\gamma\overline{\mathcal{F}}_\varphi$ by Corollary 2.1. It suffices to show that $\mathcal{T}_\varphi \subset \gamma\mathcal{F}_\varphi$ to complete the proof of (4.5).

With U, V open and nonempty, we choose $s \in N(U,V)$ and let $U_1 = U \cap f^{-s}(V)$. By (3.5):

$$g^{-s}N(U,V) = N(U, f^{-s}(V)) \supset N(U_1, U_1)$$

Then $g^{-s}N(U,V) \in \mathcal{F}_\varphi$ and $N(U,V) \in \gamma\mathcal{F}_\varphi$. ■

PROPOSITION 4.3. *Let $\mathcal{F}$ be a full family for T.*

a. *The following are equivalent:*

(1) *φ is $\mathcal{F}$ central.*

(2) $\mathcal{F}_\varphi \subset \mathcal{F}$

(3) $\overline{\mathcal{F}}_\varphi \subset \mathcal{F}$

b. *The following are equivalent:*

(1) *φ is $\mathcal{F}$ transitive.*

(2) *φ is transitive and $\gamma\mathcal{F}_\varphi \subset \mathcal{F}$*

(3) *φ is $\tilde{\gamma}\mathcal{F}$ transitive ($\tilde{\gamma}\mathcal{F} = k\gamma k\mathcal{F}$ is the largest invariant family contained in $\mathcal{F}$).*

c. *Assume that $\mathcal{F}$ is translation invariant and φ is transitive. The following are equivalent:*

(1) *φ is $\mathcal{F}$ transitive.*

(2) *φ is $\mathcal{F}$ central.*

(3) $\Omega_{\mathcal{F}}\varphi = \Omega\varphi$

(4) $\mathcal{F}_\varphi \subset \mathcal{F}$

(5) $\gamma\mathcal{F}_\varphi \subset \mathcal{F}$

PROOF. (a) (1) $\Leftrightarrow$ (2) and (3) $\Rightarrow$ (2) are clear. Because $\mathcal{F}$ is full, $\mathcal{F}_\varphi \subset \mathcal{F}$ implies $\overline{\mathcal{F}}_\varphi = \mathcal{F}_\varphi \cdot k\mathcal{B}_T \subset \mathcal{F} \cdot k\mathcal{B}_T = \mathcal{F}$, i.e., (2) $\Rightarrow$ (3).

(b) (3) $\Rightarrow$ (1) because $\tilde{\gamma}\mathcal{F} \subset \mathcal{F}$.

(1) $\Rightarrow$ (2). Since φ is $\mathcal{F}$ transitive $\mathcal{T}_\varphi \subset \mathcal{F}$, so because $\mathcal{F}$ is full, $\overline{\mathcal{T}}_\varphi \subset \mathcal{F}$ as in (a). Because $\mathcal{F}$ is full, $\mathcal{F} \subset \mathcal{B}_T$, so φ is transitive. Hence (4.5) holds, and $\gamma\mathcal{F}_\varphi = \overline{\mathcal{T}}_\varphi \subset \mathcal{F}$.

(2) $\Rightarrow$ (3). Because $\gamma\mathcal{F}_\varphi$ is translation invariant, $\gamma\mathcal{F}_\varphi \subset \mathcal{F}$ implies

$$\gamma\mathcal{F}_\varphi = \tilde{\gamma}\gamma\mathcal{F}_\varphi \subset \tilde{\gamma}\mathcal{F}$$

Because φ is transitive (4.5) holds, therefore $\overline{\mathcal{T}}_\varphi = \gamma\mathcal{F}_\varphi \subset \tilde{\gamma}\mathcal{F}$. Hence φ is $\tilde{\gamma}\mathcal{F}$ transitive.

(c) Since φ is transitive, $\Omega\varphi = X \times X$. Hence (3) $\Leftrightarrow$ (1). (1) $\Rightarrow$ (2) $\Rightarrow$ (4) are obvious.

(4) $\Rightarrow$ (5). Since $\mathcal{F}$ is translation invariant, $\gamma\mathcal{F}_\varphi \subset \gamma\mathcal{F} = \mathcal{F}$ which is (5).

(5) $\Rightarrow$ (1). From (b). ■

There is a gap between $\mathcal{P}_+$ transitivity and $\mathcal{B}_T$ transitivity. For example let $f: X \to X$ be a topologically transitive homeomorphism of a compact metric space with $e \in X$ a fixed point for f, i.e., with $T_0 = \mathbf{Z}_+$ and $f^1 = f$ we have an invertible transitive action. Assume $X \neq e$. With $T = \mathbf{Z}_+ \times \mathbf{Z}_+$, we let

$$\varphi(i,j,x) = \begin{cases} f^i(x) & \text{if } j = 0 \\ e & \text{if } j > 0 \end{cases} \tag{4.6}$$

φ is a $\mathcal{P}_+$ transitive action of T on X but it is not a dense action, so is not $\mathcal{B}_T$ transitive. However with $T = \mathbf{Z}_+$ or $\mathbf{R}_+$, this kind of phenomenon cannot occur.

PROPOSITION 4.4. *Assume that for each t in the uniform monoid T, the tail T_t is a cobounded subset of T; i.e., $T \backslash T_t$ has compact closure in T. If a uniform action $\varphi: T \times X \to X$ is $\mathcal{P}_+$ transitive, then it is $\mathcal{B}_T$ transitive.*

PROOF. By Proposition 4.2b, it suffices to show that the action is dense; i.e., given $t \in T$ we must show that $f^t(X) = f^{T_t}(X)$ is dense in X. Fix $z \in f^{T_t}(X)$ and let x be an arbitrary element of X. For each $V \in \mathcal{U}_X$, $N(V(z), V(x))$ is nonempty by $\mathcal{P}_+$ transitivity, so we can choose $z_V \in V(z)$ and $t_V \in T$ such that $f^{t_V}(z_V) \in V(x)$. The net $\{z_V\}$ indexed by the directed set $\mathcal{U}_X$ converges to z and $\{f^{t_V}(z_V)\}$ converges to x.

If for V in a cofinal subset in $\mathcal{U}_X$, $t_V \in T_t$, then the associated subnet $\{f^{t_V}(z_V)\}$ lies in $f^{T_t}(X)$ and converges to x. Then x is a limit point of $f^{T_t}(X)$. Otherwise the net $\{t_V\}$ eventually lies in $T \backslash T_t$ and thereby admits a subnet converging to some $t^* \in T$. Thus after restricting to a subnet, we obtain that (t_V, z_V) converges to (t^*, z). Then $f^{t_V}(z_V) = \varphi(t_V, z_V)$ converges to $\varphi(t^*, z) = f^{t^*}(z)$. Since $x = \text{Lim}\{f^{t_V}(z_V)\}$, $x = f^{t^*}(z)$. Because $z \in f^{T_t}(X)$, a $+$ invariant subset, $x \in f^{T_t}(X)$ in this case. ■

The common description of topological transitivity is "a dense orbit," i.e., $\overline{f^T(x)} = X$ for some $x \in X$. If T is separable, i.e., has a countable dense subset, then any orbit closure $\overline{f^T(x)}$ is separable as well. So for such monoids, a dense orbit can occur only in a separable space. As we will see, transitive $\mathbf{Z}_+$ actions can occur in spaces of arbitrarily large cardinality. On the other hand, if X consists of a single orbit and a fixed point to which the orbit converges, then the $\mathbf{Z}_+$ action has a dense orbit but it is not even $\mathcal{P}_+$ transitive. Finally, the $\mathbf{Z}_+ \times \mathbf{Z}_+$ action of (4.6) has a dense orbit, and it is $\mathcal{P}_+$ transitive but not transitive.

PROPOSITION 4.5. *Let $\varphi : T \times X \to X$ be a uniform action. Assume that for some $x_0 \in X$, $\overline{f^T(x_0)} = X$; i.e., the orbit of x_0 is dense.*

a. We have

$$\mathcal{N}^{\varphi}(u[x_0]) = \mathcal{F}_{\varphi} \tag{4.7}$$

b. For any proper family $\mathcal{F}$ for T, φ is $\mathcal{F}$ central iff $x_0 \in \Omega_{\mathcal{F}}\varphi(x_0)$. For any translation invariant family $\mathcal{F}$, φ is $\mathcal{F}$ transitive iff $x_0 \in \Omega_{\mathcal{F}}\varphi(x_0)$.

c. The following are equivalent:

(1) $x_0 \in \Omega\varphi(x_0)$

(2) φ is transitive.

(3) φ is central.

(4) φ is dense.

(5) $\omega\varphi(x_0) = X$

PROOF. (a) If U is open and nonempty then $f^t(x_0) \in U$ for some $t \in T$. Hence:

$$N(f^{-t}(U), f^{-t}(U)) \in \mathcal{N}^{\varphi}(u[x_0])$$

But by (3.6) with $h = f^t : \varphi \to \varphi$:

$$N(f^{-t}(U), f^{-t}(U)) = N(f^t(f^{-t}(U)), U) \subset N(U,U)$$

Then $N(U,U) \in \mathcal{N}^{\varphi}(u[x_0])$ by heredity. Equation (4.7) follows from (4.2).

(b) By Proposition 3.9a, $x_0 \in \Omega_{\mathcal{F}}\varphi(x_o)$ implies $\mathcal{N}^{\varphi}(u[x_0]) \subset \mathcal{F}$. Then φ is $\mathcal{F}$ central by (4.7) and Proposition 4.3a. Alternatively by (3.41), $x_0 \in \Omega_{\mathcal{F}}\varphi(x_0)$ implies $f^t(x_0) \in \Omega_{\mathcal{F}}\varphi(f^t(x_0))$ for all $t \in T$, i.e., $1_{f^T(x_0)} \subset \Omega_{\mathcal{F}}\varphi$. But $1_{f^T(x_0)}$ is dense in 1_X and $\Omega_{\mathcal{F}}\varphi$ is closed. Then $1_X \subset \Omega_{\mathcal{F}}\varphi$; i.e., φ is $\mathcal{F}$ central. If $\mathcal{F}$ is translation invariant, then by (3.42), $x_0 \in \Omega_{\mathcal{F}}\varphi(x_0)$ implies

$$f^{t_1}(x_0) \in \Omega_{\mathcal{F}}\varphi(f^{t_2}(x_0))$$

for all $t_1, t_2 \in X$. Then $f^T(x_0) \times f^T(x_0) \subset \Omega_{\mathcal{F}}\varphi$; hence $X \times X \subset \Omega_{\mathcal{F}}\varphi$; i.e., φ is $\mathcal{F}$ transitive.

(c) (1) $\Rightarrow$ (2) follows from (b) with $\mathcal{F} = \mathcal{B}_T$, which is translation invariant.

(2) $\Rightarrow$ (3) is obvious and (3) $\Rightarrow$ (4) follows from Proposition 4.2a. (5) $\Rightarrow$ (1) is obvious, so it suffices to prove (4) $\Rightarrow$ (5).

Since $f^T(x_0)$ is dense in X, $f^{T_t}(x_0) = f^t(f^T(x_0))$ is dense in $f^t(X)$. If in addition the action is dense, then $f^t(X)$ is dense in X. Then for all $t \in T$,

$$\overline{f^{T_t}(x_0)} = X$$

Intersecting over $t \in T$, we obtain $\omega\varphi(x_0) = X$. ∎

REMARK. Since $\omega\varphi(x_0) \subset \overline{f^T(x_0)}$, the condition $\omega\varphi(x_0) = X$ implies that the orbit of x_0 is dense, so φ is transitive by (c).

Notice also that the argument proving (4.7) shows more generally:

$$y \in \overline{f^T(x)} \Rightarrow \mathcal{N}(u[y]) \subset \mathcal{N}(u[x]) \tag{4.8}$$

∎

Recall that a point x is called *recurrent* for the action $\varphi : T \times X \to X$ if $x \in |\omega\varphi|$, i.e., $x \in \omega\varphi(x)$. A point x is called a *transitive point* if $\omega\varphi(x) = X$. Define

$$\text{Trans}_\varphi = \{x \in X : \quad \omega\varphi(x) = X\} \tag{4.9}$$

By Proposition 4.5b, $\text{Trans}_\varphi \neq \emptyset$ implies φ is a transitive action, and in that case Proposition 4.5c implies that Trans_φ coincides with $\{x \in X : \overline{f^T(x)} = X\}$. By Proposition 3.14a we see that φ is a minimal action exactly when $\text{Trans}_\varphi = X$. In particular a minimal action is transitive. Under suitable separability conditions, transitive points do occur.

PROPOSITION 4.6. *Let $\varphi : T \times X \to X$ be a uniform action with X a complete, separable metric space. Assume $\mathcal{F}$ is a proper family for T with a countably generated dual, e.g., $\mathcal{P}_+$.*

a. If φ is $\mathcal{F}$ central, then $|\omega_\mathcal{F}\varphi| = \{x : x \in \omega_\mathcal{F}\varphi(x)\}$ is a dense G_δ subset of X. In particular if $k\mathcal{B}_T$ is countably generated (cf. Proposition 3.16) and φ is central, then the set of recurrent points $|\omega\varphi|$ is a dense G_δ.

b. If φ is $\mathcal{F}$ transitive, then $\{x : \omega_\mathcal{F}\varphi(x) = X\}$ is a dense G_δ subset of X. In particular if $k\mathcal{B}_T$ is countably generated and φ is transitive, then the set of transitive points Trans_φ *is a dense G_δ.*

PROOF. By Proposition 3.17d, there is a residual set R such that:

$$\Omega_\mathcal{F}\varphi(x) = \omega_\mathcal{F}\varphi(x)$$

for $x \in R$. If φ is $\mathcal{F}$ central, then $x \in \omega_\mathcal{F}\varphi(x)$ for $x \in R$; if φ is $\mathcal{F}$ transitive, then $\omega_\mathcal{F}\varphi(x) = X$ for $x \in R$. In fact Trans_φ is the intersection of the family of $(f^F)^{-1}(U)$s, where U varies over a countable base for the topology of X and F

varies over a generating set for $k\mathcal{F}$. So $\{x : \omega_{\mathcal{F}}\varphi(x) = X\}$ is a G_δ. Similarly $|\omega_{\mathcal{F}}\varphi|$ is the intersection of:

$$(X\backslash\overline{V}) \cup (U \cap (f^F)^{-1}(U))$$

where U and V vary in the countable base for the topology with $\overline{V} \subset U$, and F varies in the generating set for $k\mathcal{F}$. ∎

Note that if $h : \varphi \to \varphi_1$ is a uniform action map, then by (3.29), $x \in \omega_{\mathcal{F}}\varphi(x)$ implies $h(x) \in \omega_{\mathcal{F}}\varphi_1(h(x))$, i.e.:

$$h(|\omega_{\mathcal{F}}\varphi|) \subset |\omega_{\mathcal{F}}\varphi_1| \tag{4.10}$$

In particular with $h = f^t$ it follows that $|\omega_{\mathcal{F}}\varphi|$, the set of $\mathcal{F}$ recurrent points, is + invariant as is its closure.

If in addition h is dense, then $\omega_{\mathcal{F}}\varphi(x) = X$ implies $\omega_{\mathcal{F}}\varphi_1(h(x)) = X_1$, so:

$$h(\text{Trans}_{\varphi}) \subset \text{Trans}_{\varphi_1} \tag{4.11}$$

The set of transitive points is $\pm$ invariant; for $\mathcal{F}$ any translation invariant family (3.30) and (3.31) imply

$$\omega_{\mathcal{F}}\varphi(x) = X \Leftrightarrow \omega_{\mathcal{F}}\varphi(f^t(x)) = X \tag{4.12}$$

For a + invariant subset A of X the induced subsystem is the restriction $\varphi_A : T \times A \to A$ of the map φ. Then the inclusion $i_A : A \to X$ maps φ_A to φ. A is called an $\mathcal{F}$ *central* (or $\mathcal{F}$ *transitive*) *subset* if it is + invariant and the subsystem φ_A is $\mathcal{F}$ central (resp. $\mathcal{F}$ transitive). If x is a recurrent point for φ, then $\omega\varphi(x)$ is a transitive subset (i.e., $\mathcal{F}$ transitive with $\mathcal{F} = \mathcal{B}_T$.) The closure of $|\omega\varphi|$ is a central subset called the *Birkhoff center* for φ.

Suppose for a space X we have an indexed collection of continuous maps $\{h_\alpha : X_\alpha \to Y_\alpha\}$ with $X_\alpha \subset X$ for each α in the index set, I. We say the collection $\{h_\alpha\}$ *observes the open sets* of X if each h_α is a surjection and for every U open and nonempty in X, there exists for some $\alpha \in I$ and U_α open and nonempty in Y_α so that:

$$h_\alpha^{-1}(U_\alpha) \subset U \cap X_\alpha \tag{4.13}$$

Since h_α is surjective, we also have

$$h_\alpha h_\alpha^{-1}(U_\alpha) = U_\alpha \subset Y_\alpha \tag{4.14}$$

We say that the collection $\{h_\alpha\}$ *observes pairs of open sets* in X if $\{h_\alpha \times h_\alpha : X_\alpha \times X_\alpha \to Y_\alpha \times Y_\alpha\}$ observes the open sets in $X \times X$.

Notice that $\{h_\alpha\}$ observing the open sets implies $\cup\{X_\alpha\}$ is dense in X, and conversely this condition implies $\{1_{X_\alpha} : X_\alpha \to X_\alpha\}$ observes the open sets of X. Consequently observing the pairs implies $\cup\{X_\alpha \times X_\alpha\}$ is dense in $X \times X$, and conversely this condition implies $\{1_{X_\alpha}\}$ observes the pairs.

PROPOSITION 4.7. *Let $\varphi : T \times X \to X$ be a uniform action and $\mathcal{F}$ a proper family for T. Let $\{X_\alpha : \alpha \in I\}$ be an indexed collection of $+$ invariant subsets of X, $\{\varphi_\alpha : T \times Y_\alpha \to Y_\alpha, \alpha \in I\}$ an indexed collection of uniform actions, and $\{h_\alpha : \varphi_{X_\alpha} \to \varphi_\alpha\}$ a collection of continuous action maps.*

If each φ_α is $\mathcal{F}$ central and the collection $\{h_\alpha : X_\alpha \to Y_\alpha\}$ observes open sets, then φ is $\mathcal{F}$ central.

If each φ_α is $\mathcal{F}$ transitive and the collection $\{h_\alpha\}$ observes pairs of open sets, then φ is $\mathcal{F}$ transitive.

PROOF. Given U open and nonempty in X, choose U_α to satisfy (4.12). Then by (3.6) and (4.13):

$$N(U,U) \supset N(h_\alpha^{-1}(U_\alpha), h_\alpha^{-1}(U_\alpha)) =$$
$$N(h_\alpha h_\alpha^{-1}(U_\alpha), U_\alpha) = N(U_\alpha, U_\alpha)$$

Now if φ_α is $\mathcal{F}$ central, then $N(U_\alpha, U_\alpha) \in \mathcal{F}$, so $N(U,U) \in \mathcal{F}$ by heredity.

Similarly if $\{h_\alpha\}$ observes pairs, then given U, V open and nonempty we can choose $U_\alpha \times V_\alpha$ to satisfy (4.12) for $U \times V$ in $X \times X$. Thus:

$$h_\alpha^{-1}(U_\alpha) \subset U \cap X_\alpha \qquad h_\alpha^{-1}(V_\alpha) \subset V \cap X_\alpha$$

with the same choice of α. Just as before we have $N(U,V) \supset N(U_\alpha, V_\alpha)$. If φ_α is $\mathcal{F}$ transitive $N(U,V) \in \mathcal{F}$. ∎

PROPOSITION 4.8. *Let $\varphi : T \times X \to X$ be a uniform action and $\mathcal{F}$ a proper family for T.*

a. If there exists an indexed collection $\{X_\alpha : \alpha \in I\}$ of $\mathcal{F}$ central subsets of X, with $\cup\{X_\alpha\}$ dense in X then φ is $\mathcal{F}$ central. If $\cup\{X_\alpha \times X_\alpha\}$ is dense in $X \times X$ and each X_α is $\mathcal{F}$ transitive, then φ is $\mathcal{F}$ transitive.

b. Assume φ is the surjective inverse limit of a sequence of actions; i.e., there is a sequence $\varphi_n : T \times X_n \to X_n$ of actions, surjections $k_n : \varphi_{n+1} \to \varphi_n$, and φ is the subsystem of the product action $\prod_n \varphi_n$ on the subset $X = \{\{x_n\} : k_n(x_{n+1}) = x_n,\ n = 1,2,\dots\}$. If each φ_n is $\mathcal{F}$ central (or $\mathcal{F}$ transitive) then φ is $\mathcal{F}$ central (resp. $\mathcal{F}$ transitive).

c. For any index set I let φ_* denote the product action on X^I. The following conditions are equivalent:

(1) φ_* on X^I is $\mathcal{F}$ central for every index set I.

(2) φ_* on X^I is $\mathcal{F}$ central for every finite index set I.

(3) The point 1_X of X^X satisfies $1_X \in \Omega_{\mathcal{F}}\varphi_*(1_X)$.

(4) There exists a filter $\mathcal{F}_1$ such that $\mathcal{F}_1 \subset \mathcal{F}$ and φ is $\mathcal{F}_1$ central.

d. If φ is $\mathcal{F}$ transitive, then φ_* on X^I is $\mathcal{F}$ central for any index set I.

PROOF. (a) Apply the Proposition 4.7 to the family $\{1_{X_\alpha} : \alpha \in I\}$.

(b) Let $h_n : X \to X_n$ be the restriction of projection to the nth coordinate. Since each k_n is surjective, it is easy to show by induction that h_n is surjective. Thus $\{h_n : \varphi \to \varphi_n\}$ is a sequence of surjective maps of actions. By definition of the inverse limit, the family $\{h_n\}$ observes pairs of open sets. So the result follows from the Proposition 4.7.

(c) (1) $\Rightarrow$ (3) follows by using $I = X$.

(3) $\Rightarrow$ (2). We can assume $I = \{1, \dots, n\}$. For a point $(x_1, \dots, x_n) \in X^I$ define $\text{ev}_{(x_1,\dots,x_n)} : X^X \to X^I$ by:

$$\text{ev}_{(x_1,\dots,x_n)}(h) = (h(x_1), \dots, h(x_n))$$

Because this is an action map (3.40) and (3) imply

$$(x_1, \dots, x_n) \in \Omega_{\mathcal{F}} \varphi_*(x_1, \dots, x_n)$$

(2) $\Rightarrow$ (4) $\Rightarrow$ (1). If finite products φ_* are $\mathcal{F}$ central, then by (3.7) finite intersections of elements of $\mathcal{F}_\varphi$ are in $\mathcal{F}$. Therefore these finite intersections form a filter contained in $\mathcal{F}$. If $\mathcal{F}_1$ is a filter between $\mathcal{F}_\varphi$ and $\mathcal{F}$, then applying (3.7) to a basic open set U in X^I yields $N(U,U) \in \mathcal{F}_1$. Notice that $N(U_\alpha, U_\alpha) = T$ for the all but finitely many indices α for which $U_\alpha = X$.

(d) If φ is $\mathcal{F}$ transitive, then by Proposition 4.2b, $\mathcal{F}_\varphi$ is a filter, so the result follows from (c). We present an alternative proof because the construction is useful in other contexts. We show that for any index set I, X^I is densely filled with closed $+$ invariant subsets on which φ_* is isomorphic to φ.

Fix $i_0 \in I$ and let $\tilde{I}$ be the subset of T^I consisting of maps $\alpha : I \to T$ such that $\alpha(i_0) = 0$. For each such α define $j_\alpha : X \to X^I$ by:

$$j_\alpha(x)(i) = f^{\alpha(i)}(x) \tag{4.15}$$

Because each $f^{\alpha(i)}$ is continuous, j_α is continuous to the product and the projection $\pi_0 : X^I \to X$ at the i_0 coordinate maps $j_\alpha(x)$ to x. Hence j_α is an embedding of X into X^I for each α in $\tilde{I}$. For each $t \in T$:

$$j_\alpha(f^t(x))(i) = f^{\alpha(i)+t}(x) = f^t(f^{\alpha(i)}(x)) = f^t(j_\alpha(x)(i))$$

We see that j_α maps φ to φ_* and the subset $X_\alpha = j_\alpha(X)$ is a $+$ invariant set. Since j_α is an isomorphism between φ and the subsystem $(\varphi_*)_{X_\alpha}$, it follows that each X_α is an $\mathcal{F}$ transitive, and hence $\mathcal{F}$ central, subset. The result follows once we prove $\cup\{X_\alpha\}$ is dense in X^I.

Given $z \in X^I$, a basic neighborhood for z is described by a finite list of distinct indices, which we may assume includes $i_0 : i_0, i_1, \ldots, i_k$, and a corresponding list of open sets $U_0, U_1, \ldots, U_k$ such that $z(i_j) \in U_j$ for $j = 0, 1, \ldots, k$.

Because φ is $\mathcal{P}_+$ transitive each $(f^T)^{-1}(U_j)$ is open and dense in X. Therefore we can choose inductively $t_1, t_2, \ldots t_k$ such that:

$$U_0 \cap f^{-t_1}(U_1) \cap \ldots \cap f^{-t_k}(U_k) \neq \emptyset$$

Let x be in the intersection, so that $x \in U_0$ and $f^{t_j}(x) \in U_j$ for $j = 1, 2, \ldots, k$. Let $\alpha(i_0) = 0$, $\alpha(i_j) = t_j$ for $j = 1, \ldots, k$, and extend α arbitrarily on the rest of I. Then $j_\alpha(x)$ is in the given basic neighborhood of z, so this neighborhood intersects X_α. ■

REMARK. Suppose the collection of subsets $\{X_\alpha\}$ is directed by inclusion; i.e., given α_1 and α_2 there exists α_3 so that $X_{\alpha_1} \cup X_{\alpha_2} \subset X_{\alpha_3}$ (e.g., this happens if the collection is totally ordered by inclusion). Then $\cup\{X_\alpha\}$ dense in X implies $\cup\{X_\alpha \times X_\alpha\}$ is dense in $X \times X$ because

$$\cup_\alpha\{X_\alpha \times X_\alpha\} = \cup_{\alpha_1,\alpha_2}\{X_{\alpha_1} \times X_{\alpha_2}\} = (\cup_\alpha\{X_\alpha\}) \times (\cup_\alpha\{X_\alpha\})$$

Inverse limits can be defined for any directed set. The proof works in general provided the projection maps h_α are surjective. This follows in the sequence case here and in general if the spaces X_α are all compact. ■

PROPOSITION 4.9. *Assume that $\mathcal{F}_1, \mathcal{F}_2$ and $\mathcal{F}_3$ are proper families for T such that $\mathcal{F}_1 \cdot \mathcal{F}_2 \subset \mathcal{F}_3$. For $i = 1, 2$, let $\varphi_i : T \times X_i \to X_i$ be a uniform action that is $\mathcal{F}_i$ central (or $\mathcal{F}_i$ transitive). The product action $\varphi : T \times X_1 \times X_2 \to X_1 \times X_2$ is $\mathcal{F}_3$ central (resp. $\mathcal{F}_3$ transitive).*

PROOF. With U_i, V_i open sets in X_i $(i = 1, 2)$ (3.7) implies

$$N^\varphi(U_1 \times U_2, V_1 \times V_2) = N^{\varphi_1}(U_1, V_1) \cap N^{\varphi_2}(U_2, V_2)$$

from which the result is clear. ■

COROLLARY 4.1. *Let $\{\varphi_\alpha : T \times X_\alpha \to X_\alpha\}$ be a family of uniform actions and φ_π the induced product action. If $\mathcal{F}$ is a filter of subsets of T and each φ_α is $\mathcal{F}$ central (or $\mathcal{F}$ transitive) then φ_π is $\mathcal{F}$ central (resp. $\mathcal{F}$ transitive).*

PROOF. With $\mathcal{F}_1 = \mathcal{F}_2 = \mathcal{F}_3 = \mathcal{F}$, Proposition 4.9 and induction imply the result for finite products. For every finite subset F of the index set let φ_F be the product action on X_F, the product of the factors indexed by F. Let h_F be the projection that maps φ to φ_F for all such F. The collection $\{h_F\}$ clearly observes pairs of open sets in X by definition of the product topology. So the result follows from Proposition 4.7. ■

We now prove the beautiful *Furstenberg Intersection Lemma.*

LEMMA 4.1. *Let $\varphi : T \times X \to X$ be a uniform action with $\varphi \times \varphi$ the product action on $X \times X$. Assume that for every $x \in X$*

$$\{x\} \times X \subset \Omega(\varphi \times \varphi)(x,x) \tag{4.16}$$

Then φ is transitive, and whenever U_1, V_1, U_2, V_2 are nonempty open subsets of X, there exist nonempty open subsets U_3, V_3 of X such that:

$$\begin{aligned} N^{\varphi}(U_3,V_3) &\subset N^{\varphi}(U_1,V_1) \cap N^{\varphi}(U_2,V_2) \\ &= N^{\varphi\times\varphi}(U_1 \times U_2, V_1 \times V_2) \end{aligned} \tag{4.17}$$

PROOF. By projecting to the second factor, we see from (3.40) and (4.16) that $\Omega\varphi(x) = X$ for all $x \in X$, so φ is transitive. By Proposition 4.2 the action is dense.

Because $N(U_1,V_1) \neq \emptyset$ (φ is transitive), there exists $s_1 \in T$ such that:

$$U_0 = U_1 \cap f^{-s_1}(V_1)$$

is a nonempty open set. $N(U_0,U_2) \neq \emptyset$ implies there exists $s_2 \in T$ such that:

$$U = U_1 \cap f^{-s_1}(V_1) \cap f^{-s_2}(U_2) \neq \emptyset$$

Let $x \in U$.

Because $\{x\} \times X \subset \Omega(\varphi \times \varphi)(x,x)$, we have

$$\begin{aligned} \emptyset &\neq N(U \times U, U \times f^{-s_1-s_2}(V_2)) \\ &= N(U,U) \cap N(U, f^{-s_1-s_2}(V_2)) \\ &\subset N(U_1, f^{-s_2}(U_2)) \cap N(f^{-s_1}(V_1), f^{-s_1}(f^{-s_2}(V_2))) \\ &= N(U_1, f^{-s_2}(U_2)) \cap N(f^{s_1} f^{-s_1}(V_1), f^{-s_2}(V_2)) \\ &\subset N(U_1, f^{-s_2}(U_2)) \cap N(V_1, f^{-s_2}(V_2)) \end{aligned}$$

by applying (3.6) and (3.7). Notice that $f^{-(s_1+s_2)}(V_2) \neq \emptyset$ because $f^{s_1+s_2}(X)$ is dense in X.

Fix

$$t_0 \in N(U_1, f^{-s_2}(U_2)) \cap N(V_1, f^{-s_2}(V_2))$$

With $t = t_0 + s_2$, the open sets:

$$U_3 = U_1 \cap f^{-t}(U_2) \qquad V_3 = V_1 \cap f^{-t}(V_2)$$

are nonempty. Just as before

$$\begin{aligned} N(U_3,V_3) &\subset N(U_1,V_1) \cap N(f^{-t}(U_2), f^{-t}(V_2)) \\ &\subset N(U_1,V_1) \cap N(U_2,V_2) \end{aligned}$$

■

THEOREM 4.1. *Let $\varphi : T \times X \to X$ be a uniform action.*

a. The following conditions are equivalent. When they hold we say that φ is weak mixing.

(1) The product action $\varphi \times \varphi$ on $X \times X$ is transitive.

(2) $\Omega(\varphi \times \varphi)(x,x) \supset \{x\} \times X$ for all $x \in X$

(3) $\mathcal{T}_\varphi$ is a filter contained in $\mathcal{B}_T$.

(4) $\overline{\mathcal{T}}_\varphi$ is a full filter.

(5) $\mathcal{T}_\varphi \cdot \mathcal{T}_\varphi \subset \mathcal{B}_T$

(6) φ is transitive and $\gamma\mathcal{F}_\varphi$ is a filter.

(7) φ is transitive and $\mathcal{F}_\varphi \subset \tau\mathcal{B}_T$.

(8) φ is a $\tau\mathcal{B}_T$ transitive.

b. Assume $\mathcal{F}$ is a full family for T. The following conditions are equivalent. When they hold we say that φ is $\mathcal{F}$ mixing.

(1) $\varphi \times \varphi$ is $\mathcal{F}$ transitive.

(2) φ is $\tau\mathcal{F}$ transitive.

(3) φ is $\mathcal{F}$ transitive and weak mixing.

(4) φ is $\tilde{\gamma}\mathcal{F}$ central and weak mixing.

(5) There exists a translation invariant filter $\mathcal{F}_1$ such that $\mathcal{F}_1 \subset \mathcal{F}$ and φ is $\mathcal{F}_1$ transitive.

(6) For every $x \in X$, there exists a filter $\mathcal{F}_1$ such that $\mathcal{F}_1 \subset \mathcal{F}$ and $\Omega_{\mathcal{F}_1}\varphi(x) = X$.

PROOF. (a) We prove

$$\begin{array}{ccccccccccc} (1) & \Rightarrow & (2) & \Rightarrow & (3) & \Rightarrow & (4) & \Rightarrow & (5) & \Rightarrow & (1) \\ & & & & & & \Downarrow & & & & \Uparrow \\ & & & & & & (6) & \Rightarrow & (7) & \Rightarrow & (8) \end{array}$$

(1) $\Rightarrow$ (2). This is obvious.

(2) $\Rightarrow$ (3). By the Furstenberg Intersection Lemma, $\mathcal{T}_\varphi$ is a filter and φ is transitive, i.e., $\mathcal{T}_\varphi \subset \mathcal{B}_T$.

(3) $\Rightarrow$ (4). Since $\mathcal{T}_\varphi$ is a filter contained in $\mathcal{B}_T$, $\overline{\mathcal{T}}_\varphi$ is a full filter (Proposition 2.2d).

(4) $\Rightarrow$ (5). $\mathcal{T}_\varphi \cdot \mathcal{T}_\varphi \subset \overline{\mathcal{T}}_\varphi \subset \mathcal{B}_T$ since $\overline{\mathcal{T}}_\varphi$ is a full filter.

(5) $\Rightarrow$ (1). For nonempty open sets U_1, U_2, V_1, V_2 $N^{\varphi\times\varphi}(U_1 \times U_2, V_1 \times V_2)$ is by (3.7), in $\mathcal{T}_\varphi \cdot \mathcal{T}_\varphi$ and hence in $\mathcal{B}_T$ by (5). Therefore $\varphi \times \varphi$ is transitive.

(4) $\Rightarrow$ (6). As $\mathcal{T}_\varphi \subset \overline{\mathcal{T}}_\varphi \subset \mathcal{B}_T$, φ is transitive. Therefore (4.5) holds, so $\gamma\mathcal{F}_\varphi = \overline{\mathcal{T}}_\varphi$ is a filter.

(6) $\Rightarrow$ (7). Because $\gamma\mathcal{F}_\varphi$ is proper and translation invariant, it is full (Corollary 2.1 of Proposition 2.5), and so is contained in $\mathcal{B}_T$. Because it is a translation invariant filter, it is thick (Proposition 2.7a). Therefore:

$$\mathcal{F}_\varphi \subset \gamma\mathcal{F}_\varphi = \tau\gamma\mathcal{F}_\varphi \subset \tau\mathcal{B}_T$$

(7) $\Rightarrow$ (8). By Proposition 4.3a, φ is $\tau\mathcal{B}_T$ central. Since $\tau\mathcal{B}_T$ is translation invariant Proposition 4.3c implies φ is $\tau\mathcal{B}_T$ transitive.

(8) $\Rightarrow$ (1). Because φ is transitive, $\overline{\mathcal{F}}_\varphi$ is a full filter (Proposition 4.2c). Because φ is $\tau\mathcal{B}_T$ transitive, $\overline{\mathcal{F}}_\varphi \subset \tau\mathcal{B}_T$ (Proposition 4.3c). Then by Proposition 2.6 there is a translation invariant filter $\tilde{\mathcal{F}}$ which contains $\overline{\mathcal{F}}_\varphi$. So φ is $\tilde{\mathcal{F}}$ transitive (Proposition 4.3c again) and hence $\varphi \times \varphi$ is $\tilde{\mathcal{F}}$ transitive by Corollary 4.1. Because $\tilde{\mathcal{F}}$ is translation invariant, and hence full (Corollary 2.1), $\tilde{\mathcal{F}} \subset \mathcal{B}_T$, so $\varphi \times \varphi$ is transitive.

(b) We prove

$$\begin{array}{ccccccccccc} (5) & \Rightarrow & (1) & & & & & & & & \\ \Downarrow & & \Downarrow & & & & & & & & \\ (2) & \Rightarrow & (3) & \Rightarrow & (5) & \Rightarrow & (6) & \Rightarrow & (3) & \Leftrightarrow & (4) \end{array}$$

(5) $\Rightarrow$ (1) and (2). Since $\mathcal{F}_1$ is a filter, $\varphi \times \varphi$ is $\mathcal{F}_1$ transitive by Corollary 4.1. It is $\mathcal{F}$ transitive since $\mathcal{F}_1 \subset \mathcal{F}$. Because $\mathcal{F}_1$ is a translation invariant filter, it is thick (Proposition 2.7a). Hence:

$$\mathcal{F}_1 = \tau\mathcal{F}_1 \subset \tau\mathcal{F}$$

Therefore φ is $\tau\mathcal{F}$ transitive.

(1) $\Rightarrow$ (3). As a factor of $\varphi \times \varphi$, φ is $\mathcal{F}$ transitive (Proposition 4.1d). Since $\mathcal{F} \subset \mathcal{B}_T$ (Corollary 2.1), $\varphi \times \varphi$ is transitive, and φ is weak mixing.

(2) $\Rightarrow$ (3). $\tau\mathcal{F} \subset \mathcal{F}$ and $\tau\mathcal{F} \subset \tau\mathcal{B}_T$. Thus φ is $\mathcal{F}$ transitive and $\tau\mathcal{B}_T$ transitive.

(3) $\Leftrightarrow$ (4). Proposition 4.3b and c.

(3) $\Rightarrow$ (5). Because φ is $\mathcal{F}$ transitive, $\gamma\mathcal{F}_\varphi \subset \mathcal{F}$ by Proposition 4.3b. Because φ is weak mixing, $\gamma\mathcal{F}_\varphi$ is a filter — clearly translation invariant. With $\mathcal{F}_1 = \gamma\mathcal{F}_\varphi$, φ is $\mathcal{F}_1$ transitive by Proposition 4.3b.

(5) $\Rightarrow$ (6). This is obvious.

(6) $\Rightarrow$ (3). Since $\mathcal{F}_1 \subset \mathcal{F}$, $\Omega_\mathcal{F}\varphi(x) = X$ for all x; i.e., φ is $\mathcal{F}$ transitive. By (3.44)

$$X \times X = \Omega_{\mathcal{F}_1}\varphi(x) \times \Omega_{\mathcal{F}_1}\varphi(x) \subset \Omega_{\mathcal{F}_1}(\varphi \times \varphi)(x,x)$$

since $\mathcal{F}_1$ is a filter. Because $\mathcal{F}$ is full, $\mathcal{F}_1 \subset \mathcal{B}_T$, so $\Omega(\varphi \times \varphi)(x,x) = X \times X$ for all x in X. By (a), φ is weak mixing. ■

REMARK. Clearly, $\mathcal{F}$ mixing with $\mathcal{F} = \mathcal{B}_T$ is weak mixing. For $\mathcal{F}$ a translation invariant filter, or more generally by Proposition 2.7a, for $\mathcal{F}$ a proper, translation invariant, thick family, φ is $\mathcal{F}$ mixing iff φ is $\mathcal{F}$ transitive. ■

PROPOSITION 4.10. *Let $\varphi : T \times X \to X$ be a uniform action. Let $\mathcal{F}$ be a full family for T.*

a. If φ is both $\mathcal{F}$ transitive and $k\mathcal{F}$ central, then φ is weak mixing.

b. For I any nonempty index set let φ_* be the product action on X^I induced by φ on each factor. If φ is $\mathcal{F}$ mixing, then φ_* is $\mathcal{F}$ mixing.

c. For A any nonempty, compact, zero-dimensional space let φ_* be the induced action on $C(A;X)$, the space of continuous maps with the uniformity of uniform convergence. If φ is $\mathcal{F}$ mixing, then φ_* is $\mathcal{F}$ mixing.

PROOF. (a) Since $\mathcal{F}$ is full, $k\mathcal{F} \cdot \mathcal{F} \subset \mathcal{B}_T$ by Proposition 2.2a. If φ is both $\mathcal{F}$ transitive, and $k\mathcal{F}$ transitive then $\varphi \times \varphi$ is transitive by Proposition 4.9. If φ is only $k\mathcal{F}$ central then we adapt the proof. For any nonempty open sets U, V $N(U,U) \in k\mathcal{F}$ and $N(U,V) \in \mathcal{F}$ because φ is $k\mathcal{F}$ central and $\mathcal{F}$ transitive. Hence:

$$N(U \times U, U \times V) = N(U,U) \cap N(U,V)$$

is in $\mathcal{B}_T$. This means that $(x,y) \in \Omega(\varphi \times \varphi)(x,x)$ for all $x,y \in X$. Then by Theorem 4.1a(2) φ is weak mixing.

(b) Let $\mathcal{F}_1 \subset \mathcal{F}$ be a translation invariant filter with φ $\mathcal{F}_1$ transitive (cf. Theorem 4.1b). By Corollary 4.1, φ_* is $\mathcal{F}_1$ transitive, it is $\mathcal{F}$ mixing by Theorem 4.1b again.

(c) Let $\mathcal{U}$ be a finite partition of A by clopen sets. Then $X^{\mathcal{U}}$ is a finite product of copies of X and we can regard $X^{\mathcal{U}}$ as the closed set in $C(A;X)$ of functions constant on each member of $\mathcal{U}$. For each such $\mathcal{U}$, $X^{\mathcal{U}}$ is a closed φ_* + invariant subspace on which by (b) φ_* is $\mathcal{F}$ mixing, thus it is $\tau\mathcal{F}$ transitive. Taking the union over all such finite partitions yields a dense subset of $C(A;X)$; i.e., every continuous function from A to X can be uniformly approximated by a locally constant function. Furthermore the collection $\{X^{\mathcal{U}}\}$ is directed with respect to inclusion: If $\mathcal{U}_1$ and $\mathcal{U}_2$ are partitions with common refinement $\mathcal{U}_3$ then $X^{\mathcal{U}_1} \cup X^{\mathcal{U}_2} \subset X^{\mathcal{U}_3}$; i.e., a function constant either on the elements of $\mathcal{U}_1$ or those of $\mathcal{U}_2$ is constant on those of $\mathcal{U}_3$. Hence $\cup\{X^{\mathcal{U}} \times X^{\mathcal{U}}\}$ is dense in $C(A,X) \times C(A,X)$. Thus φ_* is $\tau\mathcal{F}$ transitive by Proposition 4.9 and the remarks preceding it. ■

Notice that if A is a Cantor space, i.e., a perfect, compact, metric, zero-dimensional space, and φ is a weak mixing action on a compact metric space, then $C(A;X)$ is a complete separable metric space. By Proposition 4.6 it contains a residual set of points that are transitive with respect to φ_*. The natural map

$C(A;X) \to X^A$ is a dense injection, so we obtain a dense set of transitive points for φ_* on X^A although X^A which has cardinality 2^c.

If A is uncountable and X contains at least two points, then Trans_{φ_*} on X^A contains no nonempty G_δ. To see this observe that if $u \in C$, a G_δ subset of X^A, then there is an intersection of countably many basic sets contained in C and including u. Such an intersection restricts only a countable set of coordinates; i.e., there exists A_0 a countable subset of A such that $v|A_0 = u|A_0$ implies $v \in C$. Choose a_1, a_2 two distinct points of $A \backslash A_0$ and define v so that $v|A_0 = u|A_0$ and $v(a_1) = v(a_2) = x_0$ $v \in C$ but $v \notin \text{Trans}_{\varphi_*}$ because v projects to (x_0, x_0) in the diagonal of $X^{\{a_1,a_2\}}$, a nontransitive point for $\varphi \times \varphi$.

In fact if there exists a point $y \in X$ not recurrent, then the set of recurrent points contains no nonempty G_δ. Proceeding as before, choose

$$a_1 \in A \backslash A_0 \qquad v|A_0 = u|A_0 \qquad v(a_1) = y$$

Then v is not recurrent in X^A.

THEOREM 4.2. *Let $\varphi : T \times X \to X$ be a weak mixing uniform action with X a complete, separable metric space with at least two points and the monoid T separable metric as well. There is a Cantor space subset $A_0 \subset X$; i.e., A_0 is a compact, perfect, zero-dimensional subset, such that the family of functions $\{f^t|A_0 : t \in T\}$ is uniformly dense in $C(A_0;X)$ and pointwise dense in X^{A_0}. Such a subset is called a* Kronecker subset *for the system φ.*

PROOF. With A a Cantor space, $C(A;X)$ is a complete, separable metric space and φ induces a weak mixing action φ_* on it. By Proposition 3.16 $k\mathcal{B}_T$ is countably generated, so by Proposition 4.6 Trans_{φ_*} is a residual subset of $C(A;X)$. If $u \in \text{Trans}_{\varphi_*}$, then $u : A \to X$ is injective. If instead $u(a_0) = u(a_1)$ then u projects to a nontransitive point for the action on $X^{\{a_0,a_1\}}$. Since A is compact, u is a homeomorphism onto a subset A_0 of X. The isometry:

$$u^* : C(A_0;X) \to C(A;X)$$

is an isomorphism between actions induced by φ, and it associates the inclusion map $i : A_0 \to X$ to $u : A \to X$. Thus i is a transitive point for φ_* on $C(A_0;X)$, and so the orbit of i, $\{f^t|A_0 : t \in T\}$ is uniformly dense. Since $C(A_0;X) \to X^{A_0}$ has a dense image, this set is pointwise dense in X^{A_0}. ■

A uniform action φ is called *topological mixing* or *strong mixing* if it is $k\mathcal{B}_T$ transitive, or equivalently since $k\mathcal{B}_T$ is an invariant filter, $k\mathcal{B}_T$ mixing. The action φ is called *topologically ergodic* if it is $k\tau\mathcal{B}_T$ transitive. φ is called *ergodic mixing* if it is topologically ergodic and weak mixing.

PROPOSITION 4.11. *Let $\varphi : T \times X \to X$ and $\varphi_1 : T \times X_1 \to X_1$ be uniform actions.*

a. The following are equivalent:

(1) φ *is ergodic mixing, i.e.,* $k\tau\mathcal{B}_T$ *and* $\tau\mathcal{B}_T$ *transitive.*

(2) φ *is* $k\tau\mathcal{B}_T$ *mixing.*

(3) φ *is* $\tau k\tau\mathcal{B}_T$ *transitive.*

b. Let $\mathcal{F}$ *be a full family for* T. *If* φ *is strong mixing and* φ_1 *is* $\mathcal{F}$ *central/*$\mathcal{F}$ *transitive/*$\mathcal{F}$ *mixing then* $\varphi \times \varphi_1$ *is* $\mathcal{F}$ *central/*$\mathcal{F}$ *transitive/*$\mathcal{F}$ *mixing, respectively.*

c. If φ *is ergodic and* φ_1 *is weak mixing, then* $\varphi \times \varphi_1$ *is transitive.*

d. If φ *is ergodic mixing and* φ_1 *is ergodic/weak mixing/ergodic mixing, then* $\varphi \times \varphi_1$ *is ergodic/weak mixing/ergodic mixing, respectively.*

PROOF. (a) The equivalences follow from Theorem 4.1b with $\mathcal{F} = k\tau\mathcal{B}_T$.

(b), (c), and (d) all follow from Proposition 4.9. For (b) use $k\mathcal{B}_T \cdot \mathcal{F} = \mathcal{F}$ because $\mathcal{F}$ is full. For (c) $k\tau\mathcal{B}_T \cdot \tau\mathcal{B}_T \subset \mathcal{B}_T$. For (d) use Proposition 2.9 with $\mathcal{F} = k\mathcal{B}_T$. ■

Note: These results motivate Furstenberg's choice of $k\tau\mathcal{B}_T$ transitivity as the appropriate topological notion of ergodicity.

PROPOSITION 4.12. *Let* $T = \mathbf{Z}^*$, *the monoid of positive integers under multiplication, and* X *be the unit circle in the complex plane. The action* $T \times X \to X$ *by* $(n,z) \to z^n$ *is a strong mixing action. There exists a Kronecker subset, i.e., a Cantor set* $A_0 \subset X$, *such that the power functions* $\{f^n|A_0 : n \in T\}$ *are uniformly dense in* $C(A_0;X)$ *and pointwise dense in* X^{A_0}.

PROOF. If U is a nonempty open set, then U contains some arc of length $\varepsilon > 0$ with midpoint z. The image $f^n(U)$ contains the arc of length $n\varepsilon$ with midpoint z^n. Hence if $n > 2\pi/\varepsilon$, $f^n(U) = X$, and $N(U,V)$ contains $\{n \in T : n > 2\pi/\varepsilon\}$ for every nonempty V. Thus $N(U,V)$ is in $k\mathcal{B}_T$, and φ is strong mixing. A fortiori φ is weak mixing, so a Kronecker subset exists by Theorem 4.2. ■

For a uniform space X, we let $C(X)$ denote the set of closed subsets of X. We give $C(X)$ a uniform structure by associating the set V_C to $V \in \mathcal{U}_X$:

$$V_C = \{(A_1,A_2) \in C(X) \times C(X) : A_2 \subset V(A_1) \text{ and } A_1 \subset V^{-1}(A_2)\} \qquad (4.18)$$

Clearly $(V^{-1})_C = (V_C)^{-1}$ (so we drop the parentheses) and $V_C \circ W_C \subset (V \circ W)_C$. Thus the set of V_Cs generates a uniformity $\mathcal{U}_{C(X)}$. A closed set $A = \cap V(A)$ as V varies over $\mathcal{U}_X$, so the uniformity is Hausdorff.

If F_0 is a finite subset of X and $V = V^{-1} \in \mathcal{U}_X$ such that $V(F_0) \supset A$, we can choose for each $x_0 \in F_0$ an $x_1 \in V(x_0) \cap A$ to obtain $F_1 \subset A$ with $V \circ V(F_1) \supset A$. It follows that $C_b(X)$, the closure of the set of finite subsets of X, is exactly the set of totally bounded, closed subsets of X. Recall that a uniform space is compact iff it is complete and totally bounded. Then if X is complete, $C_b(X)$ is the set of compact subsets of X. If X is compact metric then $C_b(X) = C(X)$ is compact with the Hausdorff metric [see, e.g., Akin (1993) Chap. 7]. In general if X is compact, $C_b(X) = C(X)$ is compact by a similar argument.

That $C(X)$ is totally bounded is easy to see: If F is a finite subset of X with $V(F) = X$, $V = V^{-1}$ in $\mathcal{U}_X$, then $C(F)$ is a finite subset of $C(X)$ with $V_C(C(F)) = C(X)$. For completeness begin with a Cauchy net $\{A_\alpha\}$ in $C(X)$ and let

$$A = \limsup\{A_\alpha\} = \cap_\alpha(\overline{\cup_{\beta>\alpha} A_\alpha}) \in C(X)$$

Given $V \in \mathcal{U}_X$, there exists by compactness α_0 so that $\alpha > \alpha_0$ implies $V(A) \supset \overline{\cup_{\beta>\alpha} A_\beta}$. Choose W so that $W \circ W \subset V$. Since the net is Cauchy, there exists α_1 so that $\beta_1, \beta_2 > \alpha_1$ imply $(A_{\beta_1}, A_{\beta_2}) \in W_C$. For $\alpha > \alpha_1, W(A_\alpha) \supset \cup_{\beta>\alpha} A_\beta$, hence:

$$V(A_\alpha) \supset W \circ W(A_\alpha) \supset \overline{\cup_{\beta>\alpha} A_\beta} \supset A$$

That is, for $\alpha > \alpha_0, \alpha_1$, $(A, A_\alpha) \in V_C$, i.e., $\{A_\alpha\}$ converges to A.

If $f : X_0 \to X_1$ is uniformly continuous, then $C(f) : C(X_0) \to C(X_1)$ defined by:

$$C(f)(A) = \overline{f(A)} \tag{4.19}$$

is uniformly continuous, mapping $C_b(X_0)$ into $C_b(X_1)$. If $V_1 \in \mathcal{U}_{X_1}$, $W \circ W \subset V_1$, and $V_0 \in \mathcal{U}_{X_2}$ so that $(f \times f)(V_0) \subset W$ then since $W(f(A)) \supset \overline{f(A)}$,

$$C(f) \times C(f)(V_{0C}) \subset V_{1C}$$

It easily follows that:

$$\begin{aligned} C : C^u(X_0;X_1) &\to C^u(C(X_0);C(X_1)) \\ f &\mapsto C(f) \end{aligned} \tag{4.20}$$

is a uniformly continuous map.

There are also natural, uniformly continuous maps associating to each point the singleton set and to each n-tuple the associated subset of cardinality $\leq n$:

$$\begin{aligned} i : X &\to C(X) & i_n : X^n &\to C(X) \\ x &\mapsto \{x\} & (x_1, \dots, x_n) &\mapsto \{x_1, \dots, x_n\} \end{aligned} \tag{4.21}$$

PROPOSITION 4.13. *Let $\varphi : T \times X \to X$ be a uniform action. $C(\varphi) : T \times C(X) \to C(X)$ is a uniform action defined by:*

$$C(f)^t = C(f^t) \qquad t \in T \tag{4.22}$$

The subset $C_b(X)$ is closed and + invariant. If the action φ is uniformly reversible, then $C(\varphi)$ is uniformly reversible with $\overline{C(\varphi)} = C(\overline{\varphi})$. In the reversible case, $C_b(X)$ is invariant so the action on $C_b(X)$ is uniformly reversible as well.

Let $\mathcal{F}$ be a full family for T.

a. If φ is $\mathcal{F}$ transitive, then $C(\varphi)$ on $C_b(X)$ is $\mathcal{F}$ central. If $C(\varphi)$ is $\mathcal{F}$ transitive, then φ is $\mathcal{F}$ mixing.

b. The following conditions are equivalent:

(1) φ is $\mathcal{F}$ mixing.

(2) $C(\varphi)$ on $C_b(X)$ is $\mathcal{F}$ mixing.

(3) $C(\varphi)$ on $C_b(X)$ is $\mathcal{F}$ transitive.

PROOF. Because C is functorial, the uniformly continuous map C: $C^u(X,X) \to C^u(C(X),C(X))$ of (4.20) is a homomorphism. Hence $(C(\varphi))^{\#} : T \to C^u(C(X),C(X))$ is the composition $C \circ \varphi^{\#}$ of uniformly continuous homomorphisms. Hence the action $C(\varphi)$ is uniform. Since the image of a finite set is finite, $C_b(X)$ is clearly + invariant. Functoriality implies that in the reversible case $C(f^{-t})$ is the inverse of $C(f^t)$.

If φ is $\mathcal{F}$ transitive (or $\mathcal{F}$ mixing), then the n-fold product on X^n is $\mathcal{F}$ central (resp. $\mathcal{F}$ mixing) by Proposition 4.8d (resp. Proposition 4.10b). Hence the same is true on the factor, which is the image of i_n in $C(X)$ defined by (4.21). Then by Proposition 4.8a, φ is $\mathcal{F}$ central (resp. $\mathcal{F}$ mixing) on $C_b(X)$, since the union of the increasing sequence $\{i_n(X^n)\}$ of subsets is dense in $C_b(X)$. By the remark following Proposition 4.8, the sequence observes pairs of open sets. This proves the first assertion of (a) and (1) $\Rightarrow$ (2) of (b). (2) $\Rightarrow$ (3) is obvious. It remains to show that if either $C(\varphi)$ or its subsystem on $C_b(X)$ is $\mathcal{F}$ transitive, then φ is $\mathcal{F}$ mixing.

Let $x,x_1,x_2 \in X$. Let $V \in \mathcal{U}_X$ with $V = V^{-1}$. Let $F = N^{C(\varphi)}(V_C(\{x\}), V_C(\{x_1,x_2\}))$, which is in $\mathcal{F}$ by assumption. For $t \in F$, there exists $A \in C(X)$ such that $A \in V_C(\{x\})$ and $f^t(A) \subset V_C(\{x_1,x_2\})$, i.e., $A \subset V(x)$ and $V(\overline{f^t(A)}) \supset \{x_1,x_2\}$. Then there exist $y_1,y_2 \in A \subset V(x)$ such that $x_1 \in V(f^t(y_1))$ and $x_2 \in V(f^t(y_2))$. Hence $(y_1,y_2) \in (V(x) \times V(x)) \cap f^{-t}(V(x_1) \times V(x_2))$. It follows that:

$$F \subset N^{\varphi}(V \times V((x,x)), V \times V((x_1,x_2)))$$

Hence:

$$(x_1,x_2) \in \Omega_{\mathcal{F}}(\varphi \times \varphi)(x,x)$$

Since (x_1,x_2) was an arbitrary point of $X \times X$:

$$\Omega_{\mathcal{F}}(\varphi \times \varphi)(x,x) = X \times X$$

By projecting to either factor, we see that φ is $\mathcal{F}$ transitive. By Theorem 4.1, φ is $\mathcal{F}$ mixing. ■

PROPOSITION 4.14. *Assume $\varphi : T \times X \to X$ is a surjective uniform action with X compact and $\mathcal{F}$ is a proper translation invariant family for T. If for $x \in X$, $x \notin \Omega_{k\tau\mathcal{F}}\varphi(x)$, then there exists a closed φ invariant subset A of X with $x \notin A$ such that for every open set G containing A:*

$$J^{\varphi}(X \backslash G, G) \in \tau\mathcal{F} \tag{4.23}$$

Conversely if A is a closed subset of X such that (4.23) holds for every open set G containing A, then:

$$\Omega_{k\tau\mathcal{F}}\varphi(X \backslash A) \subset A \tag{4.24}$$

In particular if $x \notin A$, then $x \notin \Omega_{k\tau\mathcal{F}}\varphi(x)$.

PROOF. Since $x \notin \Omega_{k\tau\mathcal{F}}\varphi(x)$, there is an open set U containing x such that:

$$N(U,U) = N(\overline{U},U) \notin k\tau\mathcal{F}$$

Then its complement $F_0 = J(\overline{U}, X \backslash U)$ is in $\tau\mathcal{F}$. Applying Proposition 2.6 to $[F_0]$, the filter of sets containing F_0, in $\tau\mathcal{F} = \tau\tau\mathcal{F}$, we obtain a translation invariant filter $\mathcal{F}_0$ such that

$$J(\overline{U}, X \backslash U) \in \mathcal{F}_0 \subset \tau\mathcal{F} \tag{4.25}$$

Now define

$$B = \omega_{k\mathcal{F}_0}[\overline{U}] = \bigcap_{F \in \mathcal{F}_0} \overline{f^F(U)} \tag{4.26}$$

Since $k\mathcal{F}_0$ is a translation invariant filterdual, Proposition 3.6 implies that B is an invariant subset, nonempty by Proposition 3.5. By (4.25) $J(\overline{U}, X \backslash U) \in \mathcal{F}_0$, so $B \subset X \backslash U$. If we now define

$$U_1 = (f^T)^{-1}(U) = \bigcup_{t \in T} f^{-t}(U) \tag{4.27}$$

then U_1 is an open set containing $U = f^{-0}(U)$ and $-$ invariant; i.e., $f^t(y) \in U_1$ implies $y \in U_1$. Furthermore $y \in U_1$ implies $f^t(y) \in U$ for some $t \in T$. Because $B \subset X \backslash U$ is invariant, it follows that $y \notin B$. That is

$$A_1 = X \backslash U_1 = \bigcap_{t \in T} f^{-t}(X \backslash U) \tag{4.28}$$

is a closed $+$ invariant subset containing B. We define

$$A = \omega\varphi[A_1] = \bigcap_{t \in T} f^t(A_1) \tag{4.29}$$

Noting that A_1 + invariant implies $f^s(A_1) = f^{T_s}(A_1)$. Since $\mathcal{B}_T$ is a translation invariant filterdual, Proposition 3.6 again implies that A is an invariant subset. Moreover $B \subset A_1$ and $f^s(B) = B$ for all s imply $B \subset A$ as well. Since $A \subset A_1 \subset X\backslash U$, $x \notin A$.

Now if G is an open set containing A, then applying Proposition 3.5 to $A = \omega\varphi[A_1]$;

$$G \supset f^s(A_1) = f^{T_s}(A_1)$$

for some $s \in T$. Since $A \supset B = \omega_{k\mathcal{F}_0}\varphi[\overline{U}]$, it also implies that $f^F(\overline{U}) \subset G$ for some $F \in \mathcal{F}_0$. Thus $J(U,G) \in \mathcal{F}_0$.

Let $C = X\backslash G$. Since:

$$C \cap f^s(A_1) = \emptyset \qquad f^{-s}(C) \cap A_1 = \emptyset$$

$f^{-s}(C) \subset U_1$. By compactness of C, there is a finite subset $\{t_1, \dots, t_k\}$ of T such that:

$$f^{-s}(C) \subset \cup_{i=1}^k f^{-t_i}(U)$$

By (3.5) $f^{t_i}(f^{-t_i}(U)) = U$ implies

$$\bigcap_{i=1}^k g^{t_i}(J(U,G)) \subset \bigcap_{i=1}^k J(f^{-t_i}(U), G)$$

$$= J(\bigcup_{i=1}^k f^{-t_i}(U), G) \subset J(f^{-s}(C), G)$$

But $J(U,G) \in \mathcal{F}_0$ which is a translation invariant filter. Thus $J(f^{-s}(C), G) \in \mathcal{F}_0$. Since f^s is surjective (3.5) and (2.35) also imply

$$J(f^{-s}(C), G) \cap T_s = g^s(J(C,G))$$

Since $\mathcal{F}_0$ is a translation invariant filter, and therefore full, $J(C,G) \in \mathcal{F}_0 \subset \tau\mathcal{F}$, proving (4.23).

Conversely if (4.23) holds for every neighborhood G of a closed set A and $x, y \in X\backslash A$, we can choose G so that $x, y \notin \overline{G}$. Since $J(X\backslash G, G) \in \tau\mathcal{F}$, duality implies $N(X\backslash G, X\backslash G) \notin k\tau\mathcal{F}$. As $X\backslash G$ is a neighborhood of both x and y, we see that $y \notin \Omega_{k\tau\mathcal{F}}\varphi(x)$. Since this holds for all y, $\Omega_{k\tau\mathcal{F}}\varphi(x) \subset A$ for every $x \notin A$. ■

We call a uniform action $\tilde{\varphi} : T \times \tilde{X} \to \tilde{X}$ an $\mathcal{F}$ *eversion* (or an eversion when $\mathcal{F} = \mathcal{B}_T$) if $\tilde{X}$ is compact, the action is surjective, and there is a fixed point $e \in \tilde{X}$ of $\tilde{\varphi}$ such that for every neighborhood G of e

$$J^{\tilde{\varphi}}(\tilde{X}\backslash G, G) \in \mathcal{F} \tag{4.30}$$

the singleton case $\tilde{X} = e$ we call the *trivial eversion*.

PROPOSITION 4.15. *Let $\tilde{\varphi}: T \times \tilde{X} \to \tilde{X}$ be a surjective uniform action on a compact space $\tilde{X}$, and $\mathcal{F}$, $\mathcal{F}_1$, etc. proper families for T.*

a. If $\mathcal{F}_1 \subset \mathcal{F}_2$ and $\tilde{\varphi}$ is an $\mathcal{F}_1$ eversion, then it is an $\mathcal{F}_2$ eversion. If $\tilde{\varphi}$ is an $\mathcal{F}$ eversion then there exists a thick filter $\mathcal{F}_0 \subset \tau\mathcal{F}$ such that $\tilde{\varphi}$ is an $\mathcal{F}_0$ eversion. If moreover $\mathcal{F}$ is translation invariant, then $\mathcal{F}_0$ can be chosen translation invariant.

b. Assume $\mathcal{F}_0 \subset \mathcal{F}$ with $\mathcal{F}_0$ a full filter and $\tilde{\varphi}$ is an $\mathcal{F}_0$ eversion with fixed point e. Then:

$$\begin{aligned} &\omega_{\mathcal{F}_0}\tilde{\varphi}(\tilde{X}\backslash e) = \Omega_{\mathcal{F}_0}\tilde{\varphi}(\tilde{X}\backslash e) = \\ &\omega_{k\mathcal{F}_0}\tilde{\varphi}(\tilde{X}\backslash e) = \Omega_{k\mathcal{F}_0}\tilde{\varphi}(\tilde{X}\backslash e) = e \end{aligned} \tag{4.31}$$

$$\begin{aligned} e \subset \omega_{\mathcal{F}}\tilde{\varphi}(\tilde{X}\backslash e) \subset \Omega_{\mathcal{F}}\tilde{\varphi}(\tilde{X}\backslash e) \\ \omega_{k\mathcal{F}}\tilde{\varphi}(\tilde{X}\backslash e) \subset \Omega_{k\mathcal{F}}\tilde{\varphi}(\tilde{X}\backslash e) \subset e \end{aligned} \tag{4.32}$$

On the other hand

$$\Omega_{\mathcal{F}_0}\tilde{\varphi}(e) = \Omega_{\mathcal{F}}\tilde{\varphi}(e) = \Omega_{k\mathcal{F}_0}\varphi(e) = \tilde{X} \tag{4.33}$$

c. Assume $\mathcal{F}$ is a translation invariant filter and $e \in \tilde{X}$ satisfies $\Omega_{k\mathcal{F}}\tilde{\varphi}(\tilde{X}\backslash e) = e$. Then $\tilde{\varphi}$ is an $\mathcal{F}$ eversion with fixed point e.

d. If $\tilde{\varphi}$ is an eversion with fixed point e, then e is the unique minimal subset of $\tilde{X}$, and the point measure δ_e is the unique $\tilde{\varphi}$ invariant regular Borel measure on $\tilde{X}$.

e. Assume that $\varphi_1 : T \times X_1 \to X_1$ is a uniform action and $h : \tilde{\varphi} \to \varphi_1$ is a surjective, continuous action map. If $\tilde{\varphi}$ is an $\mathcal{F}$ eversion with fixed point e then φ_1 is an $\mathcal{F}$ eversion with fixed point $h(e)$. Conversely if φ_1 is an $\mathcal{F}$ eversion with fixed point e_1 and $h^{-1}(e_1)$ is the singleton e, then $\tilde{\varphi}$ is an $\mathcal{F}$ eversion with fixed point e.

PROOF. (a) Notice that if $U_1 \subset U$ and U_1 is a neighborhood of e, then $J(\tilde{X}\backslash U_1, U_1) \subset J(\tilde{X}\backslash U, U)$. Define

$$\tilde{\mathcal{F}} = \{F : F \supset J(\tilde{X}\backslash U, U) \text{ for some neighborhood } U \text{ of } e\} \tag{4.34}$$

The family $\tilde{\mathcal{F}}$ is proper iff $J(\tilde{X}\backslash U, U) \neq \emptyset$ for all neighborhoods U of e, in which case $\tilde{\mathcal{F}}$ is a filter. Clearly $\tilde{\varphi}$ is an $\mathcal{F}$ eversion iff e is a fixed point for $\tilde{\varphi}$ and $\tilde{\mathcal{F}} \subset \mathcal{F}$. Thus the monotonicity result is clear.

If U is a neighborhood of e and $t \in T$, then $f^t(e) = e$ implies $f^{-t}(U)$ is a neighborhood of e. Therefore if $U_1 = U \cap f^{-t}(U)$, by (3.5):

$$J(\tilde{X}\backslash U_1, U_1) \subset J(\tilde{X}\backslash U, f^{-t}(U)) = g^{-t}(J(X\backslash U, U))$$

Thus $F \in \tilde{\mathcal{F}}$ implies $g^{-t}(F) \in \tilde{\mathcal{F}}$. Since $\tilde{\mathcal{F}}$ is a filter, it follows that $\tilde{\mathcal{F}}$ is thick. Then $\tilde{\mathcal{F}} \subset \mathcal{F}$ implies $\tilde{\mathcal{F}} = \tau\tilde{\mathcal{F}} \subset \tau\mathcal{F}$. If $\mathcal{F}$ is translation invariant, then by Proposition 2.6, there is a translation invariant filter $\mathcal{F}_0$ such that $\tilde{\mathcal{F}} \subset \mathcal{F}_0 \subset \tau\mathcal{F}$.

(b) Since $\tilde{\varphi}$ is an $\mathcal{F}_0$ eversion with fixed point e, the proof that (4.23) implies (4.24) shows that for $x, y \neq e, y \notin \Omega_{k\mathcal{F}_0}\varphi(x)$, so $\Omega_{k\mathcal{F}_0}\varphi(x) \subset e$. For a filter $\mathcal{F}_0$ (4.31) then follows because $\omega_{k\mathcal{F}_0}\tilde{\varphi}(x) \neq \emptyset$ by compactness and:

$$\begin{array}{ccc} \omega_{\mathcal{F}_0}\tilde{\varphi}(x) & \subset & \Omega_{\mathcal{F}_0}\tilde{\varphi}(x) \\ \cap & & \cap \\ \omega_{k\mathcal{F}_0}\tilde{\varphi}(x) & \subset & \Omega_{k\mathcal{F}_0}\tilde{\varphi}(x) \subset e. \end{array}$$

Then by Proposition 3.8, $\omega_{\mathcal{F}_0}\tilde{\varphi}(x) = \omega_{k\mathcal{F}_0}\tilde{\varphi}(x)$ is the singleton e. (4.32) follows from $\mathcal{F}_0 \subset \mathcal{F}$ and $k\mathcal{F} \subset k\mathcal{F}_0$.

For $x \neq e$ and U a neighborhood of e we can shrink U to assume $x \notin U$. Since each f^t is onto, $f^t(\tilde{X} \setminus U) \subset U$ implies $f^t(U) \supset \tilde{X} \setminus U$ and in particular that $x \in f^t(U)$. Therefore if U_1 is any neighborhood of x:

$$N(U, U_1) \supset J(\tilde{X} \setminus U, U) \in \mathcal{F}_0$$

Consequently, $x \in \Omega_{\mathcal{F}_0}\tilde{\varphi}(e)$. Thus $\tilde{X} \setminus e \subset \Omega_{\mathcal{F}_0}\tilde{\varphi}(e)$, and since e is a fixed point, this yields $\Omega_{\mathcal{F}_0}\tilde{\varphi}(e) = \tilde{X}$. Since $\mathcal{F}_0 \subset \mathcal{F}$ and $\mathcal{F}_0 \subset k\mathcal{F}_0$, (4.33) follows.

(c) By Proposition 3.9d $\Omega_{k\mathcal{F}}\tilde{\varphi}(x)$ is $+$ invariant, so e is a fixed point. If U is an open neighborhood of e, then $C = \tilde{X} \setminus U$ is a closed subset of $\tilde{X} \setminus e$. Therefore by (3.48)

$$\omega_{k\mathcal{F}}\tilde{\varphi}[C] \subset \Omega_{k\mathcal{F}}\tilde{\varphi}(C) = e$$

By Proposition 3.5 there exists $F \in \mathcal{F}$ such that $\overline{f^F(C)} \subset U$. Thus $F \subset J(\tilde{X} \setminus U, U)$, so $\tilde{\mathcal{F}} \subset \mathcal{F}$. This means that $\tilde{\varphi}$ is an $\mathcal{F}$ eversion. Notice that translation invariance of $\mathcal{F}$ was only needed to obtain e a fixed point.

(d) By (4.32), $e \in \omega\varphi(x)$ for any $x \in \tilde{X}$, so every closed invariant subset contains e. Thus e is the only minimal set. Now if μ is an invariant Borel measure on X. then:

$$\mu(f^{-t}(e)) = (f^t_*\mu)(e) = \mu(e)$$

so $\mu(f^{-t}(e) \setminus e) = 0$. Because μ is regular, $\operatorname{Inf}\{\mu(U) : U$ open neighborhood of $e\} = \mu(e)$. Then given $\varepsilon > 0$, there exists open U such that $e \in U$ and $\mu(U \setminus e) < \varepsilon/2$. Choose $t \in J(\tilde{X} \setminus U, U)$:

$$\begin{aligned} \varepsilon &= 0 + \varepsilon/2 + \varepsilon/2 \\ &\geq \mu(f^{-t}(e) \setminus e) + \mu(f^{-t}(U \setminus e)) + \mu(U \setminus e) \\ &\geq \mu((f^{-t}(e) \setminus e) \cup (\tilde{X} \setminus (U \cup f^{-t}(e))) \cup (U \setminus e)) \\ &= \mu(\tilde{X} \setminus e) \end{aligned}$$

Thus $\mu(\tilde{X} \setminus e) = 0$, so μ is concentrated at e.

(e) If U_1 is a neighborhood of $e_1 = h(e)$ then $\tilde{U} = h^{-1}(U_1)$ is a neighborhood of e with $\tilde{X}\backslash\tilde{U} = h^{-1}(X_1\backslash U_1)$. For $x \in h^{-1}(X_1\backslash U_1)$ $f^t(x) \in \tilde{U}$ iff $hf^t(x) = f_1^t(h(x)) \in U_1$. Since h is surjective:

$$J^{\varphi}(X_1\backslash U_1, U_1) = J^{\varphi}(\tilde{X}\backslash\tilde{U}, \tilde{U}) \in \mathcal{F}$$

Conversely if U is a neighborhood of $e = h^{-1}(e_1)$, then there exists U_1 a neighborhood of e_1 such that $\tilde{U} = h^{-1}(U_1) \subset U$, because $\cap\{h^{-1}(\overline{U}_1) : U_1$ neighborhood of $e_1\} = e$. Then

$$J^{\varphi}(\tilde{X}\backslash U, U) \supset J^{\varphi}(\tilde{X}\backslash\tilde{U}, \tilde{U}) = J^{\varphi_1}(X_1\backslash U_1, U_1) \in \mathcal{F}$$

■

REMARK. If for an eversion $\tilde{\varphi}$ the fixed point e is $-$ invariant, i.e., $f^{-t}(e) = e$ for all $t \in T$, then the full filter $\tilde{\mathcal{F}} \cdot k\mathcal{B}_T$ is translation invariant. For if $g^{-s}F \supset J(X\backslash U, U)$ we can choose U_1 a neighborhood of e such that $f^{-s}(U_1) \cup U_1 \subset U$. Then:

$$g^{-s}F \supset J(X\backslash U_1, f^{-s}(U_1)) = g^{-s}(J(X\backslash U_1, U_1))$$

so mod $k\mathcal{B}_T$ F contains $J(X\backslash U_1, U_1)$. ■

THEOREM 4.3. *Let $\varphi : T \times X \to X$ be a uniform action and $\mathcal{F}$ a translation invariant family for T.*

a. If $\tilde{\varphi} : T \times \tilde{X} \to \tilde{X}$ is an $\mathcal{F}$ eversion with fixed point e and $h : \varphi \to \tilde{\varphi}$ is a continuous action map, then for $x \in X$, $x \in \Omega_{k\tau\mathcal{F}}\varphi(x)$ implies $h(x) = e$.

b. If φ is surjective, X is compact, and $x \in X$ satisfies $x \notin \Omega_{k\tau\mathcal{F}}\varphi(x)$, then there exists an $\mathcal{F}$ eversion $\tilde{\varphi}$ with fixed point e and a surjective continuous action map $h : \varphi \to \tilde{\varphi}$ such that $h(x) \neq e$. The map h can be chosen to restrict to a homeomorphism from $X\backslash h^{-1}(e) = h^{-1}(\tilde{X}\backslash e)$ onto $\tilde{X}\backslash e$. Furthermore if φ is reversible then $\tilde{\varphi}$ can be chosen reversible.

PROOF. (a) If $x \in \Omega_{k\tau\mathcal{F}}\varphi(x)$, then by (3.40) $h(x) \in \Omega_{k\tau\mathcal{F}}\tilde{\varphi}(h(x))$, so applying (4.32) to the $\tau\mathcal{F}$ eversion $\tilde{\varphi}$, $h(x) = e$.

(b). Following Proposition 4.14, choose a closed φ invariant set A with $x \notin A$ satisfying (4.23) for every open subset G containing A. Let $\tilde{X}$ be the quotient space of X obtained by identifying the points of A together to become the single point e of $\tilde{X}$, that is, the space of equivalence classes for the relation $1_X \cup (A \times A)$. Clearly the quotient map $h : X \to \tilde{X}$ is surjective and a homeomorphism of $X\backslash A$ with $\tilde{X}\backslash e$. Because A is $+$ invariant, we obtain an induced system $\tilde{\varphi}$ on $\tilde{X}$ and h maps φ to $\tilde{\varphi}$. Because A is invariant, $f^{-t}(A) = A$ for all t when φ is reversible, so $\tilde{\varphi}$ is then reversible. By (4.23) and (3.6) $\tilde{\varphi}$ is an $\mathcal{F}$ eversion. ■

COROLLARY 4.2. *If $\varphi : T \times X \to X$ is a uniform action with X compact and $\mathcal{F}$ is a translation invariant family, then φ is not $k\tau\mathcal{F}$ central iff there is a surjection of φ onto a nontrivial $\mathcal{F}$ eversion.*

COROLLARY 4.3. *Let $\varphi : T \times X \to X$ be a uniform action with X compact. If either the minimal points of X are dense, i.e., the mincenter for φ is all of X, or there exists an invariant regular Borel measure whose support is all of X, then φ is $k\tau\mathcal{B}_T$ central. If φ is transitive, i.e., $\mathcal{B}_T$ transitive, as well, then it is topologically ergodic.*

PROOF. These results are not difficult to prove directly. The minimal points are $k\tau\mathcal{B}_T$ recurrent, so if they are dense $\Omega_{k\tau\mathcal{B}_T}\varphi$ contains 1_X. If there is a measure with full support, then one can use the standard proof of the Poincaré Recurrence Theorem. However both results are immediate from Corollary 4.2 because any eversion factor of the hypothesized systems must be trivial by Proposition 4.15d. If the system is also $\mathcal{B}_T$ transitive, then it is $k\tau\mathcal{B}_T$ transitive, i.e., topologically ergodic, by Proposition 4.3c. ■

For example, suppose x is a wandering point, i.e., $x \notin \Omega\varphi(x)$. Then we can apply Theorem 4.3b with $\mathcal{F} = k\mathcal{B}_T = \tau k\mathcal{B}_T$ to obtain a surjection onto a $k\mathcal{B}_T$ eversion. A $k\mathcal{B}_T$ eversion is easy to construct; for example with $T = \mathbf{R}_+$, let $\hat{X}$ be the one-point compactification of $\mathbf{R}$ with e the point at infinity and $f^t(s) = s + t$. Equivalently on the circle with coordinate θ mod 2π use the differential equation:

$$\frac{d\theta}{dt} = \sin^2 2\theta \tag{4.35}$$

More interesting is the case of a topologically transitive system φ, where $\Omega\varphi(x) = X$ for all x, but which is not topologically ergodic, i.e., for some $x \in X$, $x \notin \Omega_{k\tau\mathcal{B}_T}\varphi(x)$. Corollary 4.2 then projects φ onto an eversion that is topologically transitive. This is a bit more difficult to imagine, but such systems do occur [see Theorem 8.3].

5

Compactifications

For a real Banach space (written B space hereafter) E we let $B(E)$ denote the unit ball:

$$B(E) = \{x \in E : |x| \leq 1\} \tag{5.1}$$

For example $B(\mathbf{R})$ is the closed interval $[-1,1]$. For a bounded linear operator $T : E_1 \to E_2$ between B spaces, the operator norm of T can be described as:

$$\| T \| = \sup_{x \in B(E_1)} |T(x)| \tag{5.2}$$

Of course by linearity $|T(x)| \leq \| T \| \, |x|$ for all $x \in E_1$. The set $L(E_1, E_2)$ of all such bounded linear operators is a B space with the operator norm, and its unit ball is the set of operators of norm at most 1. Equivalently:

$$B(L(E_1, E_2)) = \{T \in L(E_1, E_2) : T(B(E_1)) \subset B(E_2)\} \tag{5.3}$$

The *pointwise topology* on $L(E_1, E_2)$, also called the strong operator topology, is that induced via the inclusion $L(E_1, E_2) \to E_2^{E_1}$.

Define for F any subset of $B(E_1)$ the F seminorm of T:

$$\| T \|_F = \sup_{x \in F} |T(x)| \tag{5.4}$$

Since $F \subset B(E_1)$ we have $\| T \|_F \leq \| T \|$ with equality for $F = B(E_1)$. The pointwise uniformity on $E_2^{E_1}$ pulled back to $L(E_1, E_2)$ yields the topological vector space structure induced by the family of seminorms $\| \quad \|_F$ with F varying over all finite subsets of $B(E_1)$.

For $T \in L(E_1, E_2)$ and $S \in L(E_2, E_3)$, the composition $S \circ T \in L(E_1, E_3)$, and for any $F \subset B(E_1)$:

$$\| S \circ T \|_F \leq \| S \| \| T \|_F \tag{5.5}$$

For any B space E, $L(\mathbf{R},E)$ is isometrically isomorphic with E itself via the natural identification $T \mapsto T(1)$, since $\| T \| = |T(1)| = \| T \|_F$ where $F = \{1\}$.

On the other hand, for the *dual space* $L(E,\mathbf{R})$, denoted E^*, the norm and pointwise topologies disagree unless E is finite dimensional. The pointwise topology on E^* is called the weak* topology. The unit ball is compact because $B(E^*)$ is a closed subset of $B(\mathbf{R})^{B(E)}$, since linearity is a collection of pointwise closed conditions.

PROPOSITION 5.1. *Let E, E_1, E_2, etc., be B spaces and the unit balls, $B(L(E,E))$, $B(E^*)$ etc. be given the pointwise topology and associated uniformity.*

a. The composition map:

$$B(L(E_1,E_2)) \times B(L(E_2,E_3)) \to B(L(E_1,E_3))$$
$$(T,S) \mapsto S \circ T \tag{5.6}$$

is continuous (Note: observe the reversal of order). The adjoint associate:

$$B(L(E_1,E_2)) \to C^u(B(L(E_2,E_3));B(L(E_1,E_3)))$$
$$T \mapsto T^* \qquad \text{with} \qquad T^*(S) = S \circ T \tag{5.7}$$

is well-defined and uniformly continuous, where C^u is given the uniformity of uniform convergence.

b. With respect to composition defined as before, $B(L(E,E))$ is a topological monoid (usually nonabelian, of course) with a right + invariant uniformity such that each left translation is uniformly continuous. With the preceding reversed order, the adjoint map of (5.7) $B(L(E,E)) \longrightarrow C^u(B(L(E,E_3));B(L(E,E_3)))$ is a uniformly continuous homomorphism.

PROOF. With $F \subset B(E_1)$ and $T \in B(L(E_1,E_2))$ we have $T(F) \subset B(E_2)$. The usual triangle inequality argument and (5.5) imply, for (T,S), $(\tilde{T},\tilde{S}) \in B(L(E_1,E_2)) \times B(L(E_2,E_3))$:

$$\begin{aligned} \| \tilde{S} \circ \tilde{T} - S \circ T \|_F &= \| \tilde{T}^*(\tilde{S}) - T^*(S) \|_F \\ &\leq \| \tilde{T} - T \|_F + \| \tilde{S} - S \|_{T(F)} \end{aligned} \tag{5.8}$$

Thus the composition in (5.6) is continuous. With $T = \tilde{T}$, we see that T^* is a uniformly continuous function of S, i.e., $T^* \in C^u$. Furthermore with $S = \tilde{S}$, we see that the map $T \mapsto T^*$ is itself uniformly continuous. Recall that the uniformity in C^u is generated by pairs $(\tilde{H},H) \in C^u$ with

$$\sup_S \| \tilde{H}(S) - H(S) \|_F \leq \varepsilon$$

for some $\varepsilon > 0$ and $F \subset B(E_1)$.

With $E_1 = E_2 = E$, right translation is obtained by fixing $S = \tilde{S}$ and varying T, while left translation is obtained by fixing $T = \tilde{T}$ and varying S. Hence the monoid results of (b) are clear. Note that $T^* \circ S^* = (S \circ T)^*$, so by definition of composition in (5.6) $T \mapsto T^*$ is a homomorphism. ■

Our most important application of this uses $E_3 = \mathbf{R}$ in (5.7), i.e.:

$$B(L(E_1, E_2)) \to C^u(B(E_2^*); B(E_1^*))$$
$$T \mapsto T^* \qquad \text{with} \qquad T^*(p) = p \circ T \tag{5.9}$$

When the B spaces are in fact B algebras, as is $\mathbf{R}$, then an algebra map $T : E_1 \to E_2$ is a bounded linear map mapping the unit c_1 in E_1 to the unit c_2 in E_2 and such that $T(xy) = T(x)T(y)$. Our B algebras are B algebras of functions which satisfy the condition:

$$|x^2| = |x|^2 \qquad x \in E \tag{5.10}$$

If $T : E_1 \to E_2$ is an algebra map and E_1, E_2 satisfy (5.10) then $\| T \| \leq 1$, i.e., $T \in B(L(E_1, E_2))$. If not, there exists $x \in E_1$ with $|x| < 1$, but $|T(x)| > 1$. Then $|x^{2^n}| \to 0$ as $n \to \infty$, but $|T(x^{2^n})| = |T(x)|^{2^n} \to \infty$, contradicting continuity of T. In fact $T(c_1) = c_2$ then implies $\| T \| = 1$.

Algebra maps are closed under composition and $L^a(E_1, E_2)$, the set of algebra maps between B algebras satisfying (5.10), is a closed subset of $B(L(E_1, E_2))$ since the product preserving condition is pointwise closed.

For a uniform space X, we denote by $\mathcal{B}(X)$ the real B algebra of bounded, continuous real valued functions on X equipped with the sup norm, i.e., for $u \in \mathcal{B}(X)$,

$$|u| = \sup_{x \in X} |u(x)|$$

Recall that $C(X; \mathbf{R})$ is the space of all continuous, real-valued maps with the uniformity of uniform convergence. $\mathcal{B}(X)$ is a clopen subspace of $C(X; \mathbf{R})$ since any function a finite uniform distance from a bounded function is bounded. $\mathcal{B}^u(X)$ is the closed subalgebra of uniformly continuous functions. Of course, when X is compact $\mathcal{B}^u(X) = \mathcal{B}(X) = C(X; \mathbf{R})$. Clearly, the algebra $\mathcal{B}(X)$ satisfies (5.10) as does every subalgebra.

If $h : X_1 \to X_2$ is continuous, then we obtain the algebra map $h^* : \mathcal{B}(X_2) \to \mathcal{B}(X_1)$ by composition, that is:

$$C(X_1; X_2) \to L^a(\mathcal{B}(X_2), \mathcal{B}(X_1)) \subset B(L(\mathcal{B}(X_2), \mathcal{B}(X_1)))$$
$$h \mapsto h^* \qquad \text{with} \qquad h^*(u) = u \circ h \tag{5.11}$$

From the definition of the sup norm, it is easy to see that $\| h^* \| \leq 1$, i.e., $|h^*(u)| \leq |u|$. Hence, for $(\tilde{h},\tilde{u})(h,u) \in C(X_1;X_2) \times \mathcal{B}(X_2)$:

$$\begin{aligned} |\tilde{u} \circ \tilde{h} - u \circ h| &= |\tilde{h}^*(\tilde{u}) - h^*(u)| \\ &\leq |\tilde{u} - u| + |u \circ \tilde{h} - u \circ h| \end{aligned} \tag{5.12}$$

If h is uniformly continuous, e.g., if X_1 is compact, then h^* maps $\mathcal{B}^u(X_2)$ into $\mathcal{B}^u(X_1)$.

LEMMA 5.1. *With X_1 and X_2 uniform spaces the composition map of (5.11) restricts to define*

$$C^u(X_1;X_2) \to L^a(\mathcal{B}^u(X_2), \mathcal{B}^u(X_1)) \subset B(L(\mathcal{B}^u(X_2), \mathcal{B}^u(X_1)))$$

a uniformly continuous map where C^u has the uniformity of uniform convergence and L^a has the pointwise operator uniformity induced from $B(L(\mathcal{B}^u(X_2), \mathcal{B}^u(X_1)))$.

PROOF. Given a finite subset F of $B(\mathcal{B}^u(X_2))$ and $\varepsilon > 0$, we choose an ε modulus of uniform continuity for all $u \in F$, i.e., $V \in \mathcal{U}_{X_2}$ so that $(\tilde{y},y) \in V$ implies $|u(\tilde{y}) - u(y)| \leq \varepsilon$ for all $u \in F$. Suppose $(\tilde{h},h) \in V^{X_1}$, i.e., $(\tilde{h}(x),h(x)) \in V$ for all $x \in X_1$. Then:

$$|\tilde{h}^*(u) - h^*(u)| \leq \varepsilon$$

for all $u \in F$, i.e., $\| \tilde{h}^* - h^* \|_F \leq \varepsilon$ when $(\tilde{h},h) \in V^{X_1}$. ∎

LEMMA 5.2. *Let $h : X_1 \to X_2$ be a continuous map of uniform spaces.*

a. The following are equivalent:

(1) h is dense, i.e., $h(X_1)$ is a dense subset of X_2.

(2) $h^* : \mathcal{B}(X_2) \to \mathcal{B}(X_1)$ is norm preserving.

(3) $h^* : \mathcal{B}(X_2) \to \mathcal{B}(X_1)$ is injective.

(4) The restriction of h^* to $\mathcal{B}^u(X_2)$ is injective.

b. If $h^*(\mathcal{B}(X_2))$ contains $\mathcal{B}^u(X_1)$, then h is injective. If X_1 and X_2 are compact and h is injective, then $h^* : \mathcal{B}(X_2) \to \mathcal{B}(X_1)$ is surjective.

PROOF. (a) (1) $\Rightarrow$ (2) $\Rightarrow$ (3) $\Rightarrow$ (4) are obvious. Suppose $y_0 \in X_2 \backslash \overline{h(X_1)}$. There is some pseudometric d in the gage for X_2 and some $\varepsilon > 0$ so that $d(y_0,h(x)) \geq \varepsilon$ for all $x \in X_1$. Let $u(y) = \max(0, \varepsilon - d(y_0,y))$. $u \in \mathcal{B}^u(X_2)$ with $u(y_0) = \varepsilon$, but $h^*(u) = 0$. Hence not (1) $\Rightarrow$ not (4).

(b) If $h(x_1) = h(x_2)$ with $x_1 \neq x_2$, then we can choose $u \in \mathcal{B}^u(X_1)$ such that $u(x_1) \neq u(x_2)$. Clearly u is not in the image of h^*. If X_1 is compact and h is injective, then h is an embedding i.e., a homeomorphism of X_1 onto $h(X_1)$ in X_2, and $h(X_1)$ is a closed subset of X_2. If $u \in \mathcal{B}(X_1)$, we define $u \circ h^{-1}$, a bounded real-valued continuous function on $h(X_1)$. We extend to an element of $\mathcal{B}(X_2)$ by using the Tietze Extension Theorem. ■

COROLLARY 5.1. *Let $h : X_1 \to X_2$ be a continuous map of compact spaces. h is surjective iff h^* is injective, in which case h^* preserves norm. h^* is surjective iff h is injective, in which case h is an embedding onto a closed subset.*

For a uniform space X we call a continuous dense map $h : X \to \tilde{X}$ a *compactification* when $\tilde{X}$ is compact. Gelfand theory, which we now describe, shows that compactifications are classified by closed subalgebras of $\mathcal{B}(X)$. The key result is the *Stone–Weierstrass Theorem* which states that for a compact space $\tilde{X}$, the only closed subalgebra that distinguishes points of $\tilde{X}$ is the entire algebra $\mathcal{B}(\tilde{X})$.

THEOREM 5.1. *Let X be a uniform space and E a closed subalgebra of $\mathcal{B}(X)$; i.e., the closed subspace E includes constant functions, and it is closed under multiplication.*

Let X_E denote the set of continuous algebra maps from E to $\mathbf{R}$. When the dual space E^* is given the pointwise operator topology, i.e., the weak topology, then X_E is a compact subset of $B(E^*)$, the unit ball.*

Let $j_E : X \to E^*$ be defined by $j_E(x)(u) = u(x)$. The function j_E maps X continuously onto a dense subset of X_E. Then the restriction, denoted $j_E : X \to X_E$, is a compactification of X. Furthermore $j_E^* : \mathcal{B}(X_E) \to \mathcal{B}(X)$ is an isometric algebra map onto $E \subset \mathcal{B}(X)$.

The continuous map $j_E : X \to X_E$ is uniformly continuous iff $E \subset \mathcal{B}^u(X)$.

If X is compact, then $j_E : X \to X_E$ is surjective. It is then injective iff $E = \mathcal{B}(X)$, in which case it is a homeomorphism.

PROOF. Since $E \subset \mathcal{B}(X)$ satisfies (5.10), algebra maps from E to $\mathbf{R}$ have norm 1. Hence X_E is a subset of $B(E^*)$. Furthermore the algebra map conditions:

$$\begin{aligned} p(1) &= 1 \\ p(u_1 u_2) &= p(u_1)p(u_2) \end{aligned} \tag{5.13}$$

for $p \in X_E$ and $u_1, u_2 \in E$, are pointwise closed conditions. Therefore X_E is a closed, and hence compact, subset of $B(E^*)$.

For each $x \in X$, evaluation at x defines an algebra map from $\mathcal{B}(X)$ to $\mathbf{R}$, so restricts to an element $j_E(x)$ of X_E. Because $\mathcal{B}(X)$ consists of continuous functions, evaluation defines a continuous map from X to $\mathcal{B}(X)^*$ when the weak* topology is used on the dual space. Hence $j_E : X \to X_E$ is continuous.

For any B space like E, there is an isometric inclusion of E into its double dual E^{**}. That is, each $u \in E$ defines a bounded linear map from E^* to $\mathbf{R}$, namely, evaluation at u. By restricting to the subset X_E of E^*, we define the map:

$$\begin{aligned} J_E &: E \to \mathcal{B}(X_E) \\ J_E(u) &\equiv u_E \\ u_E(p) &= p(u) \end{aligned} \tag{5.14}$$

Notice that for $x \in X$:

$$u_E(j_E(x)) = j_E(x)(u) = u(x) \tag{5.15}$$

Thus for all $u \in E$:

$$j_E^*(u_E) = u \tag{5.16}$$

So the composition

$$E \xrightarrow{J_E} \mathcal{B}(X_E) \xrightarrow{j_E^*} \mathcal{B}(X)$$

is just the inclusion of E into $\mathcal{B}(X)$.

The map J_E preserves norm. For if $u \in E$:

$$\begin{aligned} |u| &= \sup_x |u(x)| = \sup_x |u_E(j_E(x))| \\ &\leq \sup_p |u_E(p)| = \sup_p |p(u)| \leq \| p \| \, |u| = |u| \end{aligned}$$

Consequently $|u| = \sup_p |u_E(p)| = |u_E|$.

Furthermore the linear map J_E is an algebra map:

$$J_E(u_1 u_2)(p) = J_E(u_1)(p) \cdot J_E(u_2)(p)$$

for every $p \in X_E$, by (5.14) and (5.13).

Thus the image $J_E(E)$ is a closed subalgebra of $\mathcal{B}(X_E)$, but it clearly distinguishes points: If p_1 and p_2 are distinct elements of X_E, then $p_1(u) \neq p_2(u)$ for some $u \in E$. Thus, by the Stone–Weierstrass Theorem $J_E(E)$ is all of $\mathcal{B}(X_E)$; that is, J_E is an isometric isomorphism of E onto $\mathcal{B}(X_E)$; as $j_E^* \circ J_E$ is the inclusion of E, it follows that j_E^* is just the inverse isomorphism J_E^{-1} from $\mathcal{B}(X_E)$ onto E followed by the inclusion into $\mathcal{B}(X)$.

Since j_E^* is injective, j_E is dense by Lemma 5.2. In particular if X is compact, then j_E is surjective. By Corollary 5.1, if X is compact, j_E is injective iff j_E^* is surjective, i.e., $E = \mathcal{B}(X)$.

If j_E is uniformly continuous, then j_E^* maps $\mathcal{B}^u(X_E) = \mathcal{B}(X_E)$ into $\mathcal{B}^u(X)$, so its image E is contained in $\mathcal{B}^u(X)$. For the reverse direction, apply Lemma 5.1 with X_1 a single point [so that $\mathcal{B}(X_1) = \mathbf{R}$] and $X_2 = X$. $C^u(X_1;X_2)$ is naturally homeomorphic to X, so we obtain from Lemma 5.2 a uniformly continuous map $j_{\mathcal{B}^u(X)}$ from X to $L^a(\mathcal{B}^u(X), \mathbf{R}) = X_{\mathcal{B}^u(X)}$. When $E \subset \mathcal{B}^u(X)$, we further restrict to $L^a(E, \mathbf{R}) = X_E$, yielding the uniformly continuous map j_E. ■

PROPOSITION 5.2. *a. Let $h : X_1 \to X_2$ be a continuous map of uniform spaces and $E_1 \subset \mathcal{B}(X_1)$, $E_2 \subset \mathcal{B}(X_2)$ closed subalgebras. There exists a continuous map $h_{E_2E_1} : X_{E_1} \to X_{E_2}$ such that the following diagram commutes:*

$$\begin{array}{ccc} X_1 & \xrightarrow{\ h\ } & X_2 \\ {\scriptstyle j_{E_1}}\downarrow & & \downarrow{\scriptstyle j_{E_2}} \\ X_{E_1} & \xrightarrow[\ h_{E_2E_1}\ ]{} & X_{E_2} \end{array} \tag{5.17}$$

iff $h^* : \mathcal{B}(X_2) \to \mathcal{B}(X_1)$ maps E_2 into E_1. If $h_{E_2E_1}$ exists, then it is uniquely defined as the continuous map satisfying (5.17).

b. Let $E \subset \mathcal{B}(X)$ be a closed subalgebra for a uniform space X and $h : X \to X_1$ continuous with X_1 compact. There exists a continuous map $h_E : X_E \to X_1$ such that the following diagram commutes:

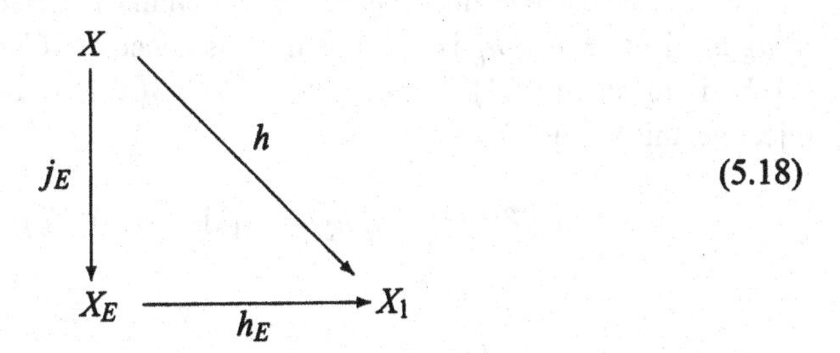

(5.18)

iff $h^*(\mathcal{B}(X_1)) \subset E$ in $\mathcal{B}(X)$. If h_E exists, then it is uniquely defined as the continuous map satisfying (5.18). h_E is surjective iff h is dense; i.e., h is a compactification. h_E is injective iff $h^*(\mathcal{B}(X_1)) = E$. Therefore h_E is a homeomorphism iff h is a compactification with $h^*(\mathcal{B}(X_1)) = E$.

PROOF. (a) If $h_{E_2E_1}$ exists then from (5.17) we have

$$h^*(E_2) = h^*(j^*_{E_2}(\mathcal{B}(X_{E_2}))) = j^*_{E_1}(h^*_{E_2E_1}(\mathcal{B}(X_{E_2}))) \subset j^*_{E_1}(\mathcal{B}(X_{E_1})) = E_1$$

Conversely if h^* maps E_2 into E_1 we can consider the dual space map of the restriction $(h^*)^* : E_1^* \to E_2^*$. Since h^* is an algebra map, $(h^*)^*$ maps X_{E_1} into X_{E_2}. Thus $h_{E_2E_1}$ is defined by

$$\begin{aligned} h_{E_2E_1}(p) &= p \circ h^* \\ h_{E_2E_1}(p)(u) &= p(u \circ h) \end{aligned} \tag{5.19}$$

In particular, $h_{E_2E_1}(j_{E_1}(x)) = j_{E_2}(h(x))$ on each $u \in E_2$, i.e., (5.17) commutes. Uniqueness follows because j_{E_1} is a dense map.

(b) $h^* = j_E^* \circ h_E^*$ shows as in (a) that $h^*(\mathcal{B}(X_1)) \subset E$ is necessary. For the converse, apply (a) with $E_1 = E$ and $E_2 = \mathcal{B}(X_1)$ to obtain a unique map $\tilde{h}$ such that the following commutes:

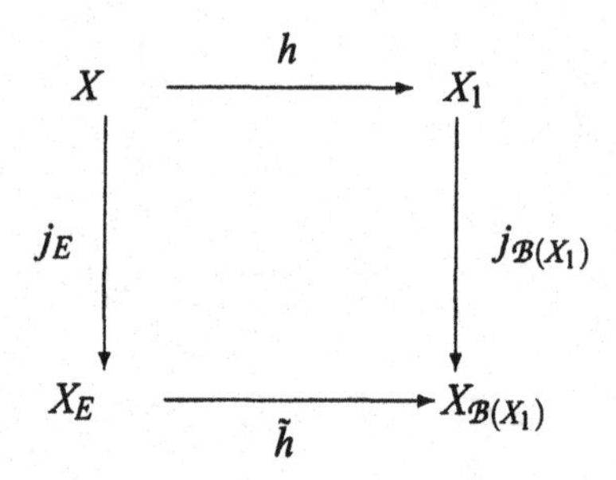

Because X_1 is compact, $j_{\mathcal{B}(X_1)}$ is a homeomorphism, so we define h_E to be $j_{\mathcal{B}(X_1)}^{-1} \circ \tilde{h}$. We have uniqueness again because j_E is dense.

Since j_E^* is injective and X_E is compact, Lemma 5.2a states that h_E is surjective iff h_E^* is injective iff $j_E^* h_E^*$ is injective iff h^* is injective iff h is dense. By Corollary 5.1 h_E is injective iff h_E^* is surjective iff $h_E^*(\mathcal{B}(X_1)) = \mathcal{B}(X_E)$, and since j_E^* is injective, this is true iff

$$h^*(\mathcal{B}(X_1)) = j_E^* h_E^*(\mathcal{B}(X_1)) = j_E^*(\mathcal{B}(X_E)) = E$$

■

In particular a compactification $h : X \to \tilde{X}$ is classified by the subalgebra $h^*(\mathcal{B}(\tilde{X}))$ in $\mathcal{B}(X)$.

COROLLARY 5.2. *Let $h_1 : X \to X_1$ and $h_2 : X \to X_2$ be compactifications of a uniform space X, i.e., dense, continuous maps. There exists a continuous map $h_{21} : X_1 \to X_2$ such that $h_{21} \circ h_1 = h_2$ and only if:*

$$h_2^*(\mathcal{B}(X_2)) \subset h_1^*(\mathcal{B}(X_1)) \tag{5.20}$$

When h_{21} exists, it is unique. It is then surjective. It is injective, iff equality holds in (5.20) in which case it is a homeomorphism.

PROOF. As usual, $h_2^* = h_1^* \circ h_{21}^*$ implies (5.20) and uniqueness follows because h_1 is dense. If (5.20) holds, we apply Proposition 5.2 twice with $E = h_1^*(\mathcal{B}(X_1))$ to obtain maps so that the following commutes:

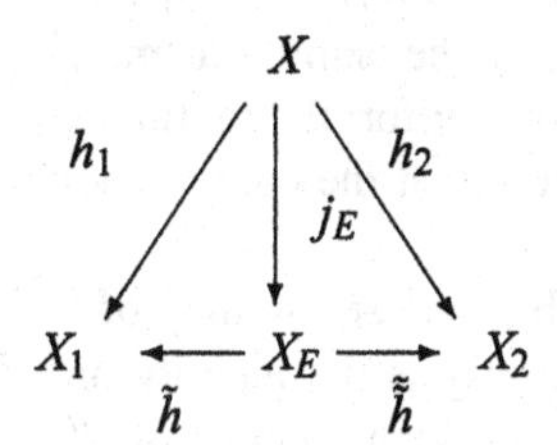

Furthermore, $\tilde{h}$ is a homeomorphism since $E = h_1^*(\mathcal{B}(X_1))$. Define h_{21} to be $\tilde{\tilde{h}} \circ \tilde{h}^{-1}$; h_{21} is a homeomorphism iff $\tilde{\tilde{h}}$ is, so iff $E = h_2^*(\mathcal{B}(X_2))$. ■

COROLLARY 5.3. *a. Let X_1 and X_2 be uniform spaces with $E_1 \subset \mathcal{B}(X_1)$ and $E_2 \subset \mathcal{B}(X_2)$ closed subalgebras. The natural map:*

$$\begin{aligned} L^a(E_2, E_1) &\longrightarrow C(X_{E_1}; X_{E_2}) \\ T &\mapsto T^* \\ T^*(p) &= p \circ T \end{aligned} \tag{5.21}$$

is a uniform isomorphism, i.e., a uniformly continuous homeomorphism with a uniformly continuous inverse. The uniformity of the pointwise operator topology is used on L^a, and uniform convergence is used on C.

b. Let $\tilde{X}_1$ and $\tilde{X}_2$ be compact spaces. The natural map:

$$\begin{aligned} C(\tilde{X}_1; \tilde{X}_2) &\to L^a(\mathcal{B}(\tilde{X}_2), \mathcal{B}(\tilde{X}_1)) \\ h &\mapsto h^* \end{aligned} \tag{5.22}$$

is a uniform isomorphism.

PROOF. By Proposition 5.1 the map of (5.7) is uniformly continuous, and with $E_3 = \mathbf{R}$ we have the commutative diagram:

$$\begin{array}{ccc} L^a(E_2, E_1) & \longrightarrow & C(X_{E_1}; X_{E_2}) \\ & & \cap \\ \cap & & C(X_{E_1}; B(E_2^*)) \\ & & \uparrow \\ B(L(E_2, E_1)) & \longrightarrow & C(B(E_1^*); B(E_2^*)) \end{array}$$

since $X_{E_\alpha} = L^a(E_\alpha, \mathbf{R}) \subset B(E_\alpha^*)$, which is compact. The map across the top is that of (5.21), so it is uniformly continuous.

By Lemma 5.1 the map of (5.22) is uniformly continuous, since $\tilde{X}_1$ and $\tilde{X}_2$ are compact.

To complete the proof of (a), we compose with the map of (5.22) using $\tilde{X}_1 = X_{E_1}$ and $\tilde{X}_2 = X_{E_2}$. We obtain the map from $L^a(E_2, E_1)$ to $L^a(\mathcal{B}(X_{E_2}), \mathcal{B}(X_{E_1}))$, which is the isomorphism induced by the isomorphisms $j^*_{E_\alpha} : \mathcal{B}(X_{E_\alpha}) \to E_\alpha$ $(\alpha = 1,2)$. With these identifications, the uniformly continuous map of (5.22) is the inverse of the map of (5.21). Observe that the composition the other way is the identity on $C(X_{E_1}; X_{E_2})$.

Similarly to complete (b), we use the map of (5.21) with $E_\alpha = \mathcal{B}(\tilde{X}_\alpha)$ $(\alpha = 1,2)$. The composition one way is the identity on $L^a(\mathcal{B}(\tilde{X}_2), \mathcal{B}(\tilde{X}_1))$, while the composition the other way is the identification of $C(\tilde{X}_1; \tilde{X}_2)$ with $C(X_{\mathcal{B}(\tilde{X}_1)}; X_{\mathcal{B}(\tilde{X}_2)})$ induced by the homeomorphisms of compacta

$$j_{\mathcal{B}(\tilde{X}_\alpha)} : \tilde{X}_\alpha \to X_{\mathcal{B}(\tilde{X}_\alpha)} \qquad (\alpha = 1,2)$$

■

PROPOSITION 5.3. *Let X be a uniform space and E a closed subalgebra of $\mathcal{B}(X)$.*

a. A subset $B \subset E$ generates *E if it satisfies the following equivalent conditions:*

(1) E is the smallest closed subalgebra containing B.

(2) The set of rational linear combinations of finite products of elements of B (including the constant 1, the empty product) is dense in E.

(3) The projection from X_E to $\mathbf{R}^B$ defined by $p \mapsto p|B$ is injective.

(4) The projection from X_E to $\mathbf{R}^B$, defined by restriction as above is an embedding, i.e., a homeomorphism onto its image.

(5) $\{u_E^{-1}(a,b) : a < b$ rational and $u \in B\}$ is a subbase for for the topology of X_E.

b. Assume now that B is any generating set for E, e.g., $B = E$. The map $j_E : X \to X_E$ is injective iff B distinguishes points of X, i.e., if x_1, x_2 are distinct points of X, there exists $u \in B$ such that $u(x_1) \neq u(x_2)$. Furthermore the following are equivalent:

(1) $j_E : X \to X_E$ is an embedding, i.e., a homeomorphism onto its image.

(2) E distinguishes points and closed sets in X, i.e., if $x \in X$ and $A \subset X$ with $x \notin \overline{A}$, there exists $u \in E$ such that $u(x) \notin \overline{u(A)}$.

(3) If $\{x_\alpha\}$ is a net in X and $x \in X$, then $\{u(x_\alpha)\} \to u(x)$ for all $u \in B$ implies $\{x_\alpha\} \to x$ in X.

(4) $\{u^{-1}(a,b) : a < b$ rational and $u \in B\}$ is a subbase for the topology of X.

c. The following are equivalent:

(1) X_E has a countable base, so is metrizable.

(2) E is separable, i.e., has a countable dense subset.

(3) E has a countable generating set.

PROOF. (a) (1) $\Leftrightarrow$ (2) is obvious. (3) $\Leftrightarrow$ (4) because X_E is compact and the projection is continuous. (4) $\Leftrightarrow$ (5) by pulling back the product topology subbase via the projection.

(1) $\Leftrightarrow$ (4). Let $\tilde{X}$ denote the image of X_E under the projection and $h_E : X_E \to \tilde{X}$ denote the projection. With $h = h_E \circ j_E$, we have the commutative diagram:

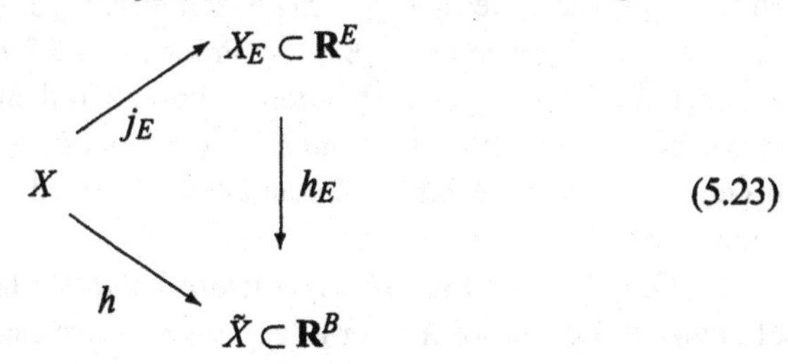

(5.23)

Since h_E is surjective by definition of $\tilde{X}$, Corollary 5.1 says that h_E is a homeomorphism iff h_E^* maps $\mathcal{B}(\tilde{X})$ onto $\mathcal{B}(X_E)$ and therefore iff $j_E^* h_E^* = h^*$ maps $\mathcal{B}(\tilde{X})$ onto E. Since projections to the B coordinates distinguish points of $\tilde{X}$, the Stone–Weierstrass Theorem implies that these projections generate $\mathcal{B}(\tilde{X})$ in the (1) sense. Regarding $u \in B$ as a coordinate index for $\mathbf{R}^B$, the u projection on $\tilde{X}$ pulls back via h to u regarded as a map on X. Thus h_E is a homeomorphism iff the smallest closed subalgebra of $B(X)$ containing B is E, i.e., (4) $\Leftrightarrow$ (1).

(b) Since B generates E, the projection h_E of (5.23) is a homeomorphism. So j_E is injective iff h is injective. Since the coordinate projections distinguish the points of $\tilde{X}$ and pull back to the elements of B, h is injective iff B distinguishes the points of X.

Similarly j_E is an embedding iff h is an embedding, from which (1) $\Leftrightarrow$ (3) and (1) $\Leftrightarrow$ (4) follow easily.

(1) $\Rightarrow$ (2). If j_E is an embedding and $x \notin A$ with A a closed subset of X then $j_E(x) \notin \overline{j_E(A)}$, so there is a continuous function $v \in \mathcal{B}(X_E)$ such that

$$v(j_E(x)) \notin v(\overline{j_E(A)}) = \overline{v(j_E(A))}$$

Let $u = j_E^* v \in j_E^*(\mathcal{B}(X_E)) = E$.

(2) $\Rightarrow$ (1). Given U open in X and $x \in U$, we show that $j_E(U)$ is a neighborhood of $j_E(x)$ in $j_E(X)$. Let $A = X \backslash U$ and choose $u \in E$ such that $u(x) \notin \overline{u(A)}$ in $\mathbf{R}$.

Using the extension u_E of u to X_E of (5.14), define $U_0 = u_E^{-1}(\mathbf{R}\backslash\overline{u(A)})$. This set is open in X_E; it contains $j_E(x)$, and pulls back under j_E to a subset of U.

(c) (2) $\Leftrightarrow$ (3) is obvious from (a). (3) $\Rightarrow$ (1) follows by using (5) of (a). Assuming (1), consider the countable set of pairs (U,V) with U,V in the countable base and $\overline{V} \subset U$. For each such pair define a function $v_{(U,V)} \in \mathcal{B}(X_E)$ that is 0 on $\overline{V}$ and 1 on $X_E\backslash U$. This countable family of functions distinguishes points of X_E, so it generates $\mathcal{B}(X_E)$ by Stone–Weierstrass. Then the pullbacks under j_E form a countable generating set for $j_E^*\mathcal{B}(X_E) = E$. Thus (1) $\Rightarrow$ (3). ■

The class of closed subalgebras of $\mathcal{B}(X)$ ordered by inclusion has a complete lattice structure, with the largest element $\mathcal{B}(X)$ itself and the smallest the algebra of constant functions. With $\{E_\alpha\}$ a family of closed subalgebras, $\wedge_\alpha E_\alpha \equiv \cap_\alpha E_\alpha$ and $\vee_\alpha E_\alpha \equiv$ the closed subalgebra generated by $\cup_\alpha E_\alpha$. Notice that if B_α generates E_α, then $\cup_\alpha B_\alpha$ generates $\cup_\alpha E_\alpha$. By Proposition 5.3 E is countably generated as an algebra iff it has a countably dense subset and therefore iff it is separable as a metric space. Thus if $E_1 \subset E$ and E is countably generated so is E_1. Furthermore if $\{E_\alpha\}$ is a countable family of countably generated subalgebras, then $\vee_\alpha\{E_\alpha\}$ is countable generated.

By Corollary 5.2 this lattice structure induces a lattice structure on the class of compactifications of X. For a family $\{E_\alpha\}$, we can explicitly characterize the compactification associated with $E = \vee_\alpha E_\alpha$. Take the product map $j = \Pi_\alpha j_{E_\alpha}$: $X \to \Pi_\alpha X_{E_\alpha}$. Since $\cup_\alpha E_\alpha$ generates E, Proposition 5.3a implies that the natural map $h_E = \Pi_\alpha h_{E_\alpha E} : X_E \to \Pi_\alpha X_{E_\alpha}$ is a homeomorphism onto the closure of $j(X)$ in the product.

The maximum compactification associated with $\mathcal{B}(X)$ is called the *Stone–Čech compactification*; it is denoted βX. We write $j_\beta : X \to \beta X$ for the map $j_{\mathcal{B}(X)}$. More important for our purposes is the *uniform Stone–Čech compactification* associated with $\mathcal{B}^u(X)$. We denote it by $\beta_u X$ and write $j_u : X \to \beta_u X$ for the map $j_{\mathcal{B}^u(X)}$. Since $\mathcal{B}^u(X)$, and a fortiori $\mathcal{B}(X)$, distinguish points and closed sets, the maps j_β and j_u are both embeddings onto dense subsets of their respective ranges, and j_u is uniformly continuous. Of course there is a natural surjection from βX to $\beta_u X$ that maps j_β to j_u.

For a separable metric space X with a metric d bounded by 1, there is an interesting compactification called the *Gromov compactification*. Define for $x \in X$ the function $d(x) \in \mathcal{B}^u(X)$ by:

$$d(x)(x_1) = d(x,x_1) \tag{5.24}$$

PROPOSITION 5.4. *The map $d : X \to \mathcal{B}^u(X)$ is an isometry from X into the unit ball of $\mathcal{B}^u(X)$, with the norm metric on the latter. The closed algebra generated by the image of d is a separable subalgebra, denoted E_d.*

PROOF. $|d(x)| \leq 1$ because the distance function is assumed bounded by 1. Furthermore:

$$\begin{aligned} d(x,y) &= |d(x)(y) - d(y)(y)| \\ &\leq \sup_{x_1} |d(x)(x_1) - d(y)(x_1)| \leq d(x,y) \end{aligned}$$

by the triangle inequality. Hence $|d(x) - d(y)| = d(x,y)$. Since X is assumed separable, the image of d has a countable dense subset, so the subalgebra E_d is separable by Proposition 5.3. ■

We thus obtain a uniformly continuous metric compactification $j_d : X \to X_{E_d}$. Observe that if $h : X \to X$ is an isometry, i.e., a surjection which preserves distance, then:

$$\begin{aligned} h^*(d(x))(x_1) &= d(x, h(x_1)) \\ &= d(h^{-1}(x), x_1) = d(h^{-1}(x))(x_1) \end{aligned} \tag{5.25}$$

i.e., $h^*(d(x)) = d(h^{-1}(x))$. Thus h^* maps E_d to E_d, and by Proposition 5.2, there is an induced homeomorphism h_d on X_{E_d} such that the following diagram commutes:

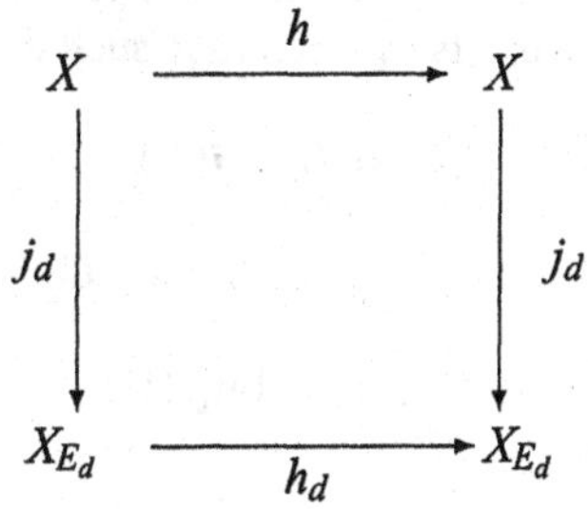

$$\begin{array}{ccc} X & \xrightarrow{\ h\ } & X \\ \downarrow{\scriptstyle j_d} & & \downarrow{\scriptstyle j_d} \\ X_{E_d} & \xrightarrow[\ h_d\ ]{} & X_{E_d} \end{array} \tag{5.26}$$

For any pseudometric d on a uniform space X, and subset A of X we define $d(A)$ by:

$$d(A)(x) = \inf\{d(x_1,x) : x_1 \in A\} \tag{5.27}$$

By the usual triangle inequality argument, we have

$$|d(A)(x_1) - d(A)(x_2)| \leq d(x_1,x_2) \tag{5.28}$$

If the pseudometric is bounded and in the gage for the uniformity on X, $d(A) \in \mathcal{B}^u(X)$ for every subset A of X.

Recall that subsets A_1 and A_2 of X are called *separated* if there exists a symmetric element V of $\mathcal{U}_X$ satisfying the equivalent conditions $V(A_1) \cap A_2 = \emptyset$,

$A_1 \cap V(A_2) = \emptyset$ and $A_1 \times A_2 \cap V = \emptyset$ [cf. (2.23)]. Choosing W symmetric in $\mathcal{U}_X$ with $W^3 \subset V$, we have

$$[W(A_1) \times W(A_2)] \cap W = \emptyset$$

There exists a bounded pseudometric d in the gage and $\varepsilon > 0$, so that $V_\varepsilon^d \subset W$. It follows that for this pseudometric and all (x_1, x_2) in $X \times X$:

$$\max(d(A_1)(x_2), d(A_2)(x_1), d(x_1, x_2)) \geq \varepsilon \tag{5.29}$$

Conversely the existence of a uniformly continuous pseudometric and an $\varepsilon > 0$ such that (5.29) holds imply that A_1 and A_2 are separated.

Using these tools, we now study the uniform Stone–Čech compactification $\beta_u X$ associated with $\mathcal{B}^u(X)$. First some preliminary results.

PROPOSITION 5.5. *a. Let D be a dense subset of a topological space X. If U_0 is an open subset of X then $U_0 \cap D$ is dense in U_0, i.e., $U_0 \subset \overline{U_0 \cap D}$.*

b. Let $j : X \to \tilde{X}$ be a dense embedding of topological spaces. If U is an open subset of X, then $j(U) \subset \mathrm{Int}[\overline{j(U)}]$.

PROOF. (a) is an easy exercise. For (b) let $D = j(X)$. Since $j(U)$ is open in the relative topology of D, there exists U_0 open in X such that:

$$j(U) = U_0 \cap j(X)$$

By (a) $j(U)$ is dense in U_0, so $U_0 \subset \overline{j(U)}$. Because U_0 is open:

$$j(U) \subset U_0 \subset \mathrm{Int}[\overline{j(U)}]$$

■

We describe some special properties of $\beta_u X$. If d is a bounded pseudometric in the gage of X and $A \subset X$ then by (5.27) $d(A) \in \mathcal{B}^u(X)$; therefore it extends to $\beta_u X$. Denote by $d_u(A)$ the unique element of $\mathcal{B}(\beta_u X)$ such that $d_u(A) \circ j_u = d(A)$.

PROPOSITION 5.6. *Let A, A_1, A_2 be subsets of a uniform space X and $j_u : X \to \beta_u X$ be its uniform Stone–Čech embedding.*

a. $p \in \overline{j_u(A)}$ iff $d_u(A)(p) = 0$ for every bounded pseudometric d in the gage of X.

b. For all $V \in \mathcal{U}_X$, we have

$$\overline{j_u(A)} \subset \mathrm{Int}\, \overline{j_u(V(A))} \tag{5.30}$$

c. A_1 and A_2 are separated subsets of X, iff $(\overline{j_u(A_1)}) \cap (\overline{j_u(A_2)}) = \emptyset$.

PROOF. (a) $\overline{d(A)(x) = 0 \text{ for all } x \in A}$, so $d_u(A)$ is 0 on $j_u(A)$. By continuity it vanishes on $\overline{j_u(A)}$. If $p \notin \overline{j_u(A)}$, then there exists a nonnegative $v \in \mathcal{B}(\beta_u X)$ such that $v(p) = 1$ and $v = 0$ on $\overline{j_u(A)}$. Because j_u is uniformly continuous:

$$d(x_1, x_2) = |v(j_u(x_1)) - v(j_u(x_2))|$$

is a pseudometric in the gage of X. Clearly $d(A) = v \circ j_u$, so $d_u(A) = v$. Then $d_u(A)(p) = 1$ with this choice of pseudometric.

(b) Choose a bounded pseudometric d in the gage and positive ε so that $d \leq \varepsilon$ on V, which we may assume is open. By (a), $d_u(A)(p) = 0$ if $p \in \overline{j_u(A)}$, so $U = d_u(A)^{-1}[0, \varepsilon)$ is an open subset of $\beta_u X$ containing $\overline{j_u(A)}$. But:

$$j_u^{-1}(U) = d(A)^{-1}[0, \varepsilon) \subset V(A)$$

Therefore we have

$$U \cap j_u(X) = j_u j_u^{-1}(U) \subset j_u(V(A))$$

So by Proposition 5.5a we get

$$\overline{j_u(A)} \subset U \subset \overline{j_u(V(A))}$$

Since U is open, (5.30) follows.

(c) There is a bounded pseudometric d in the gage and $\varepsilon > 0$ so that (5.29) holds. Restricting to the case $x_1 = x_2$ and extending to $\beta_u X$, we see that $\max(d_u(A_1)(p), d_u(A_2)(p)) \geq \varepsilon$ for all $p \in \beta_u X$. Since $d_u(A_\alpha) = 0$ on $j_u(A_\alpha)$ for $\alpha = 1, 2$, it follows that $j_u(A_1)$ and $j_u(A_2)$ have disjoint closures in $\beta_u X$. The converse is clear because $\beta_u X$ is compact and j_u is uniformly continuous. ■

Recall that for a uniform space X, a family $\mathcal{F}$ of subsets is open if $u\mathcal{F} = \mathcal{F}$, where $u\mathcal{F} = \{F : F \supset V(F_1) \text{ for some } F_1 \in \mathcal{F} \text{ and } V \in \mathcal{U}_X\}$. Dual to the operator u is $\tilde{u}\mathcal{F} = kuk\mathcal{F} = \{F : V(F) \in \mathcal{F} \text{ for all } V \in \mathcal{U}_X\}$. The properties of these operators are collected in Proposition 2.13. In particular if $\mathcal{F}$ is a filter, then $u\mathcal{F}$ is a filter. For $x \in X$, $u[x]$ is the open filter of neighborhoods of x.

A filter $\mathcal{F}$ is an ultrafilter if it is a maximal filter, that is, an element of the class of filters on X maximal with respect to inclusion (see Proposition 2.3). We call $\mathcal{F}$ a *maximal open filter* if it is a maximal element in the more restricted class of open filters.

PROPOSITION 5.7. *Let $j_u : X \to \beta_u X$ be the embedding of a uniform space into its uniform Stone–Čech compactification. For $p \in \beta_u X$, $u[p]$ is the filter of neighborhoods of p in $\beta_u X$. Let $\mathcal{F}_p$ denote the pullback of $u[p]$ by the map j_u so that:*

$$\mathcal{F}_p = j_u^{-1} u[p] = \{j_u^{-1}(U) : \quad p \in \operatorname{Int} U\} \qquad (p \in \beta_u X) \tag{5.31}$$

$\mathcal{F}_p$ is a maximal open filter of subsets of X. If $p = j_u(x)$, then $\mathcal{F}_p$ is $u[x]$, the filter of neighborhoods of x in X.

PROOF. $\mathcal{F}_p$ is defined as the family generated by $\{j_u^{-1}(U) : U \in u[p]\}$, [c.f. (2.25)]. Because j_u is injective, this collection is hereditary so is all of $\mathcal{F}_p$. Since $j_u(X)$ is dense, $p \in \overline{j_u(X)}$ and the family is a filter, i.e., proper as well as closed under intersection. Because $u[p]$ is an open family and j_u is uniformly continuous, $\mathcal{F}_p$ is an open family [cf. (2.92)]. If $p = j_u(x)$, then $\mathcal{F}_p \subset u[x]$ follows from continuity of j_u. Because j_u is an embedding, equality holds.

Suppose $\mathcal{F}$ is an open filter on X with $\mathcal{F} \supset \mathcal{F}_p$ and $F \in \mathcal{F}$. Because $\mathcal{F}$ is an open family, there exists $F_1 \in \mathcal{F}$ and $V \in \mathcal{U}_X$ such that $(V \circ V)(F_1) \subset F$. Because $\mathcal{F}$ is a filter $\mathcal{F} \subset k\mathcal{F}$, so $F_1 \in \mathcal{F}$ meets every element of $\mathcal{F}_p$. Hence $p \in \overline{j_u(F_1)}$. Let $U = \overline{j_u(V(F_1))}$. By (5.30), U is a neighborhood of p, i.e., $U \in u[p]$. Furthermore by definition of the relative topology, it follows that $U \cap j_u(X)$ is the closure of $j_u(V(F_1))$ relative to $j_u(X)$. Since j_u is an embedding, $j_u^{-1}(U)$ is $\overline{V(F_1)}$. But

$$\overline{V(F_1)} \subset V^2(F_1) \subset F$$

Since $j_u^{-1}(U) \subset F$, we have $F \in \mathcal{F}_p$ by heredity. Thus $\mathcal{F} = \mathcal{F}_p$. ■

THEOREM 5.2. *Let $j_u : X \to \beta_u X$ be the embedding of a uniform space into its uniform Stone–Čech compactification. For a filter $\mathcal{F}$ of subsets of X, the following are equivalent:*

(1) $\mathcal{F}$ is a maximal open filter.

(2) For every filter $\tilde{\mathcal{F}}$ containing $\mathcal{F}$, we have $\mathcal{F} = u\tilde{\mathcal{F}}$.

(3) There exists an ultrafilter $\tilde{\mathcal{F}}$ such that $\mathcal{F} = u\tilde{\mathcal{F}}$.

(4) $k\mathcal{F} = \tilde{u}\mathcal{F}$

(5) $\mathcal{F} = uk\mathcal{F}$

(6) $\mathcal{F}$ is open, and $\tilde{u}\mathcal{F}$ is a filterdual.

(7) $\mathcal{F}$ is open, and there exists $p \in \beta_u X$ such that $j_u\mathcal{F} \supset u[p]$.

(8) There exists $p \in \beta_u X$ such that $\mathcal{F} = \mathcal{F}_p$.

The point p of $\beta_u X$ such that $\mathcal{F} = \mathcal{F}_p$ is unique. So we have a bijection from points of $\beta_u X$ to maximal open filters given by $p \mapsto \mathcal{F}_p$.

PROOF. (1) $\Rightarrow$ (2). $\mathcal{F} \subset \tilde{\mathcal{F}}$ and $\mathcal{F}$ open imply $\mathcal{F} = u\mathcal{F} \subset u\tilde{\mathcal{F}}$. Since $u\tilde{\mathcal{F}}$ is an open filter, $\mathcal{F} = u\tilde{\mathcal{F}}$ by maximality.

(2) $\Rightarrow$ (3). By the usual Zorn's Lemma argument there exists an ultrafilter $\tilde{\mathcal{F}}$ containing $\mathcal{F}$. By (2) $\mathcal{F} = u\tilde{\mathcal{F}}$.

(3) $\Rightarrow$ (4). If $\tilde{\mathcal{F}}$ is an ultrafilter such that $\mathcal{F} = u\tilde{\mathcal{F}}$, then:

$$k\mathcal{F} = ku\tilde{\mathcal{F}} = \tilde{u}k\tilde{\mathcal{F}} = \tilde{u}\tilde{\mathcal{F}}$$

since $k\tilde{\mathcal{F}} = \tilde{\mathcal{F}}$ by Proposition 2.3. Applying (2.82) we have $k\mathcal{F} = \tilde{u}\tilde{\mathcal{F}} = \tilde{u}u\tilde{\mathcal{F}} = \tilde{u}\mathcal{F}$.

(4) $\Leftrightarrow$ (5). $k\mathcal{F} = \tilde{u}\mathcal{F}$ implies $\mathcal{F} = k\tilde{u}\mathcal{F} = uk\mathcal{F}$. Conversely $\mathcal{F} = uk\mathcal{F} = k\tilde{u}\mathcal{F}$ implies $k\mathcal{F} = \tilde{u}\mathcal{F}$.

(5) $\Rightarrow$ (6). By (5) $\mathcal{F} = uk\mathcal{F}$ is open. Since (5) $\Rightarrow$ (4), $\tilde{u}\mathcal{F} = k\mathcal{F}$ a filterdual.

(6) $\Rightarrow$ (7). Since $\mathcal{F}$ is a filter and $\beta_u X$ is compact, there exists $p \in \bigcap_{F\in\mathcal{F}} \overline{j_u(F)}$. Given U an open set in $\beta_u X$ containing p, we show that $U \in j_u\mathcal{F}$, the family generated by $\{j_u(F) : F \in \mathcal{F}\}$. Let A be the closed set $(\beta_u X)\backslash U$. There exists $v : \beta_u X \to [0,1]$ continuous with $v(p) = 0$ and $v = 1$ on A. Let $U_1 = v^{-1}[0,1/3)$ and $U_2 = v^{-1}[0,2/3)$, open subsets with $p \in U_1 \subset U_2 \subset U$. Let $\tilde{U}_1 \subset \tilde{U}_2 \subset \tilde{U}$ denote the corresponding preimages in X, via j_u^{-1}. Let $\tilde{A}_1 \supset \tilde{A}_2 \supset \tilde{A}$ denote the complements in X.

Because j_u and v are uniformly continuous:

$$V = \{(x_1,x_2) : |v(j_u(x_1)) - v(j_u(x_2))| < 1/3\} \in \mathcal{U}_X$$

Notice that $V(\tilde{U}_2) \subset \tilde{U}$ and $V(\tilde{A}_2) \subset \tilde{A}_1$. Now p lies in the open set U_1, so U_1 meets every $j_u(F)$ for $F \in \mathcal{F}$, i.e., $\tilde{U}_1 \in k\mathcal{F}$. Hence its complement $\tilde{A}_1 \notin \mathcal{F}$. Because $V(\tilde{A}_2) \subset \tilde{A}_1$, $\tilde{A}_2 \notin \tilde{u}\mathcal{F}$. Because $\tilde{u}\mathcal{F}$ is a filterdual by (6), the Ramsey Property (2.15) implies that the complement $\tilde{U}_2 \in \tilde{u}\mathcal{F}$. Since $V(\tilde{U}_2) \subset \tilde{U}$, $\tilde{U} \in \mathcal{F}$. Hence:

$$U \supset j_u j_u^{-1} U = j_u \tilde{U}$$

is in $j_u\mathcal{F}$.

(7) $\Rightarrow$ (8). $j_u\mathcal{F} \supset u[p]$ implies that $\mathcal{F} \supset \mathcal{F}_p$, since:

$$j_u(F) \subset U \text{ iff } F \subset j_u^{-1}(U)$$

$\mathcal{F}$ is an open filter. By Proposition 5.7, $\mathcal{F}_p$ is a maximal open filter, so equality holds.

(8) $\Rightarrow$ (1). Proposition 5.7.

Uniqueness of p is easy to show. If $p_1 \neq p_2$, there exist by the Hausdorff property U_1, U_2 disjoint with $U_\alpha \in u[p_\alpha]$ $(\alpha = 1,2)$. The $j_u^{-1}(U_1)$ and $j_u^{-1}(U_2)$ cannot both be in any filter. ∎

Remark. If $\mathcal{F}$ is a filter such that $\tilde{u}\mathcal{F}$ is a filterdual, then $u\mathcal{F}$ is an open filter and $\tilde{u}u\mathcal{F} = \tilde{u}\mathcal{F}$ is a filterdual. Therefore $u\mathcal{F}$ is a maximal open filter. ∎

We can describe the topology as well as the points of $\beta_u X$ by using open filters.

PROPOSITION 5.8. *Let $\mathcal{F}$ be an open filter on a uniform space X. If $F \notin \mathcal{F}$, then there exists $p \in \beta_u X$ so that $\mathcal{F} \subset \mathcal{F}_p$ and $F \notin \mathcal{F}_p$.*

PROOF. Since $F \notin \mathcal{F}$, the complement $A = X \backslash F$ lies in $k\mathcal{F}$. The family $u[A]$ is the open filter of all uniform neighborhoods of A. Since $A \in k\mathcal{F}$, $\mathcal{F}_1 \equiv \mathcal{F} \cdot u[A]$ is a proper family, so it is a filter. By (2.84) $\mathcal{F}_1$ is open. If $\mathcal{F}_1 \subset \tilde{\mathcal{F}}$ and $\tilde{\mathcal{F}}$ is an open filter, then $F \notin \tilde{\mathcal{F}}$. Suppose instead $F \in \tilde{\mathcal{F}}$. Because $\tilde{\mathcal{F}}$ is open, there exists $V \in \mathcal{U}_X$ and $F_1 \in \tilde{\mathcal{F}}$ such that $V(F_1) \subset F$. Since $\mathcal{F}_1 \subset \tilde{\mathcal{F}}$, $V^{-1}(A) \in \tilde{\mathcal{F}}$, but since F is the complement of A, F_1 and $V^{-1}(A)$ are disjoint. This contradicts the filter property of $\tilde{\mathcal{F}}$.

We now apply Zorn's Lemma to the family of open filters containing $\mathcal{F}_1$. Since the union of a chain of open filters is open as well as a filter, it follows that some maximal open filter $\tilde{\mathcal{F}}$ contains $\tilde{\mathcal{F}}_1$. By Theorem 5.2, $\tilde{\mathcal{F}} = \mathcal{F}_p$ for some $p \in \beta_u X$. ■

Now define for $F \subset X$, the *hull of* F:

$$H(F) = \{p \in \beta_u X : F \in k\mathcal{F}_p\} = \{p \in \beta_u X : F \in \tilde{u}\mathcal{F}_p\} \tag{5.32}$$

The two definitions agree, since $k\mathcal{F}_p = \tilde{u}\mathcal{F}_p$ by Theorem 5.2. For any family $\mathcal{F}$ of subsets of X, the *hull of* $\mathcal{F}$ is defined by:

$$\begin{aligned} H(\mathcal{F}) &= \{p : u\mathcal{F} \subset \mathcal{F}_p\} = \{p : \mathcal{F} \subset \tilde{u}\mathcal{F}_p\} \\ &= \{p : \tilde{u}\mathcal{F} \subset \tilde{u}\mathcal{F}_p\} = \cap_{F \in \mathcal{F}} H(F) \end{aligned} \tag{5.33}$$

Recall that by (2.83), $u\mathcal{F}_1 \subset \mathcal{F}_2$ iff $\mathcal{F}_1 \subset \tilde{u}\mathcal{F}_2$ in which case $\tilde{u}\mathcal{F}_1 \subset \tilde{u}\tilde{u}\mathcal{F}_2 = \tilde{u}\mathcal{F}_2$. If $\tilde{u}\mathcal{F}_1 \subset \tilde{u}\mathcal{F}_2$, then $\mathcal{F}_1 \subset \tilde{u}\mathcal{F}_1$ implies $\mathcal{F}_1 \subset \tilde{u}\mathcal{F}_2$. Finally $\mathcal{F} \subset \tilde{u}\mathcal{F}_p$ iff $F \in \tilde{u}\mathcal{F}_p$ for all $F \in \mathcal{F}$. On the other hand, $[F] \subset \tilde{u}\mathcal{F}_p$ iff $F \in \tilde{u}\mathcal{F}_p$ which implies:

$$H([F]) = H(F) \tag{5.34}$$

Furthermore since $uu\mathcal{F} = u\mathcal{F}$, $u\mathcal{F}_1 \subset \mathcal{F}_2$ iff $u(u\mathcal{F}_1) \subset \mathcal{F}_2$. Similarly using $\tilde{u}\tilde{u}\mathcal{F} = \tilde{u}\mathcal{F}$, we obtain

$$H(u\mathcal{F}) = H(\mathcal{F}) = H(\tilde{u}\mathcal{F}) \tag{5.35}$$

Finally we have

$$H(\mathcal{F}) = \{p : \mathcal{F} \subset \mathcal{F}_p\} \quad \text{if } \mathcal{F} \text{ is open} \tag{5.36}$$

For a single subset F of X, the dual notion is also useful:

$$\tilde{H}(F) = \{p : F \in \mathcal{F}_p\} = \beta_u X \backslash H(X \backslash F) \tag{5.37}$$

Notice that $p \notin H(X \backslash F)$ iff $X \backslash F \notin k\mathcal{F}_p$ and so iff $F \in \mathcal{F}_p$.

In the reverse direction, given $A \subset \beta_u X$, define the *kernel of A* by:

$$K(A) = u(\cap_{p \in A} \mathcal{F}_p) \tag{5.38}$$

The intersection of the filters associated with the points of A is a filter, but it need not be open unless A is finite. By applying the operator u, we obtain an open filter as the kernel of A.

We extend (5.38) by defining the kernel of the empty set to be the improper family $\mathcal{P}$ of all subsets of X. By (5.33) the hull of $\mathcal{P}$ is empty; thus:

$$K(\emptyset) = \mathcal{P} \qquad H(\mathcal{P}) = \emptyset \tag{5.39}$$

PROPOSITION 5.9. *Let $j_u : X \to \beta_u X$ be the uniform Stone–Čech embedding for a uniform space X.*

a. Let F, F_1, etc., be subsets of X. The hull of F is a closed subset of $\beta_u X$ satisfying:

$$H(F) = H(\overline{F}) = \overline{j_u(F)} \tag{5.40}$$

$\tilde{H}(F) = \tilde{H}(\text{Int} F)$ is the largest open subset U_0 of $\beta_u X$ such that $j_u^{-1}(U_0) = \text{Int} F$. Equivalently it is the union of all open subsets U_0 of $\beta_u X$ such that $j_u^{-1}(U_0) \subset F$.

$$\tilde{H}(F) \subset H(F) \qquad \tilde{H}(X) = H(X) = \beta_u X \tag{5.41}$$

$$\tilde{H}(F_1 \cap F_2) = \tilde{H}(F_1) \cap \tilde{H}(F_2) \qquad H(F_1 \cup F_2) = H(F_1) \cup H(F_2) \tag{5.42}$$

$$F_1 \subset F_2 \Rightarrow \tilde{H}(F_1) \subset \tilde{H}(F_2) \quad \text{and} \quad H(F_1) \subset H(F_2) \tag{5.43}$$

b. Let $\mathcal{F}$, $\mathcal{F}_1$, etc., be families of subsets of X. The hull $H(\mathcal{F})$ is a closed subset of $\beta_u X$:

$$H(\cup_\alpha \mathcal{F}_\alpha) = \cap_\alpha H(\mathcal{F}_\alpha) \tag{5.44}$$

$$\mathcal{F}_1 \subset \mathcal{F}_2 \Rightarrow H(\mathcal{F}_1) \supset H(\mathcal{F}_2) \tag{5.45}$$

If $\mathcal{F}$ is an open family, then:

$$H(\mathcal{F}_1) \cap H(\mathcal{F}) = H(\mathcal{F}_1 \cup \mathcal{F}) = H(\mathcal{F}_1 \cdot \mathcal{F}) \tag{5.46}$$

c. For any nonempty subset A of $\beta_u X$, $K(A) = K(\overline{A})$ is an open filter for X. If A is closed, then:

$$K(A) = \cap_{p \in A} \mathcal{F}_p \tag{5.47}$$

For subsets A_1, A_2 of $\beta_u X$:

$$K(A_1 \cup A_2) = K(A_1) \cap K(A_2) \tag{5.48}$$

$$A_1 \subset A_2 \Rightarrow K(A_1) \supset K(A_2) \tag{5.49}$$

If A_1 and A_2 are closed, then

$$K(A_1 \cap A_2) = K(A_1) \cdot K(A_2) \tag{5.50}$$

d. Let $\mathcal{F}$ be a family for X. If $u\mathcal{F}$ is not contained in any filter, e.g., if there exist $F_1, F_2 \in \mathcal{F}$ separated in X, then $H(\mathcal{F}) = \emptyset$ and $K(H(\mathcal{F})) = \mathcal{P}$. Otherwise $K(H(\mathcal{F}))$ is the smallest open filter containing $u\mathcal{F}$. For any filter $\mathcal{F}$:

$$K(H(\mathcal{F})) = \cap_{p \in H(\mathcal{F})} \mathcal{F}_p = u\mathcal{F} \tag{5.51}$$

For any subset A of $\beta_u X$:

$$H(K(A)) = \overline{A} \tag{5.52}$$

For any subset F of X:

$$K(j_u(F)) = K(H(F)) = u[F] \tag{5.53}$$

e. For any subset F of X, $j_u^{-1}(H(F))$ is the closure of F in X. Thus F is closed iff $F = j_u^{-1}(H(F))$.

Let F be a closed subset of X and A a closed subset of $\beta_u X$ such that $F = j_u^{-1}(A)$. The following conditions are equivalent:

(1) A is clopen in $\beta_u X$.

(2) $A = \tilde{H}(F)$

(3) $K(A) = [F]$

(4) $A = H(F)$ and $[F]$ is an open family for X.

(5) $A = \overline{j_u(F)}$ and $F, X \backslash F$ are separated in X.

PROOF. (a) $F \in k\mathcal{F}_p$ iff F meets $j_u^{-1}(U_0)$ for every neighborhood U_0 of p and so iff $j_u(F)$ meets every neighborhood of p. Thus $F \in k\mathcal{F}_p$ iff $p \in \overline{j_u(F)}$. Clearly $\overline{j_u(\overline{F})} = \overline{j_u(F)}$, so (5.40) follows from the definition of $H(F)$ (5.32).

$F \in \mathcal{F}_p$ iff there exists U_0 open in $\beta_u X$ with $p \in U_0$ and $j_u^{-1}(U_0) \subset F$, in which case, $j_u^{-1}(U_0) \subset \text{Int} F$. Because j_u is the embedding, there exists for any $F \subset X$, U_0 open in $\beta_u X$ so that $j_u^{-1}(U_0) = \text{Int} F$. Taking unions we see that $\tilde{H}(F)$ is the largest open set in $\beta_u X$ that pulls back to $\mathrm{Int}F$. We obtain the same result if we replace F by $\mathrm{Int}F$.

$\mathcal{F}_p \subset k\mathcal{F}_p$ for all $p \in \beta_u X$ implies (5.41). The equations of (5.42) are easy exercises using the filter property for $\mathcal{F}_p$ and the Ramsey Property for $k\mathcal{F}_p$. They are equivalent by (5.37). (5.42) implies (5.43).

(b) $H(\mathcal{F})$ is closed by (5.40) and (5.33). $\mathcal{F}_p$ contains $\cup_\alpha \mathcal{F}_\alpha$ iff $\mathcal{F}_p$ contains $\mathcal{F}_\alpha$ for all α. Hence follows (5.44). It implies (5.45) and the first equation in (5.46). For a filter $\mathcal{F}_p$,

$$(u\mathcal{F}_1) \cup (u\mathcal{F}) = u(\mathcal{F}_1 \cup \mathcal{F})$$

[cf. (2.87)] is contained in $\mathcal{F}_p$ iff $(u\mathcal{F}_1) \cdot (u\mathcal{F}) = u(\mathcal{F}_1 \cdot \mathcal{F})$ [cf. (2.84), since $\mathcal{F}$ is open] is in $\mathcal{F}_p$; (5.46) follows.

By (5.33) we have $u\mathcal{F} \subset \cap_{p \in H(\mathcal{F})} \mathcal{F}_p$; since $uu\mathcal{F} = u\mathcal{F}$, we have from monotonicity and (5.38):

$$u\mathcal{F} \subset K(H(\mathcal{F})) \subset \cap_{p \in H(\mathcal{F})} \mathcal{F}_p$$

If $\mathcal{F}_1$ is an open filter containing $u\mathcal{F}$ and $F \notin \mathcal{F}_1$, then by Proposition 5.8 there exists p such that $\mathcal{F}_1 \subset \mathcal{F}_p$ and $F \notin \mathcal{F}_p$. Since $u\mathcal{F} \subset \mathcal{F}_1 \subset \mathcal{F}_p$, $p \in H(\mathcal{F})$. Since F was arbitrary, $\cap_{p \in H(\mathcal{F})} \mathcal{F}_p \subset \mathcal{F}_1$. Since any kernel is an open filter, $K(H(\mathcal{F}))$ is the smallest open filter containing $u\mathcal{F}$. When $\mathcal{F}$ is a filter, then $\mathcal{F}_1 = u\mathcal{F}$ is an open filter, so we obtain Eq. (5.51). Clearly if $u\mathcal{F}$ is not contained in any $\mathcal{F}_p$, then $H(\mathcal{F}) = \emptyset$, so $K(H(\mathcal{F})) = \mathcal{P}$ by (5.39).

From (5.38) and (2.85) we obtain (5.48) which in turn implies the monotonicity result (5.49).

For any A, the hull $H(K(A))$ is closed, and it contains A by (5.32) and (5.38). Then:

$$A \subset \overline{A} \subset H(K(A)) \subset H(K(\overline{A}))$$

Now let A_0 be a closed subset of $\beta_u X$ containing $\overline{A}$ in its interior and choose A_1 closed so that $\overline{A} \subset \mathrm{Int}A_1$ and $A_1 \subset \mathrm{Int}A_0$. The first condition implies $j_u^{-1}(A_1) \in \mathcal{F}_p$ for all $p \in \overline{A}$. By uniform continuity of j_u, there exists $V \in \mathcal{U}_X$ so that $V(j_u^{-1}(A_1)) \subset j_u^{-1}(A_0)$;

$$j_u^{-1}(A_0) \in u(\cap_{p \in \overline{A}} \mathcal{F}_p) = K(\overline{A})$$

By (5.33) and (5.40):

$$H(K(\overline{A})) \subset H(j_u^{-1}(A_0)) = \overline{j_u j_u^{-1}(A_0)} \subset A_0$$

Intersecting over all such A_0 we obtain $\overline{A} = H(K(A)) = H(K(\overline{A}))$. Because $K(A)$ is an open filter, (5.51) then implies

$$K(A) = K(H(K(A))) = \cap_{p \in H(K(A))} \mathcal{F}_p = \cap_{p \in \overline{A}} \mathcal{F}_p$$

proving (5.47).

By (5.46) and (5.52), $H(K(A_1)\cdot K(A_2)) = \overline{A}_1 \cap \overline{A}_2$. Apply K to both sides and use (5.51) and (2.84) to obtain (5.50).

By (5.40):

$$K(j_u(F)) = K(\overline{j_u(F)}) = K(H(F))$$

Since $H(F) = H([F])$, (5.53) follows from (5.51).

(e) By (5.40), $j_u^{-1}(H(F)) = j_u^{-1}(\overline{j_u(F)})$. This is the closure of F because j_u is an embedding. In general, if A is closed and $F \subset j_u^{-1}(A)$, then by (5.40) and (5.41)

$$\tilde{H}(F) \subset H(F) \subset A \tag{5.54}$$

Now assume $F = j_u^{-1}(A)$.

(1) $\Leftrightarrow$ (2). A is closed and $\tilde{H}(F)$ is open; so (2) $\Rightarrow$ (1). On the other hand, if A is open, then $j_u^{-1}(A) \subset F$ implies $A \subset \tilde{H}(F)$ by (a). Thus by (5.54), (1) $\Rightarrow$ (2).

(3) $\Leftrightarrow$ (4). Since any kernel is open, (3) $\Rightarrow$ (4) follows from (5.52). On the other hand, by (5.51), (4) implies

$$K(A) = u[F] = [F]$$

(4) $\Leftrightarrow$ (5). By (5.40), $H(F) = \overline{j_u(F)}$. $[F]$ is an open family iff for some $V \in \mathcal{U}_X$ $F = V(F)$ and so iff F and $X\backslash F$ are separated in X.

(5) $\Leftrightarrow$ (1). If $F, X\backslash F$ are separated in X then by Proposition 5.6c, $\overline{j_u(F)}$ is disjoint from $\overline{j_u(X\backslash F)}$. Since the union of these two sets is $\beta_u X$, it follows that (5) $\Rightarrow$ (1). On the other hand, if A is clopen then by Proposition 5.5a

$$j_u(F) = A \cap j_u(X)$$

is dense in A; therefore $A = \overline{j_u(F)}$. By compactness A is a uniform neighborhood of itself. Because j_u is uniformly continuous, there exists $V \in \mathcal{U}_X$ such that $j_u(V(F)) \subset A$. Hence $V(F) = F$, so $F, X\backslash F$ are separated in X. ■

REMARK. If U_0 is a neighborhood of p in $\beta_u X$, then with $U = j_u^{-1}(U_0)$, $\tilde{H}(U)$ is an open set in $\beta_u X$ containing p and contained in $H(U) \subset \overline{U}_0$. By (5.42) it follows that $\{\tilde{H}(U) : U \text{ open in } X\}$ forms a base for the topology of $\beta_u X$. It does *not* in general suffice to allow U to vary only over a base for the topology of X. For example $\beta_u X$ need not be second countable when X is. ■

From (5.33) and (5.40) we have

$$p = H(\mathcal{F}_p) = \cap_{F \in \mathcal{F}_p} \overline{j_u(F)} \tag{5.55}$$

If X is discrete with the zero/one metric, so that $1_X \in \mathcal{U}_X$, then $\mathcal{B}^u(X) = \mathcal{B}(X)$ is the B space of all bounded real-valued functions. For every family $\mathcal{F}$, $u\mathcal{F} = \mathcal{F} = \tilde{u}\mathcal{F}$. In particular the maximum open filters are precisely the ultrafilters on X.

For a uniform space X we speak of the *set* X obtained by replacing the original uniformity with the discrete uniformity. We denote by $j_0 : X \to \beta_0 X$ the map to the Stone–Čech (= the uniform Stone–Čech) compactification of the set X. Of course the set map j_0 is not continuous with respect to the original uniform space structure. We denote by H^0 and K^0 the hull and kernel operators for $\beta_0 X$. We regard $\beta_0 X$ as the space of ultrafilters on X. We see from Proposition 5.9e applied to $A = H^0(F)$ that:

$$H^0(F) = \tilde{H}^0(F) \text{ is clopen in } \beta_0 X \text{ for all } F \subset X \tag{5.56}$$

These are all the clopen subsets of $\beta_0 X$.

The uniformly continuous identity map from the set X to the uniform space X induces a surjection from $\beta_0 X$ to $\beta_u X$.

PROPOSITION 5.10. *let* $\pi_0 : \beta_0 X \to \beta_u X$ *denote the natural surjection from the Stone–Čech compactification of the set X, considered as the space of ultrafilters on X, to the uniform Stone–Čech compactification.* π_0 *maps the ultrafilter* $\tilde{\mathcal{F}}$ *to the point p of* $\beta_u X$ *such that* $\mathcal{F}_p = u\tilde{\mathcal{F}}$.

If $\mathcal{F}$ *is any filter on X, then:*

$$\pi_0(H^0(\mathcal{F})) = H(\mathcal{F}) \tag{5.57}$$

If A_0 *is any subset of* $\beta_0 X$, *i.e., any collection of ultrafilters, then:*

$$uK^0(A_0) = K(\pi_0(A_0)) \tag{5.58}$$

PROOF. Because $\beta_0 X$ and $\beta_u X$ are compact and π_0 is continuous, we have $\pi_0(\overline{A_0}) = \overline{\pi_0(A_0)}$ for any subset A_0 of $\beta_0 X$. Since $\pi_0 \circ j_0 = j_u$, this yields for any $F \subset X$:

$$\pi_0(H^0(F)) = \pi_0(\overline{j_0(F)}) = \overline{j_u(F)} = H(F) \tag{5.59}$$

By using compactness again, we obtain (5.57) by intersecting over the filterbase of closed sets $\{H^0(F) : F \in \mathcal{F}\}$.

In particular, if $p = \pi_0(p_0)$ and $\mathcal{F}_p$, $\tilde{\mathcal{F}}_{p_0}$ are the maximal open filter associated with $p \in \beta_u X$ and the ultrafilter associated with $p_0 \in \beta_0 X$ respectively, then by (5.55) and (5.57), π_0 maps $p_0 = H^0(\tilde{\mathcal{F}}_{p_0})$ to $H(\tilde{\mathcal{F}}_{p_0})$, while $\pi_0(p_0) = p$ is also $H(\mathcal{F}_p)$. Applying (5.51) we obtain $u\tilde{\mathcal{F}}_{p_0} = u\mathcal{F}_p = \mathcal{F}_p$, the common kernel.

Consequently $K(\pi_0(A_0)) = u \bigcap_{p_0 \in A_0} u\tilde{\mathcal{F}}_{p_0}$, while $uK^0(A_0) = u \bigcap_{p \in A_0} \tilde{\mathcal{F}}_{p_0}$. These open families are equal by (2.88), proving (5.58). ■

Thus $\beta_0 X$ and $\beta_u X$ can be considered as the spaces of ultrafilters on X and maximal open filters on X, respectively. Furthermore using H^0, K^0 (and H, K) we can associate the filters on X (or the open filters on X) bijectively with the closed subsets of $\beta_0 X$ (resp. of $\beta_u X$). Recalling the space of closed subsets from Proposition 4.13 we have the commutative diagram:

$$\begin{array}{ccc} \beta_0 X & \xrightarrow{\pi_0} & \beta_u X \\ {\scriptstyle i}\downarrow & & \downarrow{\scriptstyle i} \\ C(\beta_0 X) & \xrightarrow[C(\pi_0)]{} & C(\beta_u X) \end{array} \tag{5.60}$$

Furthermore this construction relates the lattice structures by (5.44) and (5.45).

PROPOSITION 5.11. *Let $f : X_1 \to X_2$ be a uniformly continuous map of uniform spaces. There are continuous maps $\beta_u f$ and $\beta_0 f$ uniquely defined so that the following diagram commutes:*

$$\begin{array}{ccccc} \beta_0 X_1 & & \xrightarrow{\beta_0 f} & & \beta_0 X_2 \\ & \nwarrow^{j_0} & & \nearrow^{j_0} & \\ {\scriptstyle \pi_0}\downarrow & X_1 & \xrightarrow{f} & X_2 & \downarrow{\scriptstyle \pi_0} \\ & \swarrow_{j_u} & & \searrow_{j_u} & \\ \beta_u X_1 & & \xrightarrow[\beta_u f]{} & & \beta_u X_2 \end{array} \tag{5.61}$$

The operators β_0 and β_u are functors; i.e., they preserve composition.

For $\mathcal{F}$ a filter on X_1 and $f\mathcal{F}$ the filter on X_2 generated by $\{f(F) : F \in \mathcal{F}\}$, we have

$$\begin{aligned} \beta_u f(H(\mathcal{F})) &= H(f\mathcal{F}) \\ \beta_0 f(H^0(\mathcal{F})) &= H^0(f\mathcal{F}) \end{aligned} \tag{5.62}$$

For subsets $A \subset \beta_u X_1$ and $\tilde{A} \subset \beta_0 X_1$:

$$\begin{aligned} ufK(A) &= K(\beta_u f(A)) \\ fK^0(\tilde{A}) &= K^0(\beta_0 f(\tilde{A})) \end{aligned} \tag{5.63}$$

In particular for $p \in \beta_u X_1$:

$$\mathcal{F}_{\beta_u f(p)} = uf\mathcal{F}_p \tag{5.64}$$

PROOF. f^* maps $\mathcal{B}^u(X_2)$ to $\mathcal{B}^u(X_1)$, so it induces $\beta_u f$ by Proposition 5.2. Substituting the discrete uniformities yields $\beta_0 f$. The diagram (5.61) commutes because j_0 and j_u are dense maps on X_1. As usual this yields uniqueness as well, which in turn implies that β_u and β_0 preserve composition.

To prove (5.64) we note that continuity of $\beta_u f$ implies

$$u[\beta_u f(p)] \subset \beta_u f(u[p])$$

or equivalently

$$(\beta_u f)^{-1}(u[\beta_u f(p)]) \subset u[p]$$

Pulling back by j_u we obtain

$$f^{-1}\mathcal{F}_{\beta_u f(p)} \subset \mathcal{F}_p \text{ or } \mathcal{F}_{\beta_u f(p)} \subset f\mathcal{F}_p$$

Equation (5.64) follows because $\mathcal{F}_{\beta_u f(p)}$ is a maximal open filter.

For $A \subset \beta_u X_1$ (2.31), (2.88), and (2.90) imply

$$\begin{aligned} u(fK(A)) &= ufu(\cap_{p\in A}\mathcal{F}_p) = uf(\cap_{p\in A}\mathcal{F}_p) \\ &= u(\cap_{p\in A} f\mathcal{F}_p) = u(\cap_{p\in A} uf\mathcal{F}_p) \\ &= u(\cap_{p\in A}\mathcal{F}_{\beta_u f(p)}) = K(\beta_u f(A)). \end{aligned}$$

This proves (5.63) and from it we get (5.62) by applying HK (= the identity on closed subsets):

$$\begin{aligned} HK(\beta_u f(H(\mathcal{F}))) &= H(ufK(H(\mathcal{F}))) = \\ H(ufu\mathcal{F}) &= H(uf\mathcal{F}) = H(f\mathcal{F}) \end{aligned}$$

The equations (5.62) and (5.63) for β_0 follow from the β_u results applied to the discrete uniformity. ■

In particular if we interpret

$$C(\beta_0 f) : C(\beta_0 X_1) \to C(\beta_0 X_2) \qquad C(\beta_0 f) : C(\beta_u X_1) \to C(\beta_u X_2)$$

as maps of filters and open filters, respectively, then $C(\beta_0 f)$ associates the image filter $f\mathcal{F}$ to filter $\mathcal{F}$, and $C(\beta_u f)$ associates the open filter $uf\mathcal{F}$ to the open filter $\mathcal{F}$.

We now return to actions. Let $\varphi : T \times X \to X$ be a uniform action with T a uniform monoid. Let E be a closed subalgebra of $\mathcal{B}(X)$ and $j_E : X \to X_E$ be the associated compactification. We call E a φ + *invariant* subalgebra if for each $t \in T$ the algebra map $(f^t)^* : \mathcal{B}(X) \to \mathcal{B}(X)$ maps E into E, i.e., $(f^t)^*(E) \subset E$.

The function $t \mapsto (f^t)^*$ then defines a homomorphism $\varphi^* : T \to L^a(E,E)$, and each map f^t extends to define a continuous map f_E^t such that the following commutes:

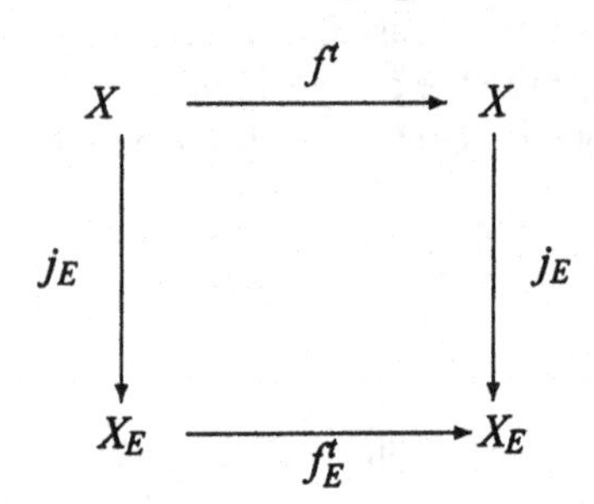

This in turn defines an action $\varphi_E : T \times X_E \to X_E$.

We say that φ *acts uniformly on* E when the homomorphism $\varphi^* : T \to L^a(E,E) \subset B(L(E,E))$ is uniformly continuous where $L^a(E,E)$ is given the pointwise operator uniformity. By Proposition 5.1b and Lemma 1.1c, it is sufficient for φ^* to be continuous at 0.

PROPOSITION 5.12. *Let $\varphi : T \times X \to X$ be a uniform action and E a closed subalgebra of $\mathcal{B}(X)$. If E is $\varphi +$ invariant and φ acts uniformly on E then $\varphi_E : T \times X_E \to X_E$ is a uniform action.*

If either T is discrete or $E \subset \mathcal{B}^u(X)$, then φ acts uniformly on E whenever E is $\varphi +$ invariant.

PROOF. By Corollary 5.3, the map from $L^a(E,E)$ to $C(X_E;X_E)$ that sends $(f^t)^*$ to extension f_E^t is a uniform isomorphism. If the homomorphism φ^* is uniformly continuous then the homomorphism $\varphi_E^\# : T \to C(X_E;X_E)$ ($C = C^u$ by compactness) is uniformly continuous which implies (1.10) and (1.11) hold; i.e., the action φ_E is uniform.

If T is discrete, then φ^* is continuous.

By Lemma 5.1, the map from $C^u(X;X)$ to $L^a(\mathcal{B}^u(X), \mathcal{B}^u(X))$, which sends f^t to $(f^t)^*$, is uniformly continuous. Because the action φ is uniform, $\varphi^\# : T \to C^u(X;X)$ is a uniformly continuous homomorphism. If $E \subset \mathcal{B}^u(X)$ and is invariant we project to $L^a(E, \mathcal{B}^u(X))$ then pull back by the isometric inclusion to $L^a(E,E)$ to obtain that $\varphi^* : T \to L^a(E,E)$ is uniformly continuous. ∎

PROPOSITION 5.13. *If $\varphi : T \times X \to X$ is a uniform action, then it extends to uniform actions $\varphi_u : T \times \beta_u X \to \beta_u X$ and $C(\varphi_u) : T \times C(\beta_u X) \to C(\beta_u X)$ by $f_u^t = \beta_u(f^t)$ and $C(f_u)^t = C(f_u^t) = C(\beta_u(f^t))$.*

Let $\mathcal{F}$ be a filter of subsets of X.

$\mathcal{F}$ is $-$ invariant iff $\mathcal{F} \subset f^t \mathcal{F}$ for all $t \in T$, in which case the open filter $u\mathcal{F}$ is $-$ invariant. $u\mathcal{F}$ is $-$ invariant iff $H(\mathcal{F}) = H(u\mathcal{F})$ is a $+$ invariant closed subset of $\beta_u X$.

$\mathcal{F}$ is invariant iff $\mathcal{F} = f^t\mathcal{F}$ for all $t \in T$, in which case the open filter $u\mathcal{F}$ satisfies

$$u\mathcal{F} = uf^t u\mathcal{F} \quad \textit{for all } t \in T \tag{5.65}$$

Condition (5.65) holds for $\mathcal{F}$ iff $H(\mathcal{F}) = H(u\mathcal{F})$ is an invariant closed subset of $\beta_u X$, i.e., a fixed point for the action $C(\varphi_u)$.

PROOF. φ acts uniformly on $\mathcal{B}^u(X)$ and induces the uniform action φ_u by Proposition 5.12. Proposition 4.13 yields the $C(\varphi_u)$ action.

$\mathcal{F}$ is $-$ invariant iff $f^{-t}\mathcal{F} \subset \mathcal{F}$, or equivalent by Proposition 2.4c, iff $\mathcal{F} \subset f^t\mathcal{F}$ for all $t \in T$. Then by (2.90):

$$u\mathcal{F} \subset uf^t\mathcal{F} = uf^t u\mathcal{F} \tag{5.66}$$

Then by (5.62),

$$\beta_u f^t(H(\mathcal{F})) = \beta_u f^t(H(u\mathcal{F})) = H(f^t u\mathcal{F}) = H(uf^t u\mathcal{F}) \subset H(u\mathcal{F}) = H(\mathcal{F})$$

For the converse, we apply the kernel operator to obtain

$$u\mathcal{F} = KH(u\mathcal{F}) \subset K(\beta_u f^t(H(\mathcal{F}))) = uf^t KH(\mathcal{F}) = uf^t u\mathcal{F} \subset f^t u\mathcal{F}$$

$\mathcal{F}$ is invariant iff in addition $f^t\mathcal{F} \subset \mathcal{F}$ for all t and so $f^t\mathcal{F} = \mathcal{F}$. By (2.90) again this implies $uf^t u\mathcal{F} = u\mathcal{F}$. Once again, (5.62) and (5.63) show that this condition is equivalent to $\beta_u f^t(H(\mathcal{F})) = H(\mathcal{F})$.

The fixed points of $C(\varphi_u)$ are exactly the closed invariant subsets of $\beta_u X$. ∎

PROPOSITION 5.14. *Let $\varphi : T \times X \to X$ be a uniform action and $\mathcal{F}$ a proper family for T. For $x_1, x_2 \in X$ suppose $x_2 \notin \Omega_{\mathcal{F}}\varphi(x_1)$. There exists a uniform action $\varphi_1 : T \times X_1 \to X_1$ with X_1 compact and a dense map of actions $h : \varphi \to \varphi_1$ such that $h(x_2) \notin \Omega_{\mathcal{F}}\varphi_1(h(x_1))$. If T is separable, then X_1 can be chosen metrizable. If φ is uniformly reversible, then φ_1 can be chosen reversible. If T is separable and φ is uniformly reversible, then X_1 can be chosen metrizable with φ_1 reversible.*

PROOF. By (3.46), $\Omega_{\mathcal{F}}\varphi(x_1)$ is the intersection of $\omega_{\mathcal{F}}\varphi[U]$s where the Us vary over neighborhoods of x_1. Furthermore $\omega_{\mathcal{F}}\varphi[U]$ is the intersection of $\overline{f^F(U)}$s, where F varies over $k\mathcal{F}$. Hence, there exists U_1 an open set containing x_1 and $F \in k\mathcal{F}$ so that $x_2 \notin \overline{f^F(U_1)}$. Let $U_2 = X \backslash \overline{f^F(U_1)}$, which is an open set containing x_2. By choosing a pseudometric d in the gage so that $d(x_2, \overline{f^F(U_1)})$ and $d(x_1, X\backslash U_1)$ are positive, we can construct uniformly continuous maps $u_\alpha : X \to [0,1]$ with $u_\alpha(x_\alpha) = 1$ and $u_\alpha = 0$ on $X\backslash U_\alpha$ $(\alpha = 1,2)$. Hence for all $t \in F$ we have that $u_1 \cdot (f^t)^* u_2 = 0$ for if $u_1(x) > 0, x \in U_1$, so $f^t(x) \notin U_2$.

Now let E be a closed, $\varphi +$ invariant subalgebra of $\mathcal{B}^u(X)$ containing u_1 and u_2. For example, use E_φ the closed subalgebra generated by $\{(f^t)^* u_\alpha : \alpha = 1,2$

and $t \in T\}$. Since u_1, u_2 and the f^ts are uniformly continuous, this is a subalgebra of $\mathcal{B}^u(X)$. By Proposition 5.11, φ acts uniformly on E and extends to φ_E on X_E so that $j_E : X \to X_E$ defines a dense map of actions. Let u_{1E} and u_{2E} denote the extensions of u_1 and u_2 to X_E. The identity $u_{1E} \cdot (f^t_E)^* u_{2E} = 0$ on X_E follows from the corresponding identity on X. Then:

$$V_2 \cap f^F_E(V_1) = \emptyset \text{ with } V_\alpha = u^{-1}_{\alpha E}(0,1] \quad (\alpha = 1,2)$$

Therefore $V_2 \cap \overline{f^F_E(V_1)} = \emptyset$. Notice that $j_E(x_\alpha) \in V_\alpha$ $(\alpha = 1,2)$, so:

$$j_E(x_2) \notin \omega_{\mathcal{F}} \varphi_E[V_1] \supset \Omega_{\mathcal{F}} \varphi_E(j_E(x_1))$$

If φ is reversible, we can use the algebra $E_{\varphi\overline{\varphi}}$ generated by $\{(f^t)^* u_\alpha, (f^{-t})^* u_\alpha : \alpha = 1,2 \text{ and } t \in T\}$ to obtain an algebra E to which both the action φ and its inverse $\overline{\varphi}$ extend so that φ_E is reversible.

If T is separable, then let T_0 be a countable dense subset of T. As $t_i \to t, f^{t_i} \to f^t$ uniformly and hence for each $v \in \mathcal{B}^u(X)$ $(f^{t_i})^*(v) \to (f^t)^*(v)$ in norm. Hence the closed algebra generated by $\{(f^t)^* u_\alpha : \alpha = 1,2 \text{ and } t \in T_0\}$ is E_φ, so E_φ is separable. Then X_E is compact and metrizable by Proposition 5.3. If φ is also uniformly reversible, then $E_{\varphi,\overline{\varphi}}$ is generated by $\{(f^t)^* u_\alpha, (f^{-t})^* u_\alpha : \alpha = 1,2 \text{ and } t \in T_0\}$, since the reverse action $\overline{\varphi}$ is uniform as well by Proposition 1.3. ∎

THEOREM 5.3. *Let $\varphi : T \times X \to X$ be a uniform action and $\mathcal{F}$ a proper family for T. Assume that T is separable. φ is $\mathcal{F}$ central (or $\mathcal{F}$ transitive) iff every compact metric dense image of φ is $\mathcal{F}$ central (resp. $\mathcal{F}$ transitive). If φ is uniformly reversible, we need only consider reversible compact metric factors.*

PROOF. If φ is $\mathcal{F}$ central or $\mathcal{F}$ transitive then every dense image satisfies the corresponding property by Proposition 4.1d. If φ is not $\mathcal{F}$ central then for some $x \in X, x \notin \Omega_{\mathcal{F}} \varphi(x)$. Apply Proposition 5.14 with $x_1 = x_2 = x$ to obtain a compact metric X_1 and a dense map $h : \varphi \to \varphi_1$ such that $h(x) \notin \Omega_{\mathcal{F}} \varphi_1(h(x))$. Hence φ_1 is not $\mathcal{F}$ central. Similarly if φ is not $\mathcal{F}$ transitive, then for some $x_1, x_2 \in X$, $x_2 \notin \Omega_{\mathcal{F}}(x_1)$. In Proposition 5.14 the image $h(x_2) \notin \Omega_{\mathcal{F}} \varphi(h(x_1))$, so φ_1 is not $\mathcal{F}$ transitive. ∎

For our last application, we relativize the concept of minimal set (cf. Proposition 3.14). Let $h : \varphi_1 \to \varphi_2$ be a continuous surjection of uniform actions on compact spaces. We call h a *minimal mapping* if for $A \subset X_1$ the conditions A closed, φ_1 + invariant and $h(A) = X_2$ taken together imply that $A = X_1$. Thus φ_1 is a minimal system iff the map to the trivial action on a point is a minimal mapping.

PROPOSITION 5.15. *Let $h : \varphi_1 \to \varphi_2$ be a continuous surjection of uniform actions on compact spaces.*

a. If h is almost injective, i.e., $\{x \in X_1 : h^{-1}h(x) = x\}$ is dense in X_1, then h is a minimal mapping.

b. There exists a closed + invariant subset A of X_1 such that the restriction of h to the subsystem on A, $h : \varphi_{1A} \to \varphi_2$ is a minimal mapping. We then call A a minimal subset relative to h.

PROOF. (a) If A is closed and $h(A) = X_2$ then $h^{-1}h(x) = x$ implies $x \in A$. This condition is dense, and A is closed so $A = X_1$.

(b) Use the usual Zorn's Lemma argument. If $\{A_\alpha\}$ is a descending chain of closed + invariant sets mapping onto X_2, then $A = \cap_\alpha \{A_\alpha\}$ is closed and + invariant. For $y \in X_2$, $\{A_\alpha \cap h^{-1}(y)\}$ is a descending chain of nonempty compact sets. Therefore the intersection $A \cap h^{-1}(y)$ is nonempty, i.e., $h(A) = X_2$. ■

PROPOSITION 5.16. *Let $h : \varphi_1 \to \varphi_2$ be a minimal mapping of uniform actions on compact spaces. Let $\mathcal{F}$ be a translation invariant filterdual of subsets of T.*

The action φ_1 is minimal iff φ_2 is; φ_1 is $\mathcal{F}$ transitive iff φ_2 is. Furthermore for $x \in X_1$, $\omega_\mathcal{F}\varphi_1(x) = X_1$ iff $\omega_\mathcal{F}\varphi_2(h(x)) = X_2$. In particular:

$$\mathrm{Trans}_{\varphi_1} = h^{-1}(\mathrm{Trans}_{\varphi_2}) \tag{5.67}$$

PROOF. If φ_1 is minimal or $\mathcal{F}$ transitive, then so is φ_2 by Propositions 3.14 and 4.1, respectively. If φ_2 is minimal and A is a closed + invariant subset of X_1, then $h(A)$ is a closed + invariant subset of X_2, by compactness, so $h(A) = X_2$ by minimality of X_2. Hence $A = X_1$ by minimality of h; thus φ_1 is minimal.

By (3.28):

$$h(\omega_\mathcal{F}\varphi_1(x)) = \omega_\mathcal{F}(\varphi_2(h(x))$$

because $\mathcal{F}$ is a filterdual and X_1 is compact. Then if $\omega_\mathcal{F}\varphi_1(x) = X_1$, $\omega_\mathcal{F}\varphi_2(h(x)) = X_2$ because h is surjective. Conversely if $\omega_\mathcal{F}\varphi_2(h(x)) = X_2$, then $\omega_\mathcal{F}\varphi_1(x)$ is a closed + invariant subset [by (3.30) and (3.31) since $\mathcal{F}$ is translation invariant] that maps onto X_2. By minimality of h, $\omega_\mathcal{F}\varphi_1(x) = X_1$. In particular, $\mathrm{Trans}_{\varphi_1} = \{x : \omega\varphi_1(x) = X_1\}$ is $h^{-1}(\mathrm{Trans}_{\varphi_2})$, proving (5.67).

For the general transitivity result, let

$$A = \{x : \Omega_\mathcal{F}\varphi_1(x) = X_1\} = \cap_{x_1 \in X}\{\Omega_\mathcal{F}\varphi_1^{-1}(x_1)\}$$

We want to show that φ_2 $\mathcal{F}$ transitive implies $A = X_1$. Observe first that by (3.42), $\Omega_\mathcal{F}\varphi(x) \subset \Omega_\mathcal{F}\varphi(f^t(x))$ for all $t \in T$, so A is + invariant. It is closed because $\Omega_\mathcal{F}\varphi_1$ is a closed relation. By minimality of h, it suffices to show h maps A onto X_2. We derive a contradiction by assuming $y \in X_2$ such that $h^{-1}(y) \cap A = \emptyset$.

For each $x \in h^{-1}(y)$, $\Omega_\mathcal{F}\varphi(x)$ is a closed + invariant set because $\mathcal{F}$ is translation invariant. Since $\Omega_\mathcal{F}\varphi(x) \neq X_1$, minimality of h implies that $h(\Omega_\mathcal{F}\varphi(x))$ is a proper, closed subset of X_2. $\{\omega_\mathcal{F}\varphi[U] : U$ a neighborhood of $x\}$ is a filterbase of compacta with intersection $\Omega_{\mathcal{F}}\varphi(x)$ by (3.46). Hence $\{h(\omega_{\mathcal{F}}\varphi[U]) : U$ a neighborhood of $x\}$ has intersection $h(\Omega_{\mathcal{F}}\varphi(x))$. Therefore we can choose a neighborhood U_x such that $h(\omega_{\mathcal{F}}\varphi[U_x])$ is a proper closed + invariant subset of X_2. Since $\mathcal{F}$ is translation invariant, φ_2 $\mathcal{F}$ transitive implies φ_2 in $\mathcal{B}_T$ transitive. Hence each $h(\omega_{\mathcal{F}}\varphi_1[U_x])$ is nowhere dense in X_2. Now select a finite subcover $U_{x_1}, U_{x_2}, \ldots, U_{x_k}$ of $h^{-1}(y)$. By compactness there exists a neighborhood V of y such that:

$$h^{-1}(V) \subset U_{x_1} \cup \cdots \cup U_{x_k}$$

Because $\mathcal{F}$ is a filterdual

$$\omega_{\mathcal{F}}\varphi_1[\cup_{i=1}^{k} U_{x_i}] = \cup_{i=1}^{k} \omega_{\mathcal{F}}\varphi_1[U_{x_i}]$$

(cf. Proposition 3.6b) and so $h(\omega_{\mathcal{F}}\varphi_1[h^{-1}(V)])$ is contained in a finite union of closed nowhere dense sets. It is therefore a proper subset of X_2. But by (3.28) this set is

$$\omega_{\mathcal{F}}\varphi_2[hh^{-1}(V)] = \omega_{\mathcal{F}}\varphi_2[V]$$

which contains $\Omega_{\mathcal{F}}\varphi_2(y) = X_2$ by $\mathcal{F}$ transitivity of φ_2. This contradiction completes the proof. ■

REMARK. Suppose conversely that $h : \varphi_1 \to \varphi_2$ is a continuous surjection of uniform actions with X_1 and X_2 compact. If for some $y \in X_2$:

$$h^{-1}(y) \subset \text{Trans}_{\varphi_1} \tag{5.68}$$

then h is a minimal mapping. ■

PROPOSITION 5.17. *Let $h : \varphi_1 \to \varphi_2$ be a continuous surjection of uniform actions with X_1 and X_2 compact. Assume $\mathcal{F}$ is a translation invariant filterdual.*

$$h(|\omega_{\mathcal{F}}\varphi_1|) = |\omega_{\mathcal{F}}\varphi_2| \tag{5.69}$$

That is, $y \in \omega_{\mathcal{F}}\varphi_2(y)$ iff there exists x such that $h(x) = y$ and $x \in \omega_{\mathcal{F}}\varphi_1(x)$.

PROOF. By (3.28), $x \in \omega_{\mathcal{F}}\varphi_1(x)$ implies $h(x) \in \omega_{\mathcal{F}}\varphi_2(h(x))$. Now suppose that $y \in \omega_{\mathcal{F}}\varphi_2(y)$.

Let $\tilde{X}_2 = \omega_{\mathcal{F}}\varphi_2(y)$ compact and φ_2 + invariant because $\mathcal{F}$ is translation invariant. Let $\tilde{\varphi}_2$ be the subsystem of φ_2 obtained by restricting to $\tilde{X}_2$. By hypothesis $y \in \tilde{X}_2$ and so $\omega_{\mathcal{F}}\tilde{\varphi}_2(y) = \tilde{X}_2$.

Let $X_0 = h^{-1}(\tilde{X}_2)$ a compact φ_1 + invariant subset of X_1. Let φ_0 be the subsystem of φ_1 obtained by restricting to X_0.

The surjection h restricts to a surjection $h : \varphi_0 \to \tilde{\varphi}_2$. By Proposition 5.15b there is a closed + invariant subset $\tilde{X}_1$ of X_0 and therefore of X_1, so that the further restriction of $h : \tilde{\varphi}_1 \to \tilde{\varphi}_2$ is a minimal mapping. Choose $x \in \tilde{X}_1$ so that $h(x) = y$. By Proposition 5.16, $\omega_{\mathcal{F}} \tilde{\varphi}_2(y) = \tilde{X}_2$ implies $\omega_{\mathcal{F}} \tilde{\varphi}_1(x) = \tilde{X}_1$, so:

$$x \in \omega_{\mathcal{F}} \tilde{\varphi}_1(x) = \omega_{\mathcal{F}} \varphi_1(x)$$

■

Assume that $\varphi : T \times X \to X$ is a transitive action with X compact and T a separable metric monoid. By Proposition 3.16 $k\mathcal{B}_T$ is countably generated. If X is metrizable, then Proposition 4.6 implies that Trans_φ is a dense G_δ. By taking X to be a large product of mixing systems, we obtained examples where Trans_φ is dense, but even the set of recurrent points contains no nonempty G_δ. By taking X to be such a large product of mixing systems, we obtain transitive examples with arbitrarily large cardinality. But if $x \in \text{Trans}_\varphi$, then $X = \overline{f^T(x)}$ is separable because T is separable. A separable space has cardinality at most 2^c (= the cardinality of the Stone–Čech compactification of a discrete countable set). Such large cardinal examples have $\text{Trans}_\varphi = \emptyset$. For large central systems, we saw that the set of recurrent points need contain no nonempty G_δ. However, following Ellis, we show that it is always dense.

PROPOSITION 5.18. *Let $\varphi : T \times X \to X$ be a uniform action with X compact and T a separable metric monoid. If φ is a central action, i.e., $x \in \Omega\varphi(x)$ for all $x \in X$, then the set of recurrent points $|\omega\varphi|$, i.e., the set of x such that $x \in \omega\varphi(x)$, is dense in X.*

PROOF. Let U be a nonempty open set in X. We show that $U \cap |\omega\varphi| \neq \emptyset$. Choose u, a nonzero function in $\mathcal{B}(X)$ that vanishes on $X \setminus U$. It suffices to find a recurrent point on which u is not 0.

Let E be the closed subalgebra of $\mathcal{B}(X)$ generated by $\{(f^t)^*u : t \in T\}$. Since T is separable, E is separable. Since $E \subset \mathcal{B}^u(X) = \mathcal{B}(X)$, we obtain the natural map $j_E : \varphi \to \varphi_E$, which is a surjection because X was already compact. Since E is separable, X_E is metrizable. Because φ is central, φ_E also is. Applying Proposition 4.6 to φ_E, $|\omega\varphi_E|$ is a dense G_δ in X_E.

Let $u_E \in \mathcal{B}(X_E)$ be the extension of u. Choose $y \in |\omega\varphi_E|$ on which u_E is nonvanishing. By Proposition 5.17 there exists $x \in |\omega\varphi|$ such that $j_E(x) = y$. Then x is a recurrent point for φ and $u(x) = u_E(j_E(x)) = u_E(y)$ is not zero. ■

6

Ellis Semigroups and Ellis Actions

A *semigroup* S is a nonempty set with an associative (usually *not* commutative) multiplication map $M : S \times S \to S$. For $p, q \in S$, we write

$$pq = M(p,q) = M^p(q) = M_q(p) \tag{6.1}$$

In terms of the *translation maps*, the associative law says

$$M^p \circ M^q = M^{pq} \qquad M_p \circ M_q = M_{qp} \tag{6.2}$$

In general, for a function $\Phi : S \times X \to X$ where S is a semigroup, for $p \in S$ and $x \in X$, we write

$$px = \Phi(p,x) = \Phi^p(x) = \Phi_x(p) \tag{6.3}$$

The map Φ is called a *semigroup action* when for all $p, q \in S$:

$$\Phi^p \circ \Phi^q = \Phi^{pq} \tag{6.4}$$

Thus M defines an action of S on itself.

For any set X, the set of maps X^X is a semigroup with map composition. The evaluation map $\mathrm{Ev} : X^X \times X \to X$ is an action of X^X on X. For $(p,x) \in X^X \times X$:

$$p(x) = \mathrm{Ev}(p,x) = \mathrm{Ev}^p(x) = \mathrm{Ev}_x(p) \tag{6.5}$$

so that Ev^p is just the map p itself.

If S_1 is a semigroup, then $g : S \to S_1$ is a *semigroup homomorphism* when $g(pq) = g(p)g(q)$ for all $p, q \in S$. A subset S_0 of S is a *subsemigroup* when it is nonempty and closed under multiplication. In that case the inclusion map $S_0 \to S$ is a semigroup homomorphism. The image of a semigroup homomorphism is a subsemigroup.

A function $\Phi : S \times X \to X$ is uniquely defined by its *adjoint associate* $\Phi^{\#} : S \to X^X$. For $(p,x) \in S \times X$:

$$\Phi^{\#}(p) = \Phi^p \qquad \Phi^{\#}(p)(x) = \Phi(p,x) \tag{6.6}$$

Thus Φ is a semigroup action iff $\Phi^{\#}$ is a semigroup homomorphism.

If $\Phi_1 : S \times X_1 \to X_1$ is a semigroup action, then $h : X \to X_1$ is an *S action map* when $h(px) = ph(x)$ for all $p \in S$ and $x \in X$.

We call a semigroup with multiplication $M : S \times S \to S$ an *Ellis semigroup* when S has a compact topology such that M is continuous in the left variable. That is, for each fixed $q \in S$, the right translation map $M_q : S \to S$, defined by (6.1), is continuous. Neither joint continuity nor even continuity of the left translation maps is assumed.

We call $\Phi : S \times X \to X$ an *Ellis action* when S is an Ellis semigroup, X is a compact space, and Φ is an action such that for each $x \in X$ $\Phi_x : S \to X$ is continuous. Again joint continuity need not hold.

The simple proofs of the following results are left to the reader.

LEMMA 6.1. *Let S be an Ellis semigroup with multiplication $M : S \times S \to S$ and let X be a compact space.*

a. M is an Ellis action of S on itself. $M^{\#} : S \to S^S$ is the continuous semigroup homomorphism associating to $p \in S$ the left translation map $M^p \in S^S$.

b. With the product topology and map composition X^X is an Ellis semigroup. $\mathrm{Ev} : X^X \times X \to X$ *is an Ellis action.*

c. Let $\Phi : S \times X \to X$ be a function with adjoint associate $\Phi^{\#} : S \to X^X$. Φ is an Ellis action iff $\Phi^{\#}$ is a continuous semigroup homomorphism. In that case, for $x \in X$:

$$\Phi_x = \mathrm{Ev}_x \circ \Phi^{\#} : S \to X \tag{6.7}$$

is a continuous S action map from M on S to Φ on X.

A closed subsemigroup of an Ellis semigroup is an Ellis semigroup in its own right. For example, the image of a continuous homomorphism between Ellis semigroups is a closed subsemigroup.

A monoid is just a semigroup with an identity element. Many of our semigroup examples, e.g., X^X, have identity elements. Whether it has an identity element or not, an Ellis semigroup is compact, it usually has only one-sided continuity, and it is usually nonabelian. We use semigroup language for such objects to emphasize the contrast with uniform monoids. Recall that a uniform monoid satisfies joint continuity, but it is usually not compact. All our uniform monoids are assumed to be abelian. The distinction is important because the Ellis semigroups we use are constructed from uniform monoids and uniform actions. In order to motivate the general theory of Ellis semigroups, we look first at the following examples.

LEMMA 6.2. *Let X be a compact space and $A \subset C(X;X)$, i.e., A is a set of continuous maps on X. If A is a subsemigroup, then its topological closure $\overline{A}$ in X^X is a subsemigroup. If in addition A is commutative, i.e., $f_1 f_2 = f_2 f_1$ for all $f_1, f_2 \in A$, then:*

$$pf = fp \qquad \text{for all} \quad f \in A,\ p \in \overline{A} \tag{6.8}$$

PROOF. Since A is a subsemigroup, $f \in A$ implies $\{q \in X^X : fq \in \overline{A}\}$ contains A. By continuity of f, this set is closed, so it contains $\overline{A}$. Hence for $q \in \overline{A}$ $\{p \in X^X : pq \in \overline{A}\}$ contains A, so it contains $\overline{A}$. Thus $\overline{A}$ is a subsemigroup. Furthermore, since $f \in A$ implies left as well as right translation by f is continuous, the set $\{p \in X^X : fp = pf\}$ is closed. If it contains A, then it contains $\overline{A}$. ■

REMARK. While it is usually not true that $\overline{A}$ is commutative, we can sharpen the results slightly to obtain (6.8) for all at $p \in \overline{A}$ and $f \in \overline{A} \cap C(X;X)$. By (6.8) this holds for all $p \in A$ and therefore for all $p \in \overline{A}$ when f is continuous.

If $\varphi : T \times X \to X$ is a uniform action with X compact, then we define the *enveloping semigroup* S_φ, of φ to be the closure in X^X of the set:

$$\{f^t : t \in T\} \subset C(X;X)$$

By Lemma 6.2, S_φ is an Ellis semigroup and:

$$pf^t = f^t p \quad \text{for all} \quad t \in T,\ p \in S_\varphi \tag{6.9}$$

Evaluation restricts to define Ev: $S_\varphi \times X \to X$ the Ellis action of S_φ on X that extends the uniform action φ of T on X.

Because φ is a uniform action, the homomorphism $\varphi^\# : T \to C(X;X)$ is uniformly continuous, so it is uniformly continuous regarded as a map to X^X. Consequently it extends to the uniform Stone–Čech compactification. We denote this extension as $\Phi^\# : \beta_u T \to X^X$. $\Phi^\#$ is the unique continuous map such that the following diagram commutes:

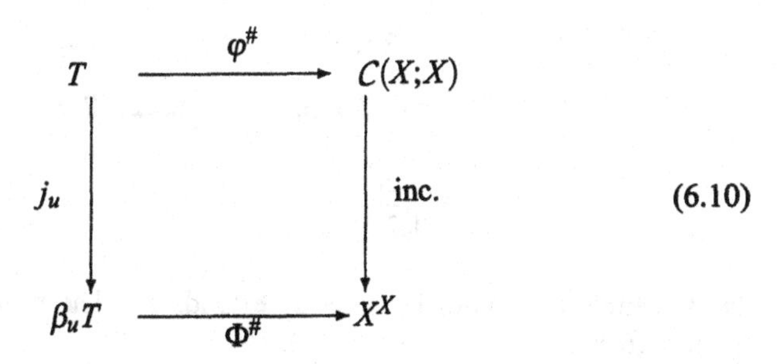

(6.10)

Following (6.6) we use $\Phi^{\#}$ to define the function $\Phi : \beta_u T \times X \to X$. For $(p,x) \in \beta_u T \times X$, we write

$$px = \Phi(p,x) = \Phi^{\#}(p)(x) \tag{6.11}$$

Diagram (6.10) implies, that for $(t,x) \in T \times X$:

$$j_u(t)x = \varphi(t,x) = f^t(x) \tag{6.12}$$

Because $j_u(T)$ is dense in $\beta_u T$ and $\beta_u T$ is compact, it follows that $\Phi^{\#}$ maps $\beta_u T$ onto S_φ, i.e.:

$$\Phi^{\#}(\beta_u T) = S_\varphi \tag{6.13}$$

Then we can rewrite (6.10) as the following commutative diagram:

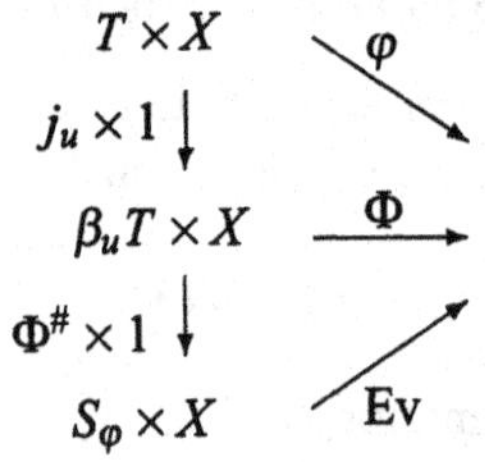

$$\begin{array}{ccc}
T \times X & \overset{\varphi}{\searrow} & \\
j_u \times 1 \downarrow & & \\
\beta_u T \times X & \xrightarrow{\Phi} & X \\
\Phi^{\#} \times 1 \downarrow & \nearrow & \\
S_\varphi \times X & \text{Ev} &
\end{array} \tag{6.14}$$

Now let $\mu : T \times T \to T$ denote addition, regarded as a uniform action of T on itself. By Proposition 5.13, μ extends to the uniform action $\mu_u : T \times \beta_u T \to \beta_u T$ with $g_u^t = \beta_u g^t$. We extend $\mu_u^{\#} : T \to \beta_u T^{\beta_u T}$ to define $M^{\#}$ as in diagram (6.10). We denote its adjoint associate $M : \beta_u T \times \beta_u T \to \beta_u T$. For $p,q \in \beta_u T$, we write

$$pq = M(p,q) = M^{\#}(p)(q) \tag{6.15}$$

The following diagram commutes.

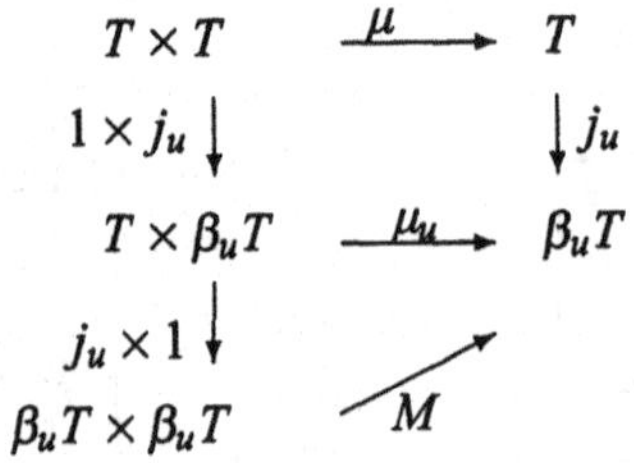

$$\begin{array}{ccc}
T \times T & \xrightarrow{\mu} & T \\
1 \times j_u \downarrow & & \downarrow j_u \\
T \times \beta_u T & \xrightarrow{\mu_u} & \beta_u T \\
j_u \times 1 \downarrow & \nearrow & \\
\beta_u T \times \beta_u T & M &
\end{array} \tag{6.16}$$

The rectangle commutes because μ_u extends μ. The triangle commutes as in diagram (6.14).

PROPOSITION 6.1. *Let T be a uniform monoid.*

a. With multiplication map $M : \beta_u T \times \beta_u T \to \beta_u T$ extending the addition map $\mu : T \times T \to T$, $\beta_u T$ is an Ellis semigroup with identity $j_u(0)$. The uniformly continuous map $j_u : T \to \beta_u T$ is a monoid homomorphism and:

$$pj_u(t) = j_u(t)p \quad \text{for} \quad p \in \beta_u T,\ t \in T \tag{6.17}$$

b. Let $\varphi : T \times X \to X$ be a uniform action with X compact. The extension $\Phi : \beta_u T \times X \to X$ is an Ellis action. The associated map $\Phi^\#$ is a continuous surjective semigroup homomorphism from $\beta_u T$ to S_φ. For every $x \in X$:

$$\Phi_x(\beta_u T) = \mathrm{Ev}_x(S_\varphi) = \overline{f^T(x)} \tag{6.18}$$

c. Let $h : \varphi \to \varphi_1$ be a continuous action map with $\varphi : T \times X \to X$ and $\varphi_1 : T \times X_1 \to X_1$ uniform actions on compact spaces. $h : \Phi \to \Phi_1$ is a $\beta_u T$ action map. If in addition h is surjective, then there is a unique map S_h such that the following diagram commutes:

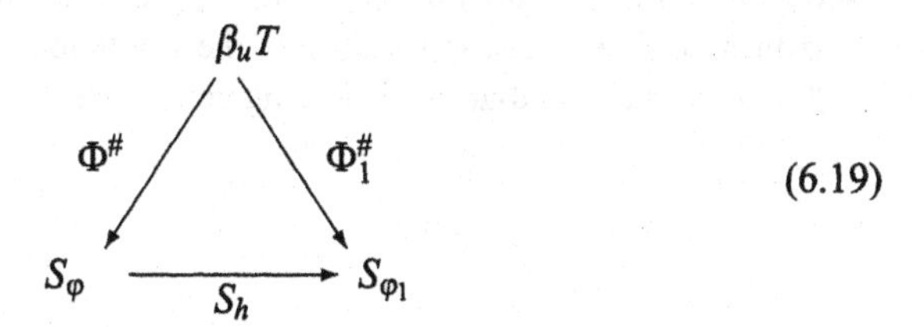

Then S_h is a surjective, continuous semigroup homomorphism.

PROOF. Because $M^\#$ is continuous, M is continuous in the left variable. For $p,q,r \in \beta_u T$:

$$p(qr) = (pq)r \tag{6.20}$$

Equation (6.20) is true for $p,q,r \in j_u(T)$ because M extends μ. With p,q fixed in $j_u(T)$ the two sides are continuous functions of $r \in \beta_u T$ because M extends μ_u. Then (6.20) holds for all such r. With $p \in j_u(T)$ and $r \in \beta_u T$, the two sides are continuous in q so (6.20) is true for all $q \in \beta_u T$. With $q,r \in \beta_u T$, the two sides are continuous in p, so the associative law is true in general. Thus $\beta_u T$ is an Ellis semigroup. Similar continuity arguments extend (6.17) and $pj_u(0) = p$ from $p \in j_u(T)$ to all of $\beta_u T$. Similarly we show in (b) that for all $p,q \in \beta_u T$ and $x \in X$:

$$p(qx) = (pq)x \tag{6.21}$$

And in (c) for all $p \in \beta_u T$ and $x \in X$:

$$h(px) = ph(x) \tag{6.22}$$

$\Phi^{\#} : \beta_u T \to S_\varphi$ is a continuous homomorphism by Lemma 6.1c. From (6.7), compactness and the density of $j_u(T)$ we obtain (6.18).

If $h : X \to X_1$ is surjective, then by (6.22) S_h is well-defined by:

$$S_h(\Phi^{\#}(p)) = \Phi_1^{\#}(p) \tag{6.23}$$

Continuity follows because surjections of compact spaces are quotient maps. Then S_h is a homomorphism because $\Phi^{\#}$ and $\Phi_1^{\#}$ are. ■

Following the construction described for Proposition 5.10, we can replace the original uniform monoid structure on T by the discrete uniformity. We refer to this as the *discrete monoid* T. Any discrete monoid is uniform, and if $\varphi : T \times X \to X$ is a uniform action, then $\varphi_0 : T \times X \to X$, the same set map but with the discrete uniformity on T is a uniform action. The set $\{f^t : t \in T\} \subset X^X$ does not depend on the topology of T, so neither does its closure, the enveloping semigroup, which we denote S_φ for φ_0 as well as for φ. On $\beta_0 T$, the Stone–Čech compactification of the discrete monoid T, we obtain an Ellis semigroup structure by applying Proposition 6.1 to discrete T. As in Proposition 5.10 the continuous surjection $\pi_0 : \beta_0 T \to \beta_u T$ is defined uniquely so that the following commutes:

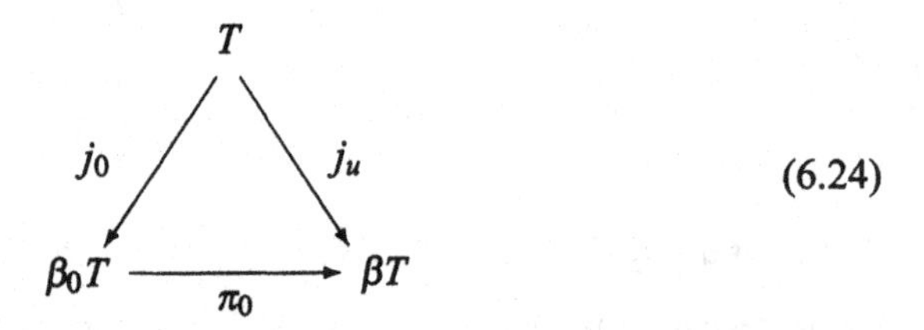

(6.24)

Note: The injection j_0 is *not* continuous with respect to the original topology on T.

PROPOSITION 6.2. *The continuous surjection $\pi_0 : \beta_0 T \to \beta_u T$ is a semigroup homomorphism. If $\varphi : T \times X \to X$ is a uniform action with X compact and $\varphi_0 : T \times X \to X$ is the same action using the discrete uniformity on T then the following diagram of continuous surjective semigroup homomorphisms commutes:*

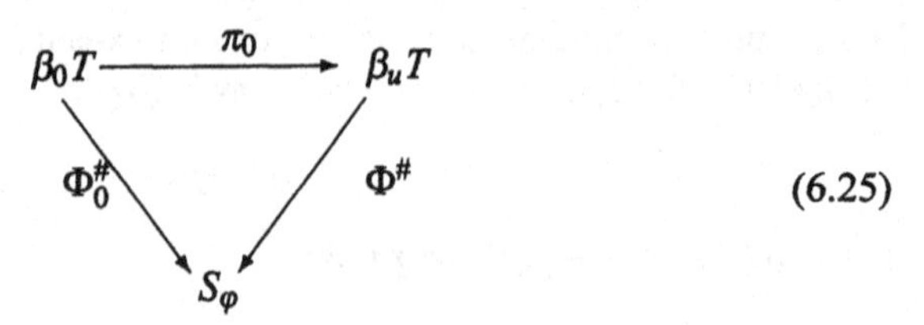

(6.25)

PROOF. To say that π_0 is a homomorphism is to say

$$\pi_0(p_0q_0) = \pi_0(p_0)\pi_0(q_0) \qquad p_0, q_0 \in \beta_0 T \tag{6.26}$$

To say that diagram (6.25) commutes is to say

$$\pi_0(p_0)x = p_0 x \qquad p_0 \in \beta_0 T \qquad x \in X \tag{6.27}$$

In each case, first check the result for p_0, q_0 in $j_0(T)$, and then use the usual continuity argument to extend it to all q_0 in $\beta_0 T$ and then to all p_0, q_0 in $\beta_0 T$. ∎

Now for the general semigroup results that we apply later. For the rest of Chapter 6, let S be a semigroup with multiplication $M : S \times S \to S$ and $\Phi : S \times X \to X$ be a semigroup action.

For a subset A of S and elements p, q of S, we write

$$\begin{aligned} Ap &\equiv M_p(A) = \{xp : x \in A\} \\ qA &\equiv M^q(A) = \{qx : x \in A\} \\ qAp &\equiv M_p(M^q(A)) = M^q(M_p(A)) = \{qxp : x \in A\} \end{aligned} \tag{6.28}$$

For subsets A, B of S, we write

$$\begin{aligned} AB &= \cup_{p \in B} M_p(A) = \cup_{q \in A} M^q(B) \\ &= \{qp : q \in A, p \in B\} \end{aligned} \tag{6.29}$$

A nonempty subset A is a *subsemigroup* when $AA \subset A$. The subsemigroup A is called an *ideal* (short for left ideal) when $SA \subset A$. A is called a *co-ideal* when it is a subsemigroup and $pq \in A$ for $q \in A$ only when p is also in A. Thus we have for a nonempty subset A:

- A is a subsemigroup iff for all $q \in A$, $M_q(A) \subset A$, i.e.,

$$p \in A \Rightarrow pq \in A$$

- A is an ideal iff for all $q \in A$, $M_q^{-1}(A) = S$, i.e.,

$$p \in S \Rightarrow pq \in A$$

- A is a co-ideal iff for all $q \in A$, $M_q^{-1}(A) = A$, i.e.,

$$p \in A \Leftrightarrow pq \in A \tag{6.30}$$

If A is an ideal (or subsemigroup with $p \in A$), then Ap is an ideal (resp. a subsemigroup). In particular Sp is an ideal, called the *principal ideal for p*.

For the action $\Phi : S \times X \to X$ and $x \in X$, we define the *isotropy set* of x

$$\text{Iso}_x \equiv (\Phi_x)^{-1}(x) = \{p \in S : px = x\} \tag{6.31}$$

Clearly if $px = x$, then $qpx = x$ iff $qx = x$. By (6.30), Iso_x is a co-ideal when it is nonempty.

In particular for any $p \in S$, we obtain the isotropy set of p:

$$\text{Iso}_p \equiv (M_p)^{-1}(p) = \{q \in S : qp = p\} \tag{6.32}$$

For $p, q \in S$, we define a partial order $\succ$ and associated equivalence relation $\sim$:

$$\begin{aligned} p \succ q &\Leftrightarrow q \in Sp \\ p \sim q &\Leftrightarrow p \succ q \text{ and } q \succ p \end{aligned} \tag{6.33}$$

Observe that A is an ideal iff $p \in A$ and $p \succ q$ implies $q \in A$. Since Sp is an ideal, $p \succ q$ implies $Sp \supset Sq$. The converse need not be true because q need not be in Sq. That is, while $\succ$ is clearly transitive, it need not be reflexive when S does not have an identity. For a subset A of S, we define

$$\begin{aligned} |A| &= \{p \in A : p \in Ap\} \\ &= \{p \in A : A \cap \text{Iso}_p \neq \emptyset\} \end{aligned} \tag{6.34}$$

In particular:

$$|S| = \{p : p \succ p\} = \{p : \text{Iso}_p \neq \emptyset\} \tag{6.35}$$

If $Sp \supset Sq$ and $q \in |S|$, then $p \succ q$. If $Sp = Sq$ and $q \in |S|$, then $p \sim q \Leftrightarrow p \in |S|$.

We call p a *minimal* element of A if $p \in A$ and for all $q \in A$, $p \succ q$ implies $p \sim q$. Let $\text{Min}(A)$ denote the (possibly empty) set of minimal elements of A.

An element p is called *idempotent* when $p^2 = p$. Let $\text{Id}(A)$ denote the (possible empty) set of idempotents in A. Clearly p is idempotent iff $p \in \text{Iso}_p$. Thus $\text{Id}(A) \subset |A|$ for any subset A.

We collect the special results which compactness assumptions yield.

PROPOSITION 6.3. *Let S be an Ellis semigroup.*

a. Let $p \in S$. If A is a closed nonempty subset of S, then Ap is closed and nonempty. In particular the principal ideal Sp is closed. The isotropy set Iso_p *is closed.*

b. If A is a closed subsemigroup of S, then $\text{Id}(A)$ *is nonempty.*

c. If S_1 is an Ellis semigroup and $h : S \to S_1$ is a surjective, continuous semigroup homomorphism, then:

$$h(\text{Id}(S)) = \text{Id}(S_1) \tag{6.36}$$

PROOF. (a) M_p is continuous and S is compact.

(b) By Zorn's Lemma and compactness, we can assume A is minimal among closed subsemigroups of S. Let $p \in A$. Ap is a closed semigroup of A. By minimality, $Ap = A$. Hence $A \cap \mathrm{Iso}_p \neq \emptyset$, so it is a closed subsemigroup of A. By minimality, $A \subset \mathrm{Iso}_p$. Thus $p^2 = p$, so $p \in \mathrm{Id}(A)$. In fact by minimality, $A = \{p\}$.

(c) If $p^2 = p$ then $h(p)^2 = h(p)$. If $p_1^2 = p_1$ in S_1 then because h is surjective $h^{-1}(p_1)$ is a closed subsemigroup of S. It contains an idempotent by (b). ■

Using Proposition 6.3b, *Namakura's Lemma*, we obtain a rich supply of idempotents. In any semigroup the idempotents and the ideals they generate play a special role, but we use compactness to prove their existence in Ellis semigroups.

For any semigroup S let $p,q \in S$. q is called a *quasi-inverse* of p when:

$$pqp = p \quad \text{and} \quad qpq = q \tag{6.37}$$

Clearly p is then a quasi-inverse of q. Notice that (6.37) implies $p,q \in |S|$.

LEMMA 6.3. *a. Assume $p,u,v \in S$ with u and v idempotents. The following equivalences hold.*

$$\begin{aligned} p \in Su, \text{ i.e., } u \succ p &\Leftrightarrow pu = p \\ p \in vS &\Leftrightarrow vp = p \\ p \in vSu &\Leftrightarrow vpu = p \Leftrightarrow vp = p \quad \text{and} \quad pu = p \end{aligned} \tag{6.38}$$

In particular, $vSu = (vS) \cap (Su)$.

The following conditions are equivalent:

(1) $p \sim u$

(2) $pu = p$ and $qp = u$ for some $q \in S$.

(3) $p \in Su$ and $qp = u$ for some $q \in uS$.

b. Assume $p,q \in S$ with q a quasi-inverse of p. Define $u,v \in S$ by:

$$qp = u \quad \text{and} \quad pq = v \tag{6.39}$$

u and v are idempotents satisfying:

$$\begin{aligned} p \in vSu \quad &\text{and} \quad p \sim u \\ q \in uSv \quad &\text{and} \quad q \sim v \end{aligned} \tag{6.40}$$

In particular if p admits a quasi-inverse, then $p \sim u$ for some idempotent u.

Conversely if $p \sim u$ for some idempotent u, then there exists $q \in uS$ such that $qp = u$. Such a q is a quasi-inverse for p.

PROOF. (a) $pu = p$ implies $p \in Su$, while $p = p_1 u$ implies $pu = p_1 u^2 = p_1 u = p$. The other equivalences of (6.38) follow similarly. The last equivalence says that $p \in vSu$ iff $p \in vS$ and $p \in Su$. Now (1) $\Leftrightarrow$ (2) is clear as is (3) $\Rightarrow$ (2). If $pu = p$ and $q_1 p = u$, then $qp = u$ with $q = uq_1$. By (6.38) $p \in Su$ and $q \in uS$. Thus, (2) $\Rightarrow$ (3).

(b) $u^2 = qpqp = qp = u$. Similarly, $vpu = p$. Then $p \sim u$ by (2) of (a). Reversing the roles of p and q, we have the remaining results.

If $p \sim u$, then by (3) of (a) there exists $q \in uS$ such that $qp = u$. Then $pqp = pu = p$ and $qpq = uq = q$. Therefore q is a quasi-inverse for p. ∎

REMARK. While an element p may admit more than one quasi-inverse, the quasi-inverse is the unique element of uSv satisfying either of the equations in (6.39). If $q_1 \in uSv$ and $q_1 p = u$, then $q_1 = q_1 v = q_1 pq = uq = q$. ∎

We now interpret these results using right translation maps on the principal ideals for idempotents. In general if $\Phi : S \times X \to X$ is an action and $x \in X$, then $\Phi_x : S \to X$ is an S action map as is its restriction to any ideal in S. In particular each right translation M_p is an S action map.

PROPOSITION 6.4. *Let u, v be idempotents in a semigroup S.*

a. Let Φ be an action of S on a set X. For $x, y \in X$, the restrictions of the S action maps Φ_x and Φ_y to the ideal Sv agree iff $vx = vy$. In particular, Φ_x and Φ_{vx} agree on Sv.

Conversely if $h : Sv \to X$ is an S action map, then $x = h(v)$ is the unique element of X such that $vx = x$ and h is the restriction of Φ_x to Sv.

In particular if $h : Sv \to S$ is an S action map, then $p = h(v)$ is the unique element of vS such that h is the restriction to Sv of M_p. Furthermore:

$$h(Sv) \subset Su \Leftrightarrow p \in Su \Leftrightarrow p \in vSu \tag{6.41}$$

b. If $h : Sv \to Su$ is a bijective S action map, then there exist unique $p \in vSu$ and $q \in uSv$ such that h is the restriction of M_p to Sv and h^{-1} is the restriction of M_q to Su.

For $p \in vSu$ and $q \in uSv$, translation maps $M_p : Sv \to Su$ and $M_q : Su \to Sv$ are inverses iff $qp = u$ and $pq = v$. In that case p and q are a quasi-inverse pair.

Conversely if p and q are a quasi-inverse pair with $qp = u$ and $pq = v$ then $M_p : Sv \to Su$ and $M_q : Su \to Sv$ are inverse bijective S action maps.

c. For $p \in vSu$ the following conditions are equivalent:

(1) The restriction $M_p : Sv \to Su$ is surjective.

(2) $p \sim u$

(3) $Sp = Su$

(4) $qp = u$ *for some* $q \in uSv$.

(5) p *has a quasi-inverse* q *such that* $qp = u$ *and* $qv = q$.

(6) There exists an idempotent $\tilde{v}$ *in* vSv *such that the restriction* $M_p : S\tilde{v} \to Su$ *is bijective with* $p \in \tilde{v}Su$*; and so*

$$v\tilde{v} = \tilde{v} = \tilde{v}v \tag{6.42}$$

If $\tilde{v}$ *is an idempotent satisfying (6.42) then* $M_{\tilde{v}}$ *maps* Sv *onto* $S\tilde{v}$ *and restricts to the identity on* $S\tilde{v}$.

If $M_p : Sv \to Su$ *is surjective and* $\tilde{v}$ *is an idempotent satisfying condition (6), then* M_p *is bijective from* Sv *to* Su *iff* $v \sim \tilde{v}$ *and iff* $v = \tilde{v}$.

PROOF. (a) Φ_x and Φ_{vx} clearly agree on Sv; so $vx = vy$ implies $\Phi_x = \Phi_y$ on Sv. Conversely if $\Phi_x = \Phi_y$ on Sv, then:

$$vx = \Phi_x(v) = \Phi_y(v) = vy$$

If h is an S action map on Sv and $q \in Sv$, then $h(q) = h(qv) = qh(v) = qp$; in particular with $q = v$, $p = h(v) = vp$. If $h = \Phi_y$ on Sv, then by the preceding argument $x = vx = vy$. Therefore if $y = vy$, then $y = x$ as well.

In particular for an S action map $h : Sv \to S$, there is a unique $p \in S$ such that $h = M_p$ on Sv and $vp = p$, i.e., $p \in vS$. Since h is an S action map its image is contained in Su iff $p = h(v)$ is in Su and so iff $p \in (Su) \cap (vS) = vSu$, i.e., (6.41) holds.

(b) By (a) there is a unique $p \in vS$ such that $h = M_p$ on Sv, and by (6.41) $p \in vSu$. Since h^{-1} is an S action map, there is similarly a unique $q \in uSv$ such that h^{-1} is M_q on Su.

In particular the identity on Su is uniquely represented in uSu by M_u. Then $M_pM_q = M_{qp}$ is the identity on Su iff $qp = u$ since $qp \in uSu$. Similarly $M_qM_p = 1_{Sv}$ iff $pq = v$. In that case, $pqp = vp = p$ and $qpq = uq = q$; therefore p, q are a quasi-inverse pair.

If p and q are quasi-inverses with $qp = u$ and $pq = v$, then by (6.40) $p \in vSu$ and $q \in uSv$. Hence M_p and M_q are inverse maps between Sv and Su by the preceding argument.

(c) (6) $\Rightarrow$ (1). This is obvious since $S\tilde{v} \subset Sv$.

(1) $\Rightarrow$ (3). $M_p(v) = vp = p$, so $M_p(Sv) = Sp$; M_p surjective implies $Sp = Su$.

(3) $\Rightarrow$ (2). Since $u^2 = u$ and $vp = p$, $u \in Su$ and $p \in Sp$, i.e., $u, p \in |S|$.

(2) $\Rightarrow$ (4). By Lemma 6.2a, $p \sim u$ implies $qp = u$ for some $q \in uS$. Since $qvp = qp = u$, we can replace q by qv if necessary to obtain $q \in uSv$.

(4) $\Rightarrow$ (5). For $p \in vSu$, $q \in uSv$ $qp = u$ implies q is a quasi-inverse: $qpq = uq = q$ and $pqp = pu = p$.

(5) $\Rightarrow$ (6). For q a quasi-inverse given by (5) let $\tilde{v} = pq$. By the last paragraph of (b), $M_p : S\tilde{v} \longrightarrow Su$ and $M_q : Su \longrightarrow S\tilde{v}$ are inverse maps. By (6.40), $p \in \tilde{v}Su$. Then $\tilde{v} \in vSv$, i.e., (6.42) holds, because $p \in vS$ and $q \in Sv$. By (6.42), $M_{\tilde{v}}(v) = \tilde{v}$, so $M_{\tilde{v}}$ maps Sv onto $S\tilde{v}$. Since $\tilde{v}$ is idempotent, $M_{\tilde{v}}$ is the identity on $S\tilde{v}$.

Clearly M_p is bijective iff $Sv = S\tilde{v}$, i.e., $v \sim \tilde{v}$. Thus $v = \tilde{v}$ is sufficient. If M_p is bijective on Sv, then $M_p(v) = p = M_p(\tilde{v})$ implies $v = \tilde{v}$. ■

Thus for a quasi-inverse pair p, q with $pq = v$ and $qp = u$, M_p maps $Sq = Sv$ bijectively to $Su = Sp$ with inverse map M_q. The map M_p takes v to p and q to u.

We call $p \in S$ *invertible* if it has a quasi-inverse q that commutes with p so that:

$$v = pq = qp = u \tag{6.43}$$

For an invertible element p, such a commuting quasi-inverse is called the *inverse* of p. The idempotent u defined by (6.43) is called the *unit* for p. Clearly if p is invertible with inverse q and unit u, then q is invertible with inverse p and the same unit.

PROPOSITION 6.5. *For an element p of a semigroup S, the following conditions are equivalent:*

(1) p is invertible.

(2) p has a quasi-inverse q such that $p \sim q$.

(3) $p \in Sp$ and the restriction $M_p : Sp \longrightarrow Sp$ is bijective.

(4) There exists an idempotent u such that $p \in uSu$ and $M_p : Su \longrightarrow Su$ is bijective.

Assume p is invertible with inverse q and unit u. The unit u is the unique idempotent in S satisfying the conditions of (4). It is the unique element u of Sp such that $up = p$. The inverse q is the unique quasi-inverse commuting with p. It is the unique element q of Sp such that $qp = u$ where u is the unit of p. q is the unique element of uSu such that M_q is the inverse of M_p on Su. Clearly:

$$p \sim u \sim q, \quad \text{i.e., } Sp = Su = Sq \tag{6.44}$$

PROOF. (4) $\Rightarrow$ (1). By Proposition 6.4b, (4) implies there is a unique $q \in uSu$ such that M_q is the inverse of M_p on Su. Furthermore q is a quasi-inverse for p. Since $v = u$, p commutes with q.

(1) $\Rightarrow$ (2). (6.43) and (6.40) imply (6.44) which yields (2).

(2) $\Rightarrow$ (3). If q is a quasi-inverse for p, with $pq = v$ and $qp = u$, then by Proposition 6.4b, M_p restricts to a bijection of $Sq = Sv$ to $Sp = Su$. Thus if $p \sim q$, (3) follows.

(3) $\Rightarrow$ (4). By (3) there is a unique $u \in Sp$ such that $M_p(u) = p$, i.e., $up = p$. Because $u^2p = up = p$ and $u^2 \in Sp$, $u^2 = u$ by uniqueness; i.e., u is an idempotent. Since $u \in Sp$, $Su \subset Sp$. But $up = p$ implies that M_p maps Su onto Sp. Since M_p is bijective on Sp, Su must be all of Sp, i.e., $Su = Sp$. Thus $p \in Su$, so $pu = p$. That is, $p \in uSu$ and since $Sp = Su$, (3) implies (4).

Now if p is invertible with inverse q and unit u, then (6.44) follows from (6.43) and (6.40) as before, so $x = u$ is the unique solution of $xp = p$ in Sp by (3). Proposition 6.4b implies that (4) holds for u. If it also holds for some idempotent u_1, then $u_1p = p$ and $Su_1 = Sp$, so that $x = u_1$ is also a solution in Sp of $xp = p$. By (6.44) and (3), $y = q$ is the unique solution in Sp of the equation $yp = u$. If q_1 is another commuting quasi-inverse for p, then $q_1 \in Sp$. By uniqueness of the unit u for p, $q_1p = u$, so that $y = q_1$ is also a solution in Sp of $yp = u$. Uniqueness of the element q in uSu such that M_q is the inverse of M_p on Su, follows from Proposition 6.4a. ■

Recalling the definition of the isotropy set for p (6.32), we see that the preceding characterization says

$$(Sp) \cap \mathrm{Iso}_p = \{u\} \tag{6.45}$$

when p is invertible with unit u.

For an idempotent u of S define the *group associated with u*, denoted G_u, to be the set of invertible elements with unit u.

PROPOSITION 6.6. *Assume u is an idempotent in a semigroup S. For an element p of S, the following properties are equivalent:*

(1) $p \in G_u$, i.e., p is invertible with unit u.

(2) $p \in uS$, $p \sim u$ and p is invertible.

(3) $p \in uSu$ and $pq = u = qp$ for some $q \in uSu$.

(4) $p \in uSu$ and $M_{p.}: Su \to Su$ is bijective.

With respect to multiplication inherited from S, G_u is a group with unit element u. For $p \in G_u$, the inverse q, i.e., the unique quasi-inverse of p that commutes with p, is the inverse element in G_u.

If v is an idempotent in S with $v \neq u$, then G_v is disjoint from G_u. Thus the set of invertible elements in S is the disjoint union of the groups G_u for u varying in $\mathrm{Id}(S)$.

PROOF. By Proposition 6.5, (1) and (4) are obviously equivalent and they imply (2) and (3) with q the inverse element for p; (3) implies q is a quasi-inverse for p. Clearly (3) then implies that p commutes with q, so q is the inverse of p and u is the unit for p. Thus (3) $\Rightarrow$ (1). Finally (2) implies $u \in Sp$ and $up = p$. By the characterization in Proposition 6.5, u is the unit for p. Thus (2) $\Rightarrow$ (1).

If p and p_1 satisfy (4) then pp_1 is in uSu, and $M_{pp_1} = M_{p_1} \circ M_p$ is a bijection on Su; thus pp_1 satisfies (4). For q the inverse of p, $q \in uSu$, and M_q is the inverse of M_p on Su; thus q satisfies (4). Since $pq = qp = u$, q is the inverse element for p in the subsemigroup G_u because u is clearly the identity element in G_u.

For any invertible element p, the associated unit is unique, (cf. Proposition 6.5). Hence $G_u \cap G_v = \emptyset$ if $u \neq v$. ■

It is helpful to think of the idempotents of S as the objects of a category $\mathcal{J}^S$ with morphisms $\mathcal{J}^S(v,u)$ the set vSu. We define composition in $\mathcal{J}^S$ by multiplication in S with the order reversed, so that for $u_1, u_2, u_3 \in \mathrm{Id}(S)$:

$$\begin{aligned} \mathcal{J}^S(u_2,u_3) \times \mathcal{J}^S(u_1,u_2) &\to \mathcal{J}^S(u_1,u_3) \\ (p,q) &\mapsto qp \end{aligned} \tag{6.46}$$

Associating to $p \in \mathcal{J}^S(v,u)$ the restriction $M_p : Sv \to Su$ defines a functor from $\mathcal{J}^S$ to the category of S actions and S action maps, $\mathcal{S}_S$. By Proposition 6.4a we see that:

$$M_\# : \mathcal{J}^S(v,u) \to \mathcal{S}_S(Sv, Su) \tag{6.47}$$

is a bijection.

The same element $p \in S$ may lie in several different $\mathcal{J}^S(v,u)$'s (strictly speaking, this violates one of the defining conditions of a category), but from Proposition 6.4b we see that p has a quasi-inverse iff it is an isomorphism from some v to some u in the category $\mathcal{J}^S$, and p is invertible when it is an isomorphism from some u to itself in $\mathcal{J}^S$. In that case, u is the unit associated with p, so it is uniquely determined. In particular we can think of the group G_u as the automorphism group of the object u in the category $\mathcal{J}^S$.

THEOREM 6.1. *Let S be an Ellis semigroup.*

a. Let H be a closed subsemigroup of S. The set of minimal idempotents of H:

$$\mathrm{Id}(\mathrm{Min}(H)) = \mathrm{Id}(H) \cap \mathrm{Min}(H) = \mathrm{Min}(\mathrm{Id}(H)) \tag{6.48}$$

is nonempty. For $p \in H$ the following conditions are equivalent:

(1) $p \in \mathrm{Min}(H)$

(2) $q \succ p$ *for some* $q \in \mathrm{Min}(H)$

(3) $u \sim p$ for some $u \in \mathrm{Id}(\mathrm{Min}(H))$

In particular $\mathrm{Min}(H)$ *is a subsemigroup of H (usually not closed), and it is contained in* $|H|$.

b. Let H be a closed subsemigroup of S. An ideal J of S is called minimal at *H when it satisfies the following equivalent conditions:*

(1) J is minimal in the family of ideals meeting H.

(2) J is minimal in the family of closed ideals meeting H.

(3) $H \cap J \neq \emptyset$ and for all $p \in H \cap J$, $J = Sp$.

(4) $J = Sp$ for some $p \in \mathrm{Min}(H)$.

(5) $J = Su$ for some $u \in \mathrm{Id}(\mathrm{Min}(H))$.

When J is an ideal of S minimal at H then $H \cap J \subset \mathrm{Min}(H)$ and:

$$p_1, p_2 \in H \cap J \;\Rightarrow\; p_1 \sim p_2 \tag{6.49}$$

Any ideal J that meets H contains ideals minimal at H, and satisfies:

$$J \cap \mathrm{Id}(\mathrm{Min}(H)) \neq \emptyset \tag{6.50}$$

Furthermore for any $p \in \mathrm{Min}(H)$, $Jp = Sp$ is an ideal minimal at H.

(c) Let H be a closed co-ideal of S and let J be an ideal of S. We have

$$p \in H \cap J \;\Rightarrow\; (M_p)^{-1}(H \cap J) = H \tag{6.51}$$

If $H \cap J \neq \emptyset$ and $p \in \mathrm{Min}(H)$, then:

$$Sp = Jp \qquad M_p(H \cap J) = H \cap (Jp) \tag{6.52}$$

Each minimal element of a closed co-ideal is invertible. In fact $\mathrm{Min}(H)$ *can be expressed as the disjoint union:*

$$\mathrm{Min}(H) = \cup\{G_u \cap H : u \in \mathrm{Id}(\mathrm{Min}(H))\} \tag{6.53}$$

Each $G_u \cap H$ is a subgroup of G_u.

PROOF. (a) (3) $\Rightarrow$ (2). This is obvious.

(2) $\Rightarrow$ (1). If $p_1 \in H$ and $p \succ p_1$, then $q \succ p$ implies $q \succ p_1$ as well. Since q is minimal, $p \sim q \sim p_1$. Thus p is minimal in H.

(1) $\Rightarrow$ (3). Hp is a closed subsemigroup of H, and so it contains some idempotent u by Namakura's Lemma, Proposition 5b. Since $p \succ u$, minimality of

p implies $p \sim u$. Since (2) $\Rightarrow$ (1) (applied to u), u is minimal. Notice that $pu = p$ and $pu \in Hp$ imply $p \in |H|$.

In (6.48) the first equality is clear. If $u \in \text{Id}(H)$ is minimal among elements of H then it is minimal among elements of the subset $\text{Id}(H)$. On the other hand, if $u \in \text{Min}(\text{Id}(H))$ then $u \in \text{Id}(H)$. If $u \succ p$ and $p \in H$, then as in the preceding argument Hp contains some idempotent u_1, so $u \succ p \succ u_1$. Since $u_1 \in \text{Id}(H)$ and $u \in \text{Min}(\text{Id}(H))$, $u \sim u_1$, so $u \sim p$, completing the proof of (6.48).

If $p,q \in \text{Min}(H)$, then $pq \in H$ and $q \succ pq$. By (3) $\Rightarrow$ (1), $pq \in \text{Min}(H)$. So $\text{Min}(H)$ is closed under multiplication. We show that $\text{Id}(\text{Min}(H))$ and a fortiori $\text{Min}(H)$ are nonempty as part of the proof of (b).

(b) (1) $\Rightarrow$ (3) and (2) $\Rightarrow$ (3). If $p \in H \cap J$, then $Hp \subset H \cap (Sp)$, so Sp is a closed ideal meeting H and contained in J. If J satisfies either (1) or (2), then $J = Sp$.

(3) $\Rightarrow$ (5). First note that if $p \in H \cap J$ and $p_1 \in H$ with $p \succ p_1$, then because J is an ideal, $p_1 \in J$. By (3) $Sp = Sp_1 = J$. Thus $p \sim p_1$, so p is minimal. In general, $p, p_1 \in H \cap J$ imply $Sp = Sp_1 = J$ by (3), so $p \sim p_1$. It follows that (3) implies (6.49). Now $H \cap J$ is a closed subsemigroup, so it contains some idempotent u. $Su = J$ by (3) and by (6.48) $u \in \text{Id}(\text{Min}(H))$.

(5) $\Rightarrow$ (4). This is obvious.

(4) $\Rightarrow$ (1) and (2). Since $J = Sp$, J is closed. If J_1 is an ideal contained in J and $p_1 \in H \cap J_1$, then:

$$Sp = J \supset J_1 \supset Sp_1$$

Therefore $p \succ p_1$. Because $p \in \text{Min}(H)$ and $p_1 \in H$, $p \sim p_1$, so $Sp = Sp_1$; hence $J = J_1$.

If J is any ideal with $p \in H \cap J$, then J contains Sp, a closed ideal meeting H. By Zorn's Lemma and compactness, Sp contains J_1 minimal among the closed ideals meeting H. Since J_1 satisfies (2) it satisfies (5) as well. Thus J contains an ideal minimal at H and meets $\text{Id}(\text{Min}(H))$, i.e., (6.50). Applied with $J = S$, we see that $\text{Id}(\text{Min}(H))$ is nonempty.

For any $p \in \text{Min}(H)$, Sp is an ideal minimal at H by (4). Let $q \in H \cap J$. Since $Sq \subset J$, we have

$$Sqp \subset Jp \subset Sp$$

Because $p,q \in H$, $qp \in H$ and $p \succ qp$. By minimality of p, $p \sim qp$, so $Sqp = Sp$; Hence $Jp = Sp$.

(c) $p \in J$ and $q \in S$ imply $qp \in J$, while $p \in H$ and $q \in H$ imply $qp \in H$. So

$$M_p(H) \subset H \cap J$$

if $p \in H \cap J$. On the other hand, $p \in H$ and $qp \in H$ imply $q \in H$ when H is a co-ideal. Thus (6.51) follows.

The first equation of (6.52) was proved in (b) for any closed subsemigroup H. From this it follows that for $p \in \mathrm{Min}(H)$:

$$M_p(H \cap J) \subset M_p(H) \subset H \cap (Sp) = H \cap (Jp) \tag{6.54}$$

But if $qp \in H$ for some $q \in J$, then because H is a co-ideal, $q \in H$, so $qp \in M_p(H \cap J)$.

Assume now that $u \in \mathrm{Id}(\mathrm{Min}(H))$. $G_u \cap H$ is a subsemigroup of G_u containing $G_u \cap \mathrm{Min}(H)$. On the other hand, if $p \in G_u \cap H$, then $p \sim u$ with $u \in \mathrm{Min}(H)$. Thus $p \in H$ implies $p \in \mathrm{Min}(H)$ by (a). Then we have

$$u \in \mathrm{Id}(\mathrm{Min}(H)) \Rightarrow G_u \cap \mathrm{Min}(H) = G_u \cap H \tag{6.55}$$

Now suppose that $p \in \mathrm{Min}(H)$. We prove that p is invertible and the associated unit u and inverse q both lie in H. Since $p, q, u \in G_u \cap H$, $p \sim u$, so by (a) $u \in \mathrm{Id}(\mathrm{Min}(H))$. Thus by (6.55) $G_u \cap H$ is a subgroup of G_u. We use the following claim to show this.

CLAIM. *Let J be an ideal minimal at H and $p \in \mathrm{Min}(H)$. $J \cap \mathrm{Iso}_p$ consists of a single element v of S. v is an idempotent and M_p restricts to a bijection from $J = Sv$ onto Sp.*

Observe first that for a co-ideal H, $p \in H$ and $qp = p$ imply $q \in H$. Hence:

$$p \in H \Rightarrow \mathrm{Iso}_p \subset H \tag{6.56}$$

Assuming the Claim we prove the preceding invertibility result. Since p is assumed minimal in H, $Sp = J$ is minimal at H, so the Claim implies that $(Sp) \cap \mathrm{Iso}_p$ consists of a unique element, an idempotent u, and M_p is a bijection from $Sp = Su$ to itself. Since $u \in \mathrm{Iso}_p$ and $p \sim u$, $p \in uSu$. By Proposition 6.6, p is invertible with unit u. By (6.56), $u \in H$; if q is the inverse of p, then $qp = u$ and $p, u \in H$ imply $q \in H$, since H is a co-ideal.

In particular (6.53) follows. By Proposition 6.6 distinct idempotents are contained in disjoint groups. Thus the union in (6.53) is disjoint.

PROOF (of Claim). By (6.52), $M_p(J) = Sp$, so $p \in |H|$ implies $J \cap \mathrm{Iso}_p$ is nonempty. Because the intersection is a closed subsemigroup, it contains some idempotent v. By (6.56), $v \in H$, so by (3), $J = Sv$. Sp is also a minimal ideal meeting H, so by (5) there exists an idempotent $u \in H$ such that $Sp = Su$. Since $v \in \mathrm{Iso}_p$ and $p \sim u$, $p \in vSu$. Clearly M_p restricts to a surjection of $J = Sv$ onto $Su = Sp$. By Proposition 6.4c, there exists an idempotent $\tilde{v} \in Sv$ such that $p \in \tilde{v}Su$ and M_p restricts to a bijection of $S\tilde{v}$ onto Su. But $\tilde{v}$ is in $Sv = J$ and in Iso_p. Then by (6.56) $\tilde{v}$, too, is in H. Again (3) implies $J = S\tilde{v}$. Thus $v \sim \tilde{v}$, and M_p is a bijection of J onto Sp, as required. Since M_p is injective on J, $x = v$ is the unique solution in J of the equation $xp = p$, i.e., $J \cap \mathrm{Iso}_p = \{v\}$. ■

From the Claim we see that any two ideals minimal at a closed co-ideal H are isomorphic as S actions.

COROLLARY 6.1. *Let H be a closed co-ideal of an Ellis semigroup S and let J_1, J_2 be ideals of S minimal at H. Given $u_2 \in J_2 \cap \mathrm{Id}(H)$ there exists a unique $u_1 \in J_1 \cap \mathrm{Id}(H)$ such that:*

$$u_1 u_2 = u_2 \qquad u_2 u_1 = u_1 \tag{6.57}$$

Then M_{u_2} restricts to a bijection of $J_1 = Su_1$ onto $J_2 = Su_2$ whose inverse is the restriction of M_{u_1}.

PROOF. Applying the Claim to $J = J_1$ and $p = u_2$, $J_1 \cap \mathrm{Iso}_{u_2}$ consists of a singleton u_1 that is an idempotent. By (6.56), u_1 is in H and hence in $J_1 \cap \mathrm{Id}(H)$. The Claim also says that M_{u_2} restricts to a bijection of $J_1 = Su_1$ onto Su_2 which equals J_2 by (3) of Theorem 6.1b. Because $u_1 \in \mathrm{Iso}_{u_2}$ and u_2 is idempotent, both $q = u_1$ and $q = u_2 u_1$ satisfy the equation $qu_2 = u_2$. Furthermore both u_1 and $u_2 u_1$ are in the ideal J_1. Thus $u_1 u_2 = u_2$, and by uniqueness of the solution, $u_2 u_1 = u_1$; i.e., (6.57) holds. Notice that (6.57) implies u_1, u_2 form a quasi-inverse pair. So the inverse of $M_{u_2} : J_1 \to J_2$ is $M_{u_1} : J_2 \to J_1$. ■

REMARK. If u_0 is any other element of $J_1 \cap \mathrm{Id}(H)$, then by (6.49), $u_0 \sim u_1$, so in contrast to (6.57), we have

$$u_0 u_1 = u_0 \qquad u_1 u_0 = u_1 \tag{6.58}$$

■

The entire semigroup S is a closed co-ideal. We call J a *minimal ideal of* S when J is an ideal of S minimal at S. From Theorem 6.1c we recover Ellis's original result.

COROLLARY 6.2. *The set of minimal elements of an Ellis semigroup S can be expressed as the disjoint union*

$$\mathrm{Min}(S) = \cup\{G_u : u \in \mathrm{Id}(\mathrm{Min}(S))\} \tag{6.59}$$

where

$$\mathrm{Id}(\mathrm{Min}(S)) = \mathrm{Id}(S) \cap \mathrm{Min}(S) = \mathit{Min}(\mathit{Id}(S)) \tag{6.60}$$

In particular every minimal element of S is invertible.

We illustrate some of these results by applying them to the semigroup $\beta_u T$. Recall that if $\varphi : T \times X \to X$ is a uniform action on a compact space, then $\Phi : \beta_u T \times X \to X$ denotes the extension of φ to an Ellis action. Because $\beta_u T$ has an identity element $j_u(0)$, it follows that $p \in \beta_u T p$ for all $p \in \beta_u T$, i.e., $|\beta_u T| = \beta_u T$. In particular, $p \succ q$ iff $\beta_u T p \supset \beta_u T q$ for all p and q. Also for any $x \in X$, $j_u(0) \in \mathrm{Iso}_x$, so the isotropy sets, being nonempty, are all closed co-ideals.

An *ambit*, denoted (φ,x), consists of a uniform action φ on a compact space X and a point x such that $\overline{f^T(x)} = X$. If (φ_1,x_1) is also an ambit, then h is a *map of ambits*, written $h : (\varphi,x) \to (\varphi_1,x_1)$, if $h : \varphi \to \varphi_1$ is a continuous action map and $h(x) = x_1$. Because h is a continuous action map and the base points have dense orbits, h is uniquely defined if it exists at all, and we say (φ,x) maps to (φ_1,x_1) if there exists some, and hence exactly one, map of ambits $h : (\varphi,x) \to (\varphi_1,x_1)$. In particular the identity is the unique ambit map from (φ,x) to itself. It follows that if $h : (\varphi,x) \to (\varphi_1,x_1)$ and $\overline{h} : (\varphi_1,x_1) \to (\varphi,x)$ exist, then they are inverse isomorphisms.

We say that $h : (\varphi,x) \to (\varphi_1,x_1)$ is *minimal* or (φ,x) *maps minimally* to (φ_1,x_1) if the action map h is minimal. For the ambit case, this says that X is the only closed + invariant subspace of X that meets $h^{-1}(x_1)$. We call h *sharp*, or say (φ,x) *maps sharply* to (φ_1,x_1) if $h^{-1}(x_1) = \{x\}$. Clearly a sharp map is minimal.

In general for any uniform action φ on a compact X and $x \in X$, we write (φ,x) for the ambit obtained by restricting φ to the closed + invariant subspace $\overline{f^T(x)}$ of X. Then $(\varphi \times \varphi_1, (x,x_1))$ denotes the restriction of the product action to the orbit closure of (x,x_1). Clearly the product ambit maps to each factor.

Using the uniform action $\mu_u : T \times \beta_u T \to \beta_u T$ on $\beta_u T$ and its closed ideals J, we write $(\beta_u T, j_u(0))$ or (J,p) for the ambits using μ_u or its restriction to the closed ideal $J = \beta_u T p$. $(\beta_u T, j_u(0))$ is the *universal ambit*. That is, if (φ,x) is an ambit, then Φ_x maps $(\beta_u T, j_u(0))$ to (φ,x) and $\Phi_x^{-1}(x)$ is the isotropy co-ideal Iso_x. Recall that J is a closed + invariant subspace of $\beta_u T$ exactly when it is a closed ideal. For any closed co-ideal H, e.g., Iso_x, we call J an H ideal if J meets H, i.e., $H \cap J \neq \emptyset$.

PROPOSITION 6.7. *Let (φ,x) and (φ_1,x_1) be ambits with Ellis action extensions Φ and Φ_1, respectively.*

a. Let J be a closed ideal in $\beta_u T$. Φ_x maps J onto X iff J is an Iso_x *ideal, i.e.,* $\mathrm{Iso}_x \cap J \neq \emptyset$. *$\Phi_x$ is a minimal action map from the restriction of μ_u to J to φ iff J is minimal for* Iso_x, *i.e., minimal in the set of* Iso_x *ideals. In that case, for any $p \in J \cap \mathrm{Iso}_x$, (J,p) is an ambit, and Φ_x maps (J,p) to (φ,x) minimally.*

b. Assume h maps (φ,x) to (φ_1,x_1). $\mathrm{Iso}_x \subset \mathrm{Iso}_{x_1}$ *and h is sharp iff* $\mathrm{Iso}_x = \mathrm{Iso}_{x_1}$. *The following are equivalent:*

(1) h is minimal.

(2) The set of Iso_x *ideals equals the set of* Iso_{x_1} *ideals, i.e., for every ideal J,* $J \cap \mathrm{Iso}_{x_1} \neq \emptyset$ *implies* $J \cap \mathrm{Iso}_x \neq \emptyset$.

(3) There exists an ideal J minimal for Iso_{x_1} *that meets* Iso_x.

c. Let $\pi : (\varphi \times \varphi_1, (x,x_1)) \to (\varphi, x)$ *and* $\pi_1 : (\varphi \times \varphi_1, (x,x_1)) \to (\varphi_1, x_1)$ *be projection maps.* $\mathrm{Iso}_x \subset \mathrm{Iso}_{x_1}$ *iff* π *is sharp. In that case,* π_1 *is also sharp iff* $\mathrm{Iso}_x = \mathrm{Iso}_{x_1}$ *while* π_1 *is minimal iff every* Iso_{x_1} *ideal is an* Iso_x *ideal.*

d. Let H be a closed co-ideal in $\beta_u T$. *There exists an ambit* $(\mu_H, 0_H)$ *such that* $H \subset \mathrm{Iso}_{0_H}$ *and* $(\mu_H, 0_H)$ *maps to* (φ, x) *iff* $H \subset \mathrm{Iso}_x$. *In particular,* $H \subset \mathrm{Iso}_x$ *implies* $H \subset \mathrm{Iso}_{0_H} \subset \mathrm{Iso}_x$.

PROOF. (a) is an easy exercise. For (b), $h(px) = ph(x) = px_1$ implies $p \in \mathrm{Iso}_x \Rightarrow p \in \mathrm{Iso}_{x_1}$. On the other hand, $p \in \mathrm{Iso}_{x_1}$ implies $px \in h^{-1}(x_1)$. So if h is sharp the isotropy sets agree. For any $y \in h^{-1}(x_1)$, there exists $p \in \beta_u T$ such that px = y by (6.18). Thus if $y \neq x$, $p \in \mathrm{Iso}_{x_1} \setminus \mathrm{Iso}_x$.

(1) $\Rightarrow$ (2). $\Phi_{x_1}(J) = X_1$ if J is an Iso_{x_1} ideal. Since $\Phi_{x_1} = h \circ \Phi_x$, $\Phi_x(J) = X$ by minimality, so J is an Iso_x ideal.

(2) $\Rightarrow$ (3). This is obvious since minimal Iso_{x_1} ideals exist.

(3) $\Rightarrow$ (1). J meets Iso_x so that $\Phi_x(J) = X$. If A is a nonempty, closed + invariant subspace of X, then $\Phi_x^{-1}(A) \cap J$ is a closed ideal. If $h(A) = X_1$ then it is an Iso_{x_1} ideal, so it equals J by minimality. Since $J \subset \Phi_x^{-1}(A)$, $\Phi_x(J) \subset A$. So $X = A$.

(c) In general, $p(x,x_1) = (px, px_1)$ implies

$$\mathrm{Iso}_{(x,x_1)} = \mathrm{Iso}_x \cap \mathrm{Iso}_{x_1} \tag{6.61}$$

Thus $\mathrm{Iso}_{(x,x_1)} = \mathrm{Iso}_x$ iff $\mathrm{Iso}_x \subset \mathrm{Iso}_{x_1}$. The remaining assertions follow from (b).

(d) Define the closed subalgebra E_H of $\mathcal{B}(\beta_u T) = \mathcal{B}^u(\beta_u T)$:

$$E_H = \{v \in \mathcal{B}(\beta_u T) : v = (M_p)^* v \quad \text{for all} \quad p \in H\} \tag{6.62}$$

(6.17) implies that $(M_p)^*(g_u^t)^* = (g_u^t)^*(M_p)^*$. So E_H is μ_u + invariant. We obtain a compactification map, i.e., a continuous surjection $j_H : \beta_u T \to X_H$, and by Proposition 5.12 we obtain a uniform action μ_H of T on X_H such that j_H is a continuous action map. Let 0_H denote $j_H(j_u(0))$. Since j_H is surjective, it maps $(\beta_u T, j_u(0))$ onto the ambit $(\mu_H, 0_H)$. Since $j_H^*(\mathcal{B}(X_H)) = E_H$ by Theorem 5.1 it follows that $v(pz) = v(z)$ for every $v \in \mathcal{B}(X_H)$, $z \in X_H$ and $p \in H$, so $H \subset \mathrm{Iso}_{0_H}$.

If h maps $(\mu_H, 0_H)$ to (φ, x), then by (b) $\mathrm{Iso}_{0_H} \subset \mathrm{Iso}_x$. On the other hand, if $H \subset \mathrm{Iso}_x$, then $\Phi_x^*(\mathcal{B}(X)) \subset E_H$. That is, for $\tilde{v} \in \mathcal{B}(X)$ define $v \in \mathcal{B}(\beta_u T)$ by $v(q) = \tilde{v}(qx)$. Hence for $p \in H \subset \mathrm{Iso}_x$,

$$v(qp) = \tilde{v}(qpx) = \tilde{v}(qx) = v(q)$$

By Proposition 5.2 the action map Φ_x factors through j_H to define the ambit map from $(\mu_H, 0_H)$ to (φ, x). ∎

For an ambit (φ,x), we can perform the construction in (d) with $H = \mathrm{Iso}_x$. Denoting the result $(\mu_x, 0_x)$, we have an ambit map to (φ,x) so that:

$$\mathrm{Iso}_x = H \subset \mathrm{Iso}_{0_x} \subset \mathrm{Iso}_x$$

That is, $\mathrm{Iso}_{0_x} = \mathrm{Iso}_x$, so the map from $(\mu_x, 0_x)$ to (φ,x) is sharp by (b). This is the universal ambit in the sharp class of (φ,x). That is, if (φ_1,x_1) has $\mathrm{Iso}_{x_1} = \mathrm{Iso}_x$, then there is a map from $(\mu_x, 0_x)$ to (φ_1,x_1) and it is sharp because $\mathrm{Iso}_{0_x} = \mathrm{Iso}_x = \mathrm{Iso}_{x_1}$.

For minimal maps the results are less canonical. If J is a minimal Iso_x ideal, then for any $p \in J \cap \mathrm{Iso}_x$, (J,p) is an ambit, and Φ_x maps (J,p) minimally to (φ,x). If J meets Iso_{x_1} and it is minimal for Iso_{x_1}, then Φ_{x_1} maps J to φ_1, but we may have to choose a different base point $p_1 \in J \cap \mathrm{Iso}_{x_1}$. Since $\beta_u T p = J = \beta_u T p_1$ we do have $p \sim p_1$, but while this condition is necessary it is not sufficient for isomorphism of the ambits (J,p) and (J,p_1). For example, if $p = u$ is an idempotent in $J \cap \mathrm{Iso}_x$ then Propositions 6.4 and 6.5 show that exactly for $p \in J \cap \mathrm{Iso}_x$ invertible with unit u, (J,u) is isomorphic to (J,p), that is, for elements of the group $G_u \cap \mathrm{Iso}_x$. In particular if u_1 is an idempotent in $J \cap \mathrm{Iso}_x$ and $u \neq u_1$, then (J,u) is not isomorphic to (J,u_1).

Which ideals are minimal may depend on the base point. We now construct an example of a transitive action $\varphi : T \times X \to X$ with X compact metric and points $x, x_1 \in \mathrm{Trans}_\varphi$ such that Iso_x is properly contained in Iso_{x_1} and there exist Iso_{x_1} ideals that do not meet Iso_x. It follows that no minimal Iso_x ideal is minimal for Iso_{x_1}.

We need a nontrivial, weak mixing system $\varphi : T \times X \to X$ that has a fixed point x_0 and such that there exists a continuous action isomorphism $h : \varphi \times \varphi \to \varphi$. To obtain such an example, we begin with a nontrivial, weak mixing system with a fixed point, defined on a compact metric space X_0, e.g., the stopped torus map from Akin (1993), Chapt. 9. For a countably infinite index set I, let $X = X_0^I$ and let φ be the action induced on the product. Let (x_1,x_2) be a transitive point for $\varphi \times \varphi$ and let $x = h(x_1,x_2)$ be the associated transitive point for φ.

$$\mathrm{Iso}_x = \mathrm{Iso}_{(x_1,x_2)} \subset \mathrm{Iso}_{x_1}$$

Thus $x, x_1, x_2 \in \mathrm{Trans}_\varphi$. Define:

$$A := \{p \in \beta_u T : px_1 = x_1 \text{ and } px_2 = x_0\}$$

Since (x_1,x_0) is in the orbit closure of (x_1,x_2), A is a closed, nonempty set. Since x_0 is a fixed point, $qx_0 = x_0$ for all $q \in \beta_u T$, so A is a subsemigroup, and it is contained in Iso_{x_1}. But $p \in \mathrm{Iso}_x$ iff $px_1 = x_1$ and $px_2 = x_2$. Thus $A \subset \mathrm{Iso}_{x_1} \setminus \mathrm{Iso}_x$. In fact if $p \in A$, then for every $q \in \beta_u T p$, $qx_2 = x_0$, so the ideal $\beta_u T p$ is disjoint from Iso_x. Notice that the elements of A fix some transitive points but map others to nontransitive points.

7

Semigroups and Families

In this chapter we consider a uniform action $\varphi : T \times X \to X$ with X compact, and compare the semigroup and family viewpoints. Recall that by Lemma 1.2 any topological action of a uniform monoid on a compact space is a uniform action. The focus of our comparison is the uniform Stone–Čech compactification of the uniform monoid T. In itself $\beta_u T$ combines three different phenomena. First T acts uniformly on the compact space $\beta_u T$, which is the orbit closure of $j_u(0)$ in $\beta_u T$. Furthermore using the maps Φ_x we see that this action is the universal compact T action ambit [see (6.18)]. Next $\beta_u T$ is an Ellis semigroup mapping onto the enveloping semigroup, S_φ, by the homomorphism $\Phi^{\#}$ [see (6.13)]. Finally we recall that $\beta_u T$ can be regarded as the space of maximal open filters on T as in Theorem 5.2. This connects $\beta_u T$ with all of the family constructions in Chapters 3 and 4.

We begin with the family interpretation of the Ellis action $\Phi : \beta_u T \times X \to X$ extending φ.

PROPOSITION 7.1. *Let $\varphi : T \times X \to X$ be a uniform action with X compact and let $\Phi : \beta_u T \times X \to X$ be the Ellis action extension. For $p \in \beta_u T$ let $\mathcal{F}_p$ denote the maximal open filter $j_u^{-1}u[p]$ on T.*

For $x \in X$:

$$px = \omega_{\mathcal{F}_p}\varphi(x) = \omega_{k\mathcal{F}_p}\varphi(x) \tag{7.1}$$

If $\mathcal{F}$ is a filter of subsets of T, then:

$$\{px : p \in H(\mathcal{F})\} = \omega_{k\mathcal{F}}\varphi(x) \tag{7.2}$$

and $y \in \omega_{\mathcal{F}}\varphi(x)$ iff $px = y$ for all $p \in H(\mathcal{F})$, in which case $\omega_{\mathcal{F}}\varphi(x) = \omega_{k\mathcal{F}}\varphi(x)$ is the singleton y.

For any subset A of $\beta_u T$:

$$\overline{\{px : p \in A\}} = \omega_{k\mathcal{F}}\varphi(x) \tag{7.3}$$

where $\mathcal{F} = K(A)$.

For $x, y \in X$:

$$\mathcal{N}^{\varphi}(x, u[y]) = K(\{p \in \beta_u T : px = y\}) \tag{7.4}$$

PROOF. By definition $px = \Phi_x(p)$. From (5.55) we have $p = \cap_{F \in \mathcal{F}_p} \overline{j_u(F)}$. From compactness of $\beta_u T$ and continuity of Φ_x, px is the intersection of the filterbase $\{\Phi_x(\overline{j_u(F)}) : F \in \mathcal{F}_p\}$. But:

$$\Phi_x(\overline{j_u(F)}) = \overline{\Phi_x(j_u(F))} = \overline{f^F(x)}$$

Therefore the intersection is $\cap_{F \in \mathcal{F}_p} \overline{f^F(x)}$, which is $\omega_{k\mathcal{F}_p}\varphi(x)$. By Theorem 5.2, $k\mathcal{F}_p = \tilde{u}\mathcal{F}_p$, so by Proposition 3.6b:

$$\omega_{k\mathcal{F}_p}\varphi(x) = \omega_{\tilde{u}\mathcal{F}_p}\varphi(x) = \omega_{\mathcal{F}_p}\varphi(x)$$

This proves (7.1).

For a filter $\mathcal{F}$, $y \in \omega_{k\mathcal{F}}\varphi(x)$ iff $y \in \omega_{\tilde{u}k\mathcal{F}}\varphi(x)$ iff $\mathcal{N}(x, u[y]) \subset \tilde{u}k\mathcal{F} = ku\mathcal{F}$ iff $\mathcal{N}(x, u[y]) \cdot u\mathcal{F}$ is a proper family iff $\mathcal{N}(x, u[y]) \cdot u\mathcal{F}$ is an open filter iff (Proposition 5.8) there exists $p \in \beta_u T$ such that $\mathcal{N}(x, u[y]) \cdot u\mathcal{F} \subset \mathcal{F}_p$ iff there exists $p \in \beta_u T$ such that $\mathcal{N}(x, u[y]) \subset \mathcal{F}_p$ and $u\mathcal{F} \subset \mathcal{F}_p$, i.e., $p \in H(\mathcal{F})$, iff there exists $p \in H(\mathcal{F})$ such that $y \in \omega_{\mathcal{F}_p}\varphi(x)$ iff there exists $p \in H(\mathcal{F})$ such that $y = px$. This proves (7.2).

Since X is compact and $\mathcal{F}$ is a filter, Proposition 3.8 says that $\omega_{\mathcal{F}}\varphi(x)$ is nonempty iff $\omega_{k\mathcal{F}}\varphi(x)$ is a singleton, in which case $\omega_{k\mathcal{F}}\varphi(x) = \omega_{\mathcal{F}}\varphi(x)$. Thus $\omega_{\mathcal{F}}\varphi(x)$ is empty unless $\{px : p \in H(\mathcal{F})\}$ is a singleton in which case $\omega_{\mathcal{F}}\varphi(x)$ consists of that point.

By continuity of Φ_x:

$$\overline{\{px : p \in A\}} = \{px : p \in \overline{A}\} = \omega_{k\mathcal{F}}\varphi(x)$$

when $H(\mathcal{F}) = \overline{A}$. By (5.52), $H(\mathcal{F}) = \overline{A}$ when $\mathcal{F} = K(A)$.

Recall that $\mathcal{N}^{\varphi}(x, u[y])$ is an open filter. Then:

$$H(\mathcal{N}^{\varphi}(x, u[y])) = \{p : \mathcal{N}^{\varphi}(x, u[y]) \subset \mathcal{F}_p\}$$

by (5.36). Thus:

$$H(\mathcal{N}^{\varphi}(x, u[y])) = \{p : y \in \omega_{\mathcal{F}_p}\varphi(x)\}$$

so by (7.1):

$$H(\mathcal{N}^{\varphi}(x,u[y])) = \{p : y = px\}$$

Because $\mathcal{N}^{\varphi}(x,u[y])$ is an open filter:

$$\mathcal{N}^{\varphi}(x,u[y]) = KH(\mathcal{N}^{\varphi}(x,u[y]))$$

proving (7.4). ∎

Recall that addition on T, $\mu : T \times T \to T$, extends to the uniform action $\mu_u : T \times \beta_u T \to \beta_u T$ and then to the semigroup multiplication $M : \beta_u T \times \beta_u T \to \beta_u T$ [see diagram (6.16)].

COROLLARY 7.1. *Let T be a uniform monoid. For $p, q \in \beta_u T$:*

$$pq = \omega_{\mathcal{F}_p}\mu_u(q) = \omega_{k\mathcal{F}_p}\mu_u(q) \tag{7.5}$$

Define the filter on T:

$$\mathcal{F}_p * \mathcal{F}_q = \{F : N^{\mu_u}(q, \tilde{H}(F)) \in \mathcal{F}_p\}$$

The following conditions are equivalent for $z \in \beta_u T$:

(1) $z = pq$

(2) $\mathcal{F}_z \subset \mathcal{F}_p * \mathcal{F}_q$

(3) $\mathcal{F}_z = u(\mathcal{F}_p * \mathcal{F}_q)$

(4) $\mathcal{F}_p * \mathcal{F}_q \subset \tilde{u}\mathcal{F}_z$

In particular for the discrete monoid T, if $p_0, q_0 \in \beta_0 T$, then $\mathcal{F}_{p_0 q_0} = \mathcal{F}_{p_0} * \mathcal{F}_{q_0}$.

PROOF. Equations (7.5) are special cases of (7.2) applied to the action μ_u.

For $F = \emptyset$, $\tilde{H}(F) = \emptyset$ and $N(q, \tilde{H}(F)) = \emptyset$. For $F = T$, $\tilde{H}(F) = T$ and $N(q, \tilde{H}(F)) = T$. Thus the family $\mathcal{F}_p * \mathcal{F}_q$ is proper because $\mathcal{F}_p$ is proper. It is a filter because $\tilde{H}(F_1 \cap F_2) = \tilde{H}(F_1) \cap \tilde{H}(F_2)$. Hence $N(q, \tilde{H}(F_1 \cap F_2)) = N(q, \tilde{H}(F_1)) \cap N(q, \tilde{H}(F_2))$.

(1) $\Rightarrow$ (2). If $z = pq$, then $z \in \omega_{\mathcal{F}_p}\mu_u(q)$, so $\mathcal{N}(q, u[z]) \subset \mathcal{F}_p$, i.e., for U any neighborhood of z, $N(q, U) \in \mathcal{F}_p$. If $F \in \mathcal{F}_z$ then z is an element of the open set $\tilde{H}(F)$ [cf. Proposition 5.9], so $N(q, \tilde{H}(F)) \in \mathcal{F}_p$. Thus (2) holds.

(2) $\Rightarrow$ (3). Since $\mathcal{F}_p * \mathcal{F}_q$ is a filter this follows from Theorem 5.2.

(3) $\Rightarrow$ (4). Apply (2.82).

(4) $\Rightarrow$ (1). By (1) $\Rightarrow$ (2), $\mathcal{F}_{pq} \subset \mathcal{F}_p * \mathcal{F}_q$. If $\mathcal{F}_p * \mathcal{F}_q \subset \tilde{u}\mathcal{F}_z$, then $\mathcal{F}_{pq} \subset \tilde{u}\mathcal{F}_z$, so:

$$\mathcal{F}_{pq} = u\mathcal{F}_{pq} \subset u\tilde{u}\mathcal{F}_z = \mathcal{F}_z$$

By maximality and uniqueness, $\mathcal{F}_{pq} = \mathcal{F}_z$ and $pq = z$.

In the discrete case:

$$u(\mathcal{F}_{p0} * \mathcal{F}_{q0}) = \mathcal{F}_{p0} * \mathcal{F}_{q0}$$

since every family is open, so $\mathcal{F}_{p0q0} = \mathcal{F}_{p0} * \mathcal{F}_{q0}$ by (1) $\Rightarrow$ (3). ■

COROLLARY 7.2. *Let T be a uniform monoid. If $\mathcal{F}$ is a filter for T, then:*

$$H(\mathcal{F}) = \omega_{k\mathcal{F}}\mu_u(j_u(0)) \tag{7.6}$$

If $\mathcal{F}$ is translation $-$ invariant (or translation invariant), then the closed set $H(\mathcal{F})$ is a $+$ invariant (resp. an invariant) subset of $\beta_u T$.

If A is a $+$ invariant (or invariant) subset of $\beta_u T$ then the open filter $K(A)$ is translation $-$ invariant (resp. translation invariant). If A is a closed subset of $\beta_u T$, then it is $+$ invariant iff it is an ideal of the semigroup $\beta_u T$.

PROOF. Equation (7.6) is the special case of (7.2) applied to $x = j_u(0)$, the identity element of the Ellis semigroup $\beta_u T$.

If $\mathcal{F}$ is translation $-$ invariant (or invariant), then the dual $k\mathcal{F}$ is $+$ invariant (resp. invariant). Then $\omega_{k\mathcal{F}}\mu(j_u(0))$ is $+$ invariant (resp. invariant) by Proposition 3.6c,d.

The $+$ invariance condition $g_u^t(A) \subset A$ implies $K(g_u^t(A)) \supset K(A)$, so by (5.63) $ug^t K(A) \supset K(A)$. Since $K(A)$ is open, this is equivalent to $g^t K(A) \supset K(A)$. So $g_u^t(A) \subset A$ for all t implies $g^t K(A) \supset K(A)$, so $K(A) \supset g^{-t}K(A)$ for all t (cf. Proposition 2.4c). Thus, $K(A)$ is $-$ invariant.

Similarly $g_u^t(A) = A$ for all t is equivalent to $ug^t K(A) = K(A)$ for all t. In particular, $V(T_t) \in K(A)$ for all $t \in T$ and $V \in \mathcal{U}_T$. By the Interior Condition (1.6), $k\mathcal{B}_T \subset K(A)$. Hence, $K(A)$ is full because it is a filter. Because it is open as well, Eq. (2.101) implies that $g^t K(A)$ is open, so

$$g^t K(A) = ug^t K(A) = K(A)$$

for all $t \in T$. Thus, $K(A)$ is invariant.

Finally $+$ invariance $j_u(T)A \subset A$ is equivalent to the ideal condition $(\beta_u T)A \subset A$, when A is closed. ■

PROPOSITION 7.2. *Let T be a uniform monoid. With μ_u the uniform action of T on $\beta_u T$, let μ_0 denote the uniform action of discrete T on $\beta_0 T$. We define the closed subsets $\beta_u^* T \subset \beta_u T$ and $\beta_0^* T \subset \beta_0 T$ by:*

$$\begin{aligned} \beta_u^* T &= H(k\mathcal{B}_T) = \cap_{t \in T} g_u^t(\beta_u T) = \omega\mu_u(j_u(0)) \\ \beta_0^* T &= H^0(k\mathcal{B}_T) = \cap_{t \in T} g_0^t(\beta_0 T) = \omega\mu_0(j_0(0)) \end{aligned} \tag{7.7}$$

$\beta_u^* T$ is a closed invariant subset of $\beta_u T$ and if A is any invariant subset of $\beta_u T$, then $A \subset \beta_u^* T$. T acts reversibly on $\beta_u^* T$. Similarly $\beta_0^* T$ is the maximum closed invariant subset of $\beta_0 T$, and discrete T acts reversibly on $\beta_0^* T$.

A filter $\mathcal{F}$ for T is full iff $H(\mathcal{F}) \subset \beta_u^* T$ iff $H^0(\mathcal{F}) \subset \beta_0^* T$. In particular $p \in \beta_u^* T$ iff $\mathcal{F}_p$ is full.

The projection π_0 of $\beta_0 T$ onto $\beta_u T$ maps $\beta_0^* T$ onto $\beta_u^* T$. In fact:

$$\beta_0^* T = \pi_0^{-1}(\beta_u^* T) \tag{7.8}$$

For any full filter $\mathcal{F}$, $H(\tilde{\gamma}\mathcal{F})$ is the smallest closed invariant subset of $\beta_u T$ containing $H(\mathcal{F})$. In fact:

$$\begin{aligned} H(\tilde{\gamma}\mathcal{F}) &= H(\tilde{\gamma}u\mathcal{F}) = \overline{(\beta_u T)H(\mathcal{F})} \\ H^0(\tilde{\gamma}\mathcal{F}) &= \overline{(\beta_0 T)H^0(\mathcal{F})} \end{aligned} \tag{7.9}$$

In particular for $p \in \beta_u^* T$ and $p_0 \in \beta_0^* T$:

$$\begin{aligned} H(\tilde{\gamma}\mathcal{F}_p) &= \beta_u T p \\ H^0(\tilde{\gamma}\mathcal{F}_{p_0}) &= \beta_0 T p_0 \end{aligned} \tag{7.10}$$

the principal ideals associated with p and p_0, respectively.

PROOF. By (5.40), for each $t \in T$:

$$H(T_t) = \overline{j_u(T_t)} = \overline{g_u^t(j_u(T))} = g_u^t(\beta_u T) \tag{7.11}$$

The equivalence of alternative definitions in (7.7) follows from (7.6). By Corollary 7.2 $\beta_u^* T$ is an invariant subset. If A is any invariant subset of $\beta_u T$, then $A = g_u^t(A)$ is contained in $g_u^t(\beta_u T)$, so $A \subset \beta_u^* T$.

A filter $\mathcal{F}$ is full iff it contains $k\mathcal{B}_T$ and so iff the open filter $u\mathcal{F}$ contains $k\mathcal{B}_T$, since the latter is open. Thus, $\mathcal{F}$ is full iff $H(\mathcal{F}) \subset \beta_u^* T$. In particular we can regard $\beta_u^* T$ as the space of full maximal open filters.

Each $g_u^t : \beta_u^* T \to \beta_u^* T$ is surjective. In fact for each $t \in T$:

$$\beta_u^* T = g_u^{-t}(\beta_u^* T) \tag{7.12}$$

because $\mathcal{F}$ is full iff $g^t \mathcal{F}$ is full, i.e., by Proposition 2.4c, $k\mathcal{B}_T \subset g^t \mathcal{F}$ iff $k\mathcal{B}_T = g^{-t} k\mathcal{B}_T \subset \mathcal{F}$. By Proposition 2.12 the T action on full filters is reversible. Hence each g_u^t is bijective on $\beta_u^* T$. By compactness each g_u^t is a homeomorphism on $\beta_u^* T$ and so by Proposition 1.3 the T action on $\beta_u^* T$ is uniformly reversible.

By (5.35), $H(u\mathcal{F}) = H(\mathcal{F})$ for any family $\mathcal{F}$. By (2.102), $u\tilde{\gamma}u\mathcal{F} = u\tilde{\gamma}\mathcal{F}$. Then replacing the filter $\mathcal{F}$ by $u\mathcal{F}$, we can reduce to the case of open filters which correspond to closed subsets of $\beta_u T$. By Corollary 7.2, translation invariant open filters correspond to invariant closed subsets. Thus $u\tilde{\gamma}\mathcal{F}$ corresponds to the smallest closed invariant set containing $H(\mathcal{F})$. If $\mathcal{F} = \mathcal{F}_p$, then the principal ideal $\beta_u T p$ is closed.

The results for $\beta_0 T$ are analogous or follow directly from these applied to discrete T. (7.8) follows because an ultrafilter $\mathcal{F}_{p_0}$ is full iff $u\mathcal{F}_{p_0} = \mathcal{F}_{\pi(p_0)}$ is full. ■

REMARK. Since $\beta_u^* T$ is an invariant subset of $\beta_u T$, the product set $\beta_u^* T q$ is invariant for any $q \in \beta_u T$; i.e., not only is $\beta_u^* T$ an ideal (a left ideal), but also on the right we have $\beta_u^* T \beta_u T = \beta_u^* T$. In particular, $\omega\mu_u(q) \subset \beta_u^* T$ for any $q \in \beta_u T$. ■

We can simplify the description of the hitting time sets for points in $\beta_u^* T$. In general we have

$$\begin{aligned} N^{\mu_u}(q, \tilde{H}(F)) &= \{t : g_u^t q \in \tilde{H}(F)\} \\ &= \{t : F \in u g^t \mathcal{F}_q\} \qquad q \in \beta_u T,\ F \subset T \end{aligned} \tag{7.13}$$

by (5.64). If $q \in \beta_u^* T$ then $\mathcal{F}_q$ is full, so by (2.101) $u g^t \mathcal{F}_q = g^t u \mathcal{F}_q = g^t \mathcal{F}_q$. Hence,

$$\begin{aligned} N^{\mu_u}(q, \tilde{H}(F)) &= \{t : F \in g^t \mathcal{F}_q\} \\ &= \{t : g^{-t}(F) \in \mathcal{F}_q\} \qquad q \in \beta_u^* T,\ F \subset T \end{aligned} \tag{7.14}$$

Alternatively, since $\beta_u^* T$ is $\pm$ invariant, we have

$$g_u^{-t}(\tilde{H}(F) \cap \beta_u^* T) = \tilde{H}(g^{-t}(F)) \cap \beta_u^* T \tag{7.15}$$

PROPOSITION 7.3. *Let T be a uniform monoid. If T is a group then $\beta_u T = \beta_u^* T$. If T is not a group then $\beta_u^* T \subset (\beta_u T) \backslash j_u(T)$. If T_t is cobounded for each t and T is not a group then $\beta_u^* T = (\beta_u T) \backslash j_u(T)$.*

PROOF. If T is a group, then $T_t = T$ for all $t \in T$. $k\mathcal{B}_T = k\mathcal{P}_+ = \{T\}$. All families are full and $\beta_u T = \beta_u^* T$.

Suppose for some $s \in T$, $j_u(s) \in \beta_u^* T$. Let $t \in T$. Since s is in every element of $k\mathcal{B}_T$, $s \in T_{s+t}$. Then there exists $\bar{t} \in T$ such that $s + t + \bar{t} = s$. By cancellation $t + \bar{t} = 0$. Thus, every element of T has an inverse, and T is a group.

If for every $t \in T$ there exists a compact subset F_t of T such that $T = F_t \cup T_t$ then:

$$\beta_u T = \overline{j_u(F_t)} \cup \overline{j_u(T_t)}$$

But $j_u(F_t)$ is compact and hence closed. Thus:

$$(\beta_u T) \backslash j_u(T) \subset (\beta_u T) \backslash (j_u(F_t)) \subset \overline{j_u(T_t)}$$

Intersecting over $t \in T$, $(\beta_u T) \backslash j_u(T) \subset \beta_u^* T$. The reverse inclusion holds when T is not a group. ■

Using Corollary 7.2 we see that minimal sets, i.e., minimal elements among closed invariant subsets of $\beta_u T$, correspond to maximal elements in the class of open translation invariant filters for T.

PROPOSITION 7.4. *Let $\mathcal{F}$ be a filter for the uniform monoid T. The following are equivalent:*

(1) $\mathcal{F}$ is maximal in the collection of open, translation invariant filters.

(2) For every filter $\tilde{\mathcal{F}}$ containing $\mathcal{F}$, $\mathcal{F} = u\tilde{\gamma}\tilde{\mathcal{F}}$.

(3) For every $p \in H(\mathcal{F})$, $\mathcal{F} = u\tilde{\gamma}\mathcal{F}_p$.

(4) For every $p \in H(\mathcal{F}), k\mathcal{F} = \tilde{u}\gamma\mathcal{F}_p$.

(5) $\mathcal{F}$ is open and $H(\mathcal{F})$ is a minimal subset of $\beta_u^* T$.

(6) $\mathcal{F} = K(A)$ for A some minimal subset of $\beta_u T$.

PROOF. (1) $\Rightarrow$ (2). If $\mathcal{F} \subset \tilde{\mathcal{F}}$ and $\mathcal{F}$ is translation invariant, then $\mathcal{F} \subset \tilde{\gamma}\tilde{\mathcal{F}}$. If $\mathcal{F}$ is also open then $\mathcal{F} \subset u\tilde{\gamma}\tilde{\mathcal{F}}$. Equality follows from maximality, since $u\tilde{\gamma}\tilde{\mathcal{F}}$ is a filter by Propositions 2.10 and 2.13b.

(2) $\Rightarrow$ (3). Applying (2) to $\tilde{\mathcal{F}} = \mathcal{F}$, we see that $\mathcal{F} = u\tilde{\gamma}\mathcal{F}$ is open. Then $p \in H(\mathcal{F})$ implies $\mathcal{F}_p$ is a filter containing $\mathcal{F}$. We apply (2) to such $\mathcal{F}_p$s.

(3) $\Rightarrow$ (4):

$$\mathcal{F} = u\tilde{\gamma}\mathcal{F}_p \quad \text{iff} \quad k\mathcal{F} = \tilde{u}\gamma k\mathcal{F}_p = \tilde{u}\gamma\tilde{u}\mathcal{F}_p$$

by Theorem 5.2. Since $\mathcal{F}$ is translation invariant, it is full, so $p \in \beta_u^* T$. Hence by (2.82) and (2.101):

$$\tilde{u}\gamma\tilde{u}\mathcal{F}_p = \tilde{u}u\gamma\tilde{u}\mathcal{F}_p = \tilde{u}\gamma u\tilde{u}\mathcal{F}_p = \tilde{u}\gamma u\mathcal{F}_p = \tilde{u}\gamma\mathcal{F}_p$$

(4) $\Rightarrow$ (3):

$$k\mathcal{F} = \tilde{u}\gamma\mathcal{F}_p \quad \text{iff} \quad \mathcal{F} = u\tilde{\gamma}k\mathcal{F}_p = u\tilde{\gamma}\tilde{u}\mathcal{F}_p = u\tilde{u}\tilde{\gamma}\mathcal{F}_p = u\tilde{\gamma}\mathcal{F}_p$$

by (2.82) and (2.101) because $k\mathcal{F}$ translation invariant again implies $p \in \beta_u^* T$.

(3) $\Rightarrow$ (1). If $\mathcal{F}_1$ is an open invariant filter containing $\mathcal{F}$, then $H(\mathcal{F}_1)$ is nonempty and contained in $H(\mathcal{F})$. For $p \in H(\mathcal{F}_1)$, $\mathcal{F} \subset \mathcal{F}_1 \subset \mathcal{F}_p$, so $\mathcal{F} \subset \mathcal{F}_1 \subset u\tilde{\gamma}\mathcal{F}_p$ because $\mathcal{F}_1$ is open and invariant. (3) applied to p implies $\mathcal{F} = \mathcal{F}_1$. Applying (3) to any $p \in H(\mathcal{F}) \neq \emptyset$, we see that $\mathcal{F}$ itself is open and translation invariant.

(1) $\Leftrightarrow$ (5). For an open filter $\mathcal{F}$, Corollary 7.2 says $\mathcal{F}$ is translation invariant iff $H(\mathcal{F})$ is invariant. Furthermore

$$\mathcal{F} \subset \mathcal{F}_1 \quad \text{iff} \quad H(\mathcal{F}) \supset H(\mathcal{F}_1)$$

Then maximal elements among open invariant filters correspond by H to minimal elements among closed invariant subsets.

(5) $\Leftrightarrow$ (6). $\mathcal{F} = K(A)$ for a closed set A iff $\mathcal{F}$ is open and $H(\mathcal{F}) = A$.

Notice that minimal subsets are invariant by compactness, so these are contained in $\beta_u^* T$. ∎

PROPOSITION 7.5. *For a nonempty closed subset B of $\beta_u T$ the following are equivalent:*

(1) B contains a nonempty + invariant subset of $\beta_u T$.

(2) B contains a minimal subset of $\beta_u T$.

(3) $K(B)$ is contained in some translation invariant filter.

(4) $K(B) \subset \tau\mathcal{B}_T$.

Furthermore:

$$\begin{aligned} u\tau\mathcal{B}_T &= \bigcup\{\mathcal{F} : \mathcal{F} \text{ is an open translation invariant filter}\} \\ &= \bigcup\{\mathcal{F} : \mathcal{F} \text{ is a maximal open translation invariant filter}\} \end{aligned} \tag{7.16}$$

$$\begin{aligned} u\tau k\tau\mathcal{B}_T &= u\bigcap\{\mathcal{F} : \mathcal{F} \text{ is a maximal open translation invariant filter}\} \\ &= u\bigcap\{\mathcal{F} : \mathcal{F} \text{ is a maximal translation invariant filter}\} \\ &= K(\bigcup\{A : A \text{ is a minimal subset of } \beta_u T\}) \end{aligned} \tag{7.17}$$

$$H(\tau k\tau\mathcal{B}_T) = \overline{\bigcup\{A : A \text{ is a minimal subset of } \beta_u T\}} \tag{7.18}$$

that is, $H(\tau k\tau\mathcal{B}_T)$ is the mincenter of $\beta_u T$.

PROOF. (1) $\Leftrightarrow$ (2). This follows from Proposition 3.14.

(2) $\Rightarrow$ (3). A closed subset A is contained in B iff $K(B) \subset K(A)$. If A is a minimal subset of B, then $K(A)$ is an invariant filter containing $K(B)$.

(3) $\Rightarrow$ (4) and (1). If $K(B)$ is contained in the invariant filter $\mathcal{F}$, then $\mathcal{F} \subset \tau\mathcal{B}_T$ by Proposition 2.7 and $B = H(K(B))$ contains the invariant set $H(\mathcal{F})$, proving (4) and (1).

(4) $\Rightarrow$ (3). Proposition 2.6 implies that the filter $K(B)$ is contained in some translation invariant filter.

It clearly follows that the union of all maximal open invariant filters is contained in the union of open invariant filters and the latter is an open family contained in $\tau\mathcal{B}_T$ and hence in $u\tau\mathcal{B}_T$. If $F \in \tau\mathcal{B}_T$, then:

$$K(H(F)) = u[F] \subset \tau\mathcal{B}_T$$

therefore so $H(F)$ contains some minimal set A. Thus $u[F]$ is contained in the maximal open invariant filter $K(A)$. If $F_1 \in u\tau\mathcal{B}_T$, then there exists $F \in \tau\mathcal{B}_T$ and $V \in \mathcal{U}_T$ such that $F_1 \supset V(F)$. Hence $F_1 \in u[F]$ is in the union of maximal open translation invariant filters. So (7.16) holds.

Applying Theorem 3.1 to the filter $\mathcal{F} = \tau k\tau\mathcal{B}_T$, $\omega_{k\mathcal{F}}\mu_u(j_u(0))$ is the mincenter of $\omega\mu_u(j_u(0))$, i.e., the closure of the union of the minimal subsets of the latter. Then by (7.6) and (7.7), $H(\mathcal{F})$ is the mincenter of β_u^*T which contains all minimal subsets of β_uT. This is (7.18).

Thus

$$u\tau k\tau\mathcal{B}_T = KH(\tau k\tau\mathcal{B}_T) = K(\text{mincenter}) = K(\cup\{A : A \text{ minimal}\})$$

by Proposition 5.9. The latter is $u\cap\{K(A) : A \text{ minimal}\}$. Then (7.17) follows from Proposition 7.4 [see also (2.88)]. ∎

Remark. Applying these results to discrete T we see that $\tau\mathcal{B}_T$ and $\tau k\tau\mathcal{B}_T$ are the union and intersection, respectively, of the collection of all maximal translation invariant filters. ∎

A closed subset A of β_uT is $+$ invariant iff it is an ideal. So A is a minimal subset for β_uT iff A is a minimal closed ideal. By Theorem 6.1 A is a minimal ideal and $A = \beta_uTe$ for some minimal idempotent e in β_uT. The idempotents of β_uT and β_0T are associated with filters satisfying special conditions. Notice that with $\pi_0 : \beta_oT \to \beta_uT$, the canonical projection, (6.36) implies

$$\begin{aligned} \pi_0(\text{Id}(\beta_0T)) &= \text{Id}(\beta_uT) \\ \pi_0(\text{Id}(\beta_0^*T)) &= \text{Id}(\beta_u^*T) \end{aligned} \tag{7.19}$$

Just as ideals are associated with thick families, subsemigroups are associated with a special class of families. This class includes the filters associated with idempotents.

For $F \subset T$ and $t_1, \ldots, t_l \in T$ define $F_{\{t_1,\ldots t_l\}} = \{t \in T : t, t+t_1, \ldots, t+t_l \in F\}$, or equivalently:

$$F_{\{t_1,\ldots t_l\}} = \cap_{i=0}^{l} g^{-t_i}(F) \qquad (t_0 \equiv 0) \tag{7.20}$$

Thus $F \in \tau\mathcal{F}$ iff for every finite subset $\{t_1,\dots t_l\}$ of T, $F_{\{t_1,\dots t_l\}} \in \mathcal{F}$. Define the family generated by the sets obtained by varying only over finite subsets of F itself:

$$\mathcal{S}(F) = [\{F_{\{t_1,\dots,t_l\}} : \{t_1,\dots t_l\} \subset F\}] \tag{7.21}$$

If this family is proper, then it is a filter. For a proper family $\mathcal{F}$, F is called *$\mathcal{F}$ semiadditive* if $\mathcal{S}(F) \subset \mathcal{F}$. F is called semiadditive if $\mathcal{S}(F)$ is proper, i.e., $\mathcal{S}(F) \subset \mathcal{P}_+$, so that F is $\mathcal{P}_+$ semiadditive.

A family $\mathcal{F}$ is called *semiadditive* if it is proper and generated by $\mathcal{F}$ semiadditive sets.

For any proper family $\mathcal{F}$ define

$$\tau_0\mathcal{F} = \{F : \text{For some } F_1 \subset F,\ F_1 \text{ is } \mathcal{F} \text{ semiadditive}\} \tag{7.22}$$

PROPOSITION 7.6. *Let $\mathcal{F}$ be a proper family for a uniform monoid T.*

a. If F is $\mathcal{F}$ semiadditive, then for $\{t_1,\dots,t_l\} \subset F$, $F_{\{t_1,\dots,t_l\}}$ is $\mathcal{F}$ semiadditive. The family $\mathcal{S}(F)$ is semiadditive when it is proper.

b. The family $\tau_0\mathcal{F}$ is proper and it is the largest semiadditive family contained in $\mathcal{F}$. That is, $\tau_0\mathcal{F}$ is semiadditive, and if $\mathcal{F}_1$ is a semiadditive family contained in $\mathcal{F}$, then $\mathcal{F}_1 \subset \tau_0\mathcal{F}$. Thus $\mathcal{F}$ is semiadditive, iff $\tau_0\mathcal{F} = \mathcal{F}$. In particular, $\tau_0\tau_0\mathcal{F} = \tau_0\mathcal{F}$ for any proper family $\mathcal{F}$. Any thick family is semiadditive and:

$$\tau\mathcal{F} \subset \tau_0\mathcal{F} \subset \mathcal{F} \tag{7.23}$$

$$\tau_0\mathcal{F} = \cup\{\mathcal{S}(F) : \mathcal{S}(F) \subset \mathcal{F}\} \tag{7.24}$$

The operator τ_0 preserves inclusions.

c. If F_α is $\mathcal{F}_\alpha$ semiadditive for $\alpha = 1,2$ and $\mathcal{F}_1 \cdot \mathcal{F}_2$ is a proper family, then $F_1 \cap F_2$ is $\mathcal{F}_1 \cdot \mathcal{F}_2$ semiadditive. If $\mathcal{F}_1$ and $\mathcal{F}_2$ are semiadditive and $\mathcal{F}_1 \cdot \mathcal{F}_2$ is proper, then $\mathcal{F}_1 \cdot \mathcal{F}_2$ is semiadditive. In general if $\mathcal{F}_1 \cdot \mathcal{F}_2$ is proper, then:

$$(\tau_0\mathcal{F}_1)\cdot(\tau_0\mathcal{F}_2) = \tau_0((\tau_0\mathcal{F}_1)\cdot(\tau_0\mathcal{F}_2)) \subset \tau_0(\mathcal{F}_1 \cdot \mathcal{F}_2) \tag{7.25}$$

d. F is an additive subset of T, i.e., closed under $+$, iff $F_{\{t_1,\dots,t_l\}} = F$ for every finite subset $\{t_1,\dots,t_l\}$ of F, or equivalently, iff $\mathcal{S}(F) = [F]$. If F is additive, it is $\mathcal{F}$ semiadditive whenever $F \in \mathcal{F}$. The family $k\mathcal{B}_T$ is semiadditive.

e. If $\{\mathcal{F}_\alpha\}$ is a collection of semiadditive families, then $\cup_\alpha \mathcal{F}_\alpha$ is semiadditive.

f. For any family $\mathcal{F} \subset \mathcal{B}_T$:

$$\tau_0\gamma\tau_0\gamma\mathcal{F} = \tau_0\gamma\mathcal{F} \qquad \gamma\tau_0\gamma\tau_0\mathcal{F} = \gamma\tau_0\mathcal{F} \tag{7.26}$$

g. If $\mathcal{F}$ is a filter, then F is $\mathcal{F}$ semiadditive iff $g^{-t}F \in \mathcal{F}$ for all $t \in F$. If $\mathcal{F}$ is a filter, then $\tau_0\mathcal{F}$ is a filter. Furthermore if $k\mathcal{B}_T \subset \mathcal{F}$, then:

$$\tilde{\gamma}\tau\mathcal{F} = \tilde{\gamma}\tau_0\mathcal{F} = \tilde{\gamma}\mathcal{F} \qquad (\mathcal{F} \text{ a filter}) \tag{7.27}$$

h. If F is $\mathcal{F}$ semiadditive and V is an open invariant element of $\mathcal{U}_T$, then $V(F)$ is $u\mathcal{F}$ semiadditive. If $\mathcal{F}$ is semiadditive, then $u\mathcal{F}$ is semiadditive. In general:

$$u\tau_0 u\mathcal{F} = u\tau_0\mathcal{F} = \tau_0 u\tau_0\mathcal{F} \tag{7.28}$$

PROOF. (a) If $\{s_1,\dots,s_k\} \in F_{\{t_1,\dots,t_l\}}$, then with $s_0 = 0$, we have $s_0 + t_0 = 0$ and $s_i + t_j \in F$ for $i = 0,\dots,k$, $j = 0,\dots,l$, and $(i,j) \neq (0,0)$. Hence:

$$\cap_{i=0}^{k} g^{-s_i}(F_{\{t_1,\dots,t_l\}}) = \cap_{i=0,j=0}^{k\ \ l} g^{-(s_i+t_j)}(F) \tag{7.29}$$

is in $\mathcal{S}(F)$. Therefore:

$$\mathcal{S}(F_{\{t_1,\dots,t_l\}}) \subset \mathcal{S}(F) \qquad \{t_1,\dots,t_l\} \subset F \tag{7.30}$$

Thus $F_{\{t_1,\dots,t_l\}}$ is $\mathcal{F}$ semiadditive when F is. In particular $\mathcal{S}(F)$ proper implies $F_{\{t_1,\dots,t_l\}}$ is $\mathcal{S}(F)$ semiadditive for $\{t_1,\dots,t_l\} \subset F$.

(b) The inclusions of (7.23) are clear as is $\tau_0\mathcal{F}_1 \subset \tau_0\mathcal{F}_2$ when $\mathcal{F}_1 \subset \mathcal{F}_2$. Also $\tau_0\mathcal{F} = \mathcal{F}$ iff $\mathcal{F}$ is semiadditive. From (a) we see that F $\mathcal{F}$ semiadditive implies $\mathcal{S}(F) \subset \tau_0\mathcal{F}$, so (7.24) is clear. In particular F is $\mathcal{F}$ semiadditive iff it is $\tau_0\mathcal{F}$ semiadditive, so $\tau_0\tau_0\mathcal{F} = \tau_0\mathcal{F}$. Thus the family $\tau_0\mathcal{F}$ is semiadditive. If $\mathcal{F}_1$ is semiadditive and contained in $\mathcal{F}$, then it is generated by sets that are $\mathcal{F}_1$ and hence $\mathcal{F}$ semiadditive. Therefore $\mathcal{F}_1 \subset \tau_0\mathcal{F}$.

(c) For any set $\{t_1,\dots,t_l\}$ we have

$$(F_1 \cap F_2)_{\{t_1,\dots,t_l\}} = F_{1\{t_1,\dots,t_l\}} \cap F_{2\{t_1,\dots,t_l\}} \tag{7.31}$$

Also: $\{t_1,\dots,t_l\} \subset F_1 \cap F_2$ implies $\{t_1,\dots,t_l\} \subset F_\alpha$ for $\alpha = 1,2$. Hence we have

$$\mathcal{S}(F_1 \cap F_2) \subset \mathcal{S}(F_1) \cdot \mathcal{S}(F_2) \tag{7.32}$$

whence if F_α is $\mathcal{F}_\alpha$ semiadditive $(\alpha = 1,2)$, then $F_1 \cap F_2$ is $\mathcal{F}_1 \cdot \mathcal{F}_2$ semiadditive; so if $\mathcal{F}_\alpha$ is generated by $\mathcal{F}_\alpha$ semiadditive sets $(\alpha = 1,2)$, then $\mathcal{F}_1 \cdot \mathcal{F}_2$ is generated by $\mathcal{F}_1 \cdot \mathcal{F}_2$ semiadditive sets (assuming $\mathcal{F}_1 \cdot \mathcal{F}_2$ is proper). In general if $\mathcal{F}_1 \cdot \mathcal{F}_2$ is proper, then $(\tau_0\mathcal{F}_1) \cdot (\tau_0\mathcal{F}_2)$ is proper, so it is semiadditive, hence the equality in (7.25). The inclusion follows by monotonicity.

(d) Results for additives are clear. In particular $k\mathcal{B}_T$, generated by the additive sets T_t, is a semi-additive family.

(e) The union result follows from (7.24) when $\tau_0\mathcal{F}_\alpha = \mathcal{F}_\alpha$ for all α.

(f) $\tau_0\gamma = \tau_0\tau_0\gamma \subset \tau_0\gamma\tau_0\gamma \subset \tau_0\gamma\gamma = \tau_0\gamma$; similarly $\gamma\tau_0\gamma\tau_0 = \gamma\tau_0$.

(g) If $F \in \mathcal{F}$ and $t \in F$, then

$$F_{\{t\}} = (F) \cap g^{-t}(F) \subset g^{-t}(F)$$

and so $g^{-t}(F) \in \mathcal{F}$ if F is $\mathcal{F}$ semiadditive. The converse follows from the filter property and (7.20) the definition of $F_{\{t_1,\dots,t_l\}}$. That $\tau_0\mathcal{F}$ is a filter when $\mathcal{F}$ is follows from (7.25). If $k\mathcal{B}_T \subset \mathcal{F}$, then $\tilde{\gamma}\tau\mathcal{F}$ is proper, and by monotonicity $\tilde{\gamma}\tau \subset \tilde{\gamma}\tau_0 \subset \tilde{\gamma}$. But $\tilde{\gamma}\mathcal{F}$ is a translation invariant filter, so it is thick. Thus $\tilde{\gamma}\mathcal{F} \subset \tau\mathcal{F}$, so $\tilde{\gamma}\mathcal{F} \subset \tilde{\gamma}\tau\mathcal{F}$.

(h) Recall that $V \in \mathcal{U}_T$ is invariant if $(g^{-t} \times g^{-t})(V) = V$ for all $t \in T$. Such invariant elements can be constructed by using neighborhoods of 0 in the group G_T. By using open neighborhoods, we obtain a base of open invariant elements of $\mathcal{U}_T$.

Now assume F is $\mathcal{F}$ semiadditive and $V \in \mathcal{U}_T$ is open and invariant. Let $\{t_1,\dots,t_l\} \subset V(F)$. Then there exists $\{\tilde{t}_1,\dots,\tilde{t}_l\} \subset F$ such that $t_i \in V(\tilde{t}_i)$ for $i = 1,\dots,l$. Because each $V(\tilde{t}_i)$ is open, there exists an invariant $W \in \mathcal{U}_T$ such that $W \subset V$ and $W(t_i) \subset V(\tilde{t}_i)$ for $i = 1,\dots,l$. Note that W depends on the particular finite sets at hand. By looking at the group G_T where $g^{\tilde{s}}$ is invertible, we see that $W(\tilde{s}+t_i) \subset V(\tilde{s}+\tilde{t}_i)$ for all $\tilde{s} \in T$ by invariance. We prove that:

$$W(F_{\{\tilde{t}_1,\dots,\tilde{t}_l\}}) \subset (V(F))_{\{t_1,\dots,t_l\}} \tag{7.33}$$

If $s \in W(\tilde{s})$ with $\tilde{s} \in F_{\{\tilde{t}_1,\dots,\tilde{t}_l\}}$ then for $i = 0,\dots,l$; $s+t_i \in W(\tilde{s}+t_i)$ by invariance of W. Then $W(\tilde{s}+t_i) \subset V(\tilde{s}+\tilde{t}_i)$ as above by invariance of W and V. Finally $\tilde{s}+\tilde{t}_i \in F$ by definition of $F_{\{\tilde{t}_1,\dots,\tilde{t}_l\}}$.

Because $\tilde{F} \in \tilde{\mathcal{F}}$ implies $W(\tilde{F}) \in u\tilde{\mathcal{F}}$, the inclusion (7.33) shows that $V(F)$ is $u\mathcal{S}(F)$ semiadditive, and so is $u\mathcal{F}$ semiadditive when F is $\mathcal{F}$ semiadditive. In particular if a collection of $\mathcal{F}$ semiadditive sets $\{F_\alpha\}$ generates $\mathcal{F}$ then $\{V(F_\alpha)\}$, with V varying over open invariant elements of $\mathcal{U}_T$, generates $u\mathcal{F}$. Therefore $u\mathcal{F}$ is semiadditive. Then $u\tau_0\mathcal{F}$ is semiadditive, so $u\tau_0 = \tau_0 u\tau_0$. Hence, $u\tau_0 u\tau_0 = uu\tau_0 = u\tau_0$ but $u\tau_0 u\tau_0 \subset u\tau_0 u \subset u\tau_0$. Therefore (7.28) follows. ∎

REMARK. The singleton $\{0\}$ is additive so it is semiadditive. In fact any set that contains 0 is semiadditive, i.e., $\mathcal{P}_+$ semiadditive. Using $\mathcal{B}_T$ semiadditivity eliminates this special role for 0. Notice that if $\mathcal{F}$ is semiadditive and $\mathcal{F} \subset \mathcal{B}_T$, then $\mathcal{F} \cdot k\mathcal{B}_T$ is semiadditive by (c) and (d). ∎

The most important examples of semiadditive sets come from dynamics.

PROPOSITION 7.7. *Let $\varphi : T \times X \to X$ be a uniform action with X compact and let $\mathcal{F}$ be a proper family for T.*

For A, a closed subset of X, and U any open neighborhood of A, $J^\varphi(A,U) = \{t : f^t(A) \subset U\}$ is a semiadditive set. If $\mathcal{F}$ is a filter, then:

$$\omega_{k\mathcal{F}}\varphi[A] \subset A \tag{7.34}$$

iff for every open neighborhood U of A, $J^{\varphi}(A,U) \in \mathcal{F}$. In that case, each such $J^{\varphi}(A,U)$ is $\mathcal{F}$ semiadditive.

For x, a point in X, and U any open neighborhood of x, each $N^{\varphi}(x,U) = J^{\varphi}(x,U)$ is $\mathcal{N}^{\varphi}(x,u[x])$ semiadditive, and so is $\mathcal{F}$ semiadditive when $x \in \omega_{\mathcal{F}}\varphi(x)$. In particular, $\mathcal{N}^{\varphi}(x,u[x])$ is a semiadditive family for all $x \in X$.

PROOF. If $\{t_1,\dots,t_l\} \subset J^{\varphi}(A,U)$ then let:

$$U_{\{t_1,\dots,t_l\}} = \cap_{i=0}^{l} f^{-t_i}(U) \qquad (t_0 = 0) \tag{7.35}$$

By definition of $J^{\varphi}(A,U)$, this is an open neighborhood of A, and by (3.5):

$$J^{\varphi}(A,U_{\{t_1,\dots,t_l\}}) \subset J^{\varphi}(A,U)_{\{t_1,\dots,t_l\}} \tag{7.36}$$

If $J^{\varphi}(A,U) \in \mathcal{F}$ for all open U containing A, then each such set is $\mathcal{F}$ semiadditive. Since X is compact, Proposition 3.5 and definition (3.27) shows that this condition is equivalent to the condition $\omega_{k\mathcal{F}}\varphi[A] \subset A$. Notice that $0 \in J^{\varphi}(A,U)$ for all U containing A, so these sets are semiadditive.

For A a singleton x, we have $J^{\varphi}(A,U) = N^{\varphi}(x,U)$. These sets are semiadditive. By (7.36) we have, for any open set U containing x:

$$\mathcal{S}(N^{\varphi}(x,U)) \subset \mathcal{N}(x,u[x]) \tag{7.37}$$

Thus, if $x \in \omega_{\mathcal{F}}\varphi(x)$, i.e., $\mathcal{N}^{\varphi}(x,u[x]) \subset \mathcal{F}$, these sets are $\mathcal{F}$ semiadditive. ∎

We digress to present an important class of algebraic examples of semiadditive sets.

Define $\mathbf{Z}_{\infty} = \mathbf{Z}_{+} \cup \{\infty\}$, the nonnegative integers with the point at infinity. Extend the usual ordering and arithmetic by $n \leq \infty$ and $\infty \pm n = \infty$ for all $n \in \mathbf{Z}_{\infty}$. Also define $n_1 - n_2 = 0$, when $n_2 > n_1$ in $\mathbf{Z}_{\infty}$. Let $\mathcal{Z}$ denote the set of functions from T to $\mathbf{Z}_{\infty}$, i.e., $\mathbf{Z}_{\infty}^{T}$. On $\mathcal{Z}$ we define addition and subtraction pointwise, and for $z, z_1, z_2 \in \mathcal{Z}$, we define:

$$\begin{aligned} \operatorname{supp} z &= \{t \in T : z(t) > 0\} \\ z_1 \leq z_2 &\Leftrightarrow z_1(t) \leq z_1(t) \quad \text{for all } t \in T \end{aligned} \tag{7.38}$$

Call z a *finite* element of $\mathcal{Z}$ if $\operatorname{supp} z$ is a finite subset of T and $z(t) \neq \infty$ for all $t \in T$. Let $\mathcal{Z}^0$ denote the set of finite elements of $\mathcal{Z}$, a monoid with respect to addition. Define the monoid homomorphism:

$$\begin{aligned} \sigma : \mathcal{Z}^0 &\longrightarrow T \\ \sigma_z &= \sum z(t)t \end{aligned} \tag{7.39}$$

where the sum is taken over the support of z or over all T when we allow the formal addition of infinitely many 0s.

For $z \in \mathcal{Z}$ we define the set

$$F_z = \{\sigma_{z_1} : z_1 \in \mathcal{Z}^0 \backslash 0 \text{ such that } z_1 \leq z\} \tag{7.40}$$

F_z is nonempty unless $z = 0$. In fact:

$$\text{supp}(z) \subset F_z \tag{7.41}$$

Letting $\delta_t(s) = 0$ for $s \neq t$ and $\delta_t(t) = 1$, $\sigma_{\delta_t} = t$, $\delta_t \in \mathcal{Z}^0 \backslash 0$ and, when $t \in \text{supp}(z)$, $\delta_t \leq z$. If z is finite, then F_z is a finite set.

For $z \in \mathcal{Z} \backslash \mathcal{Z}^0$, i.e., for nonfinite z, define

$$\mathcal{F}_z = \{F : \text{ For some } z_1 \in \mathcal{Z}^0 \text{ with } z_1 \leq z,\ F_{z-z_1} \subset F\} \tag{7.42}$$

That is, $F \in \mathcal{F}_z$ iff there exists $z_1 \in \mathcal{Z}^0$ with $z_1 \leq z$ such that:

$$z_2 \in \mathcal{Z}^0 \backslash 0 \qquad \text{and} \qquad z_1 + z_2 \leq z \Rightarrow \sigma_{z_2} \in F \tag{7.43}$$

LEMMA 7.1. *a. If $z, \tilde{z} \in \mathcal{Z} \backslash \mathcal{Z}^0$ with $z \leq \tilde{z}$, then $\mathcal{F}_{\tilde{z}} \subset \mathcal{F}_z$.*

b. If $z \in \mathcal{Z} \backslash \mathcal{Z}^0$ and $z_1 \in \mathcal{Z}^0$ with $z_1 \leq z$, then F_{z-z_1} is $\mathcal{F}_z$ semiadditive.

c. For $z \in \mathcal{Z} \backslash \mathcal{Z}^0$, $\mathcal{F}_z$ is a semiadditive filter.

d. If $\mathcal{F}_z \subset \mathcal{F}$, then $\mathcal{F}_z \subset \tau_0 \mathcal{F}$.

PROOF. (a) If $F \in \mathcal{F}_{\tilde{z}}$ then there exists $\tilde{z}_1 \leq \tilde{z}$ with $\tilde{z}_1 \in \mathcal{Z}^0$ and such that $F_{\tilde{z}-\tilde{z}_1} \subset F$. Let $z_1(t) = \min(z(t), \tilde{z}_1(t))$. Then $z_1 \leq \tilde{z}_1 \leq \tilde{z}$, so z_1 is finite. Also $z_1 \leq z$. If $z_1(t) = z(t)$, then:

$$(z - z_1)(t) = 0 \leq (\tilde{z} - \tilde{z}_1)(t)$$

Otherwise $z_1(t) = \tilde{z}_1(t)$, so:

$$(z - z_1)(t) \leq (\tilde{z} - \tilde{z}_1)(t)$$

because $z \leq \tilde{z}$. Hence $z - z_1 \leq \tilde{z} - \tilde{z}_1$, so $F_{z-z_1} \subset F_{\tilde{z}-\tilde{z}_1} \subset F$. Thus $F \in \mathcal{F}_z$.

(b)–(c) Because $z \in \mathcal{Z} \backslash \mathcal{Z}^0$ $\text{supp}(z - z_1) \neq \emptyset$ whenever $z_1 \in \mathcal{Z}^0$ and $z_1 \leq z$, so $\mathcal{F}_z$ is proper by (7.41). If $z_1, z_2 \in \mathcal{Z}^0$ with $z_1, z_2 \leq z$, we define $z_3(t) = \max(z_1(t), z_2(t))$. This is still finite and $z_1, z_2 \leq z_3$. Then $z - z_1$, $z - z_2 \geq z - z_3$. Hence:

$$F_{z-z_3} \subset (F_{z-z_1}) \cap (F_{z-z_2})$$

Thus $\mathcal{F}_z$ is a filter.

We now prove F_{z-z_1} is $\mathcal{F}_z$ semiadditive. Since $\mathcal{F}_z$ is a filter, it suffices to show $g^{-t} F_{z-z_1} \in \mathcal{F}_z$ for $t \in F_{z-z_1}$. By definition $t = \sigma_{z_2}$ with $z_2 \in \mathcal{Z}^0 \backslash 0$ and $z_2 \leq z - z_1$. Then $z_3 = z_1 + z_2$ is finite and $z_3 \leq z$. We show that $F_{z-z_3} \subset g^{-t} F_{z-z_1}$.

If $z_4 \in Z^0\backslash 0$ and $z_4 \leq z - z_3$, then $t + \sigma_{z_4} = \sigma_{z_2+z_4}$. But:

$$z \geq z_3 + z_4 = z_1 + z_2 + z_4$$

Thus:

$$z_2 + z_4 \leq z - z_1 \qquad z_2 + z_4 \neq 0$$

Then $\sigma_{z_2+z_4} \in F_{z-z_1}$ as required.

Since these F_{z-z_1}'s are $\mathcal{F}_z$ semiadditive and generate $\mathcal{F}_z$, it follows that $\mathcal{F}_z$ is semiadditive.

(d) $\tau_0\mathcal{F}$ is the largest semiadditive family contained in $\mathcal{F}$. ■

PROPOSITION 7.8 (Galvan). *Let $\mathcal{F}$ be a proper family such that $k\mathcal{F}$ is countably generated and let F be $\mathcal{F}$ semiadditive. There exists $z \in Z\backslash Z^0$ such that $F_z \subset F$ and for all $z_1 \in Z^0$ with $z_1 \leq z$,* $\text{supp}(z - z_1) \in \mathcal{F}$*, a fortiori $\mathcal{F}_z \subset \mathcal{F}$.*

PROOF. Let $\{A_1, A_2, \dots\}$ be a sequence in $k\mathcal{F}$ such that for each n, $\{A_n, A_{n+1}, \dots\}$ generates $k\mathcal{F}$, e.g., partition $\{1,2,\dots\}$ into countably many disjoint infinite sets and associate the elements of each partition member with a generating sequence.

Initially let $F_1 = F$ and choose $t_1 \in F_1 \cap A_1$; let $F_2 = F_1 \cap g^{-t_1}F_1$ and choose $t_2 \in F_2 \cap A_2$. Inductively define the decreasing sequence of $\mathcal{F}$ semiadditive sets $F_{n+1} = F_n \cap g^{-t_n}F_n$ and choose $t_{n+1} \in F_{n+1} \cap A_{n+1}$. For each n, the set $\{t_n, t_{n+1}, \dots\}$ meets a generating sequence for $k\mathcal{F}$, so it lies in $\mathcal{F}$.

Now define z_n for $n \in \mathbf{Z}_\infty$. Let $z_0 = 0$, and for $n \geq 1$, let

$$z_n = \sum_{i=1}^{n} \delta_{t_i} \tag{7.44}$$

So for $n = 1, 2, \dots$ $z_n \in Z^0\backslash 0$, and for $n = \infty$, $z_\infty \in Z\backslash Z^0$. Next we prove that for $n = 0, 1, \dots$

$$F_{z_\infty - z_n} \subset F_{n+1} \tag{7.45}$$

Since any finite $\tilde{z} \leq z_\infty - z_n$ satisfies $\tilde{z} \leq z_m - z_n$ for some $m > n$, it suffices to prove

$$F_{z_{m+1}-z_{m-j}} \subset F_{m-j+1} \qquad j = 0, \dots, m \tag{7.46}$$

We prove this by induction on j. Observe that:

$$z_{m+1} - z_{m-j} = \sum_{i=m-j+1}^{m+1} \delta_{t_i}$$

Then the set on the left of (7.46) consists of the points:

$$\sum_{i=m-j+1}^{m+1} \varepsilon_i t_i$$

where each ε_i is either 0 or 1 and not all $\varepsilon_i = 0$. In particular with $j = 0$, the set is the singleton $t_{m+1} \in F_{m+1}$. Moving inductively from j to $j+1$, we obtain a set of points in F_{m-j+1}, and then adjoin new points by adding t_{m-j}. Thus the inductive step follows from:

$$F_{m-j+1} \cup g^{t_{m-j}}(F_{m-j+1}) \subset F_{m-j}$$

With $n = 0$ and $z = z_\infty$, (7.45) says that $F_z \subset F_1 = F$. If $\tilde{z}_1$ is finite and $\tilde{z}_1 \leq z$, then $\tilde{z}_1 \leq z_n$ for some n, so:

$$\text{supp}(z - \tilde{z}_1) \supset \{t_{n+1}, t_{n+2}, \dots\} \in \mathcal{F}$$

for some n. By (7.41) and heredity, $F_{z-\tilde{z}_1} \in \mathcal{F}$, thus, $\mathcal{F}_z \subset \mathcal{F}$. ■

The sets of the form F_z for $z \in Z \backslash Z^0$ are called *IP sets*.

COROLLARY 7.3. *Let T be a uniform monoid that is separable. The family $\tau_0 \mathcal{B}_T$ is generated by the IP sets in $\mathcal{B}_T$.*

PROOF. Each IP set F_z in $\mathcal{B}_T$ is $\mathcal{F}_z$ semiadditive by Lemma 7.1. Then $F_z \in \tau_0 \mathcal{B}_T$ follows once we prove $\mathcal{F}_z \subset \mathcal{B}_T$, i.e., $F_z \in \mathcal{B}_T$ for some $z \in Z \backslash Z^0$ implies $F_{z-z_1} \in \mathcal{B}_T$ for $z_1 \in Z^0$, $z_1 \leq z$. Let $\{t_1, \dots, t_l\}$ be the finite set F_{z_1} and $t_0 = 0$. Clearly:

$$F_z = (\cup_{i=1}^{l} \{t_i\}) \cup (\cup_{i=0}^{l} g^{t_i}(F_{z-z_1}))$$

$\mathcal{B}_T$ is a filterdual, so it satisfies the Ramsey Property (2.15). Since $F_z \in \mathcal{B}_T$, it follows that either $\{t_i\}$ or $g^{t_i} F_{z-z_1}$ is in $\mathcal{B}_T$ for some $i = 0, 1, \dots, l$. If the former is true, then by invariance every singleton set, so every nonempty set is in $\mathcal{B}_T$, i.e., $\mathcal{B}_T = \mathcal{P}_+$ and $F_{z-z_1} \in \mathcal{B}_T$. If $g^{t_i} F_{z-z_1}$, then by invariance $F_{z-z_1} \in \mathcal{B}_T$. Thus, $F_z \in \mathcal{B}_T$ implies $\mathcal{F}_z \subset \mathcal{B}_T$, so F_z is $\mathcal{B}_T$ semiadditive.

On the other hand, $\tau_0 \mathcal{B}_T$ is generated by the $\mathcal{B}_T$ semiadditive sets. If F is $\mathcal{B}_T$ semiadditive, then we can apply Proposition 7.8 to $\mathcal{B}_T$ since T separable implies $k\mathcal{B}_T$ is countably generated. By Proposition 7.8 there exists $z \in Z \backslash Z^0$ such that $F_z \subset F$ and $\mathcal{F}_z \subset \mathcal{B}_T$. Hence F_z is an IP set in $\mathcal{B}_T$, so these sets generate $\tau_0 \mathcal{B}_T$. ■

We now return to the main line of our argument, relating semiadditivity to dynamic conditions.

PROPOSITION 7.9. *Let T be a uniform monoid and $\mu_u : T \times \beta_u T \to \beta_u T$ the extension of the translation action $\mu : T \times T \to T$. For an open filter $\mathcal{F}$ for T, the following conditions are equivalent:*

(1) $\mathcal{F}$ is a semiadditive family, i.e., $\mathcal{F}$ is generated by $\mathcal{F}$ semiadditive sets.

(2) The hull $H(\mathcal{F}) \subset \beta_u T$ satisfies

$$\omega_{k\mathcal{F}}\mu_u[H(\mathcal{F})] \subset H(\mathcal{F}) \tag{7.47}$$

(3) For all $F \in \mathcal{F}$, $F \cap J^{\mu_u}(H(\mathcal{F}), \tilde{H}(F))$ is an $\mathcal{F}$ semiadditive set.

(4) For all $F \in \mathcal{F}$, $\{t : g^{-t}(F) \in \mathcal{F}\} \in \mathcal{F}$

PROOF. We show (2) $\Rightarrow$ (3) $\Rightarrow$ (1) $\Rightarrow$ (4) $\Rightarrow$ (2).

(2) $\Rightarrow$ (3). By (5.36), $p \in H(\mathcal{F})$ implies $\mathcal{F} \subset \mathcal{F}_p$ because $\mathcal{F}$ is open, so $p \in \tilde{H}(F)$. That is, the open set $\tilde{H}(F)$ in $\beta_u T$ is a neighborhood of $H(\mathcal{F})$. By Proposition 7.7, (7.47) implies $J(H(\mathcal{F}), \tilde{H}(F))$ is a $\mathcal{F}$ semiadditive set. Moreover, if $t \in J(H(\mathcal{F}), \tilde{H}(F))$ and $p \in H(\mathcal{F})$, then $t \in N(p, \tilde{H}(F))$. By (7.13),

$$F \in ug^t \mathcal{F}_p \subset g^t \mathcal{F}_p$$

Hence $g^{-t}F \in \cap_{p \in H(\mathcal{F})} \mathcal{F}_p$ which is $\mathcal{F}$ by (5.51). A fortiori if $t \in F \cap J(H(\mathcal{F}), \tilde{H}(F))$ then $g^{-t}F \in \mathcal{F}$ and $g^{-t}J(H(\mathcal{F}), \tilde{H}(F)) \in \mathcal{F}$. Because $\mathcal{F}$ is a filter, this implies $F \cap J(H(\mathcal{F}), \tilde{H}(F))$ is $\mathcal{F}$ semiadditive.

(3) $\Rightarrow$ (1). $F \in \mathcal{F}$ contains the $\mathcal{F}$ semiadditive set described in (3).

(1) $\Rightarrow$ (4). If $F \in \mathcal{F}$, then $\mathcal{F}$ semiadditive implies $F_1 \subset F$ for some $\mathcal{F}$ semiadditive set F_1. Then $F_1 \subset \{t : g^{-t}F_1 \in \mathcal{F}\} \subset \{t : g^{-t}F \in \mathcal{F}\}$, so the latter is $\mathcal{F}$.

(4) $\Rightarrow$ (2). Given $F \in \mathcal{F}$ choose $F_1 \in \mathcal{F}$ and $V \in \mathcal{U}_T$ so that $V(F_1) \subset F$. By (4), $F_2 = \{t : g^{-t}F_1 \in \mathcal{F}\}$ is in $\mathcal{F}$. If $t \in F_2$, $F_1 \in g^t\mathcal{F}$, so for $p \in H(\mathcal{F})$, $F_1 \in g^t \mathcal{F}_p$. Hence $F \in ug^t \mathcal{F}_p = \mathcal{F}_{g_u^t(p)}$ [cf. (5.64)]. Thus for all $t \in F_2$, $g_u^t(H(\mathcal{F}))$ is contained in the open set $\tilde{H}(F)$, which is in turn contained in the closed set $H(F)$. Thus:

$$\omega_{k\mathcal{F}}\mu[H(\mathcal{F})] \subset \overline{g_u^{F_2}(H(\mathcal{F}))} \subset H(F)$$

Intersecting over all $F \in \mathcal{F}$ and applying (5.33) we get (7.47) as required. ∎

REMARK. Applying this result for discrete T, we see that an arbitrary filter $\mathcal{F}$ is semiadditive iff condition (4) holds, in which case for all $F \in \mathcal{F}$, the set:

$$F \cap J^{\mu_0}(H^0(\mathcal{F}), H^0(F)) = \{t : t \in F \text{ and } g^{-t}F \in \mathcal{F}\}$$

is $\mathcal{F}$ semiadditive. ∎

COROLLARY 7.4. *Let T be a uniform monoid.*

a. If a filter $\mathcal{F}$ is semiadditive then $H(\mathcal{F})$ is a closed subsemigroup of $\beta_u T$, so it contains idempotents.

b. For $p \in \beta_u T$ the maximal open filter $\mathcal{F}_p$ is semiadditive iff p is an idempotent, i.e., $p^2 = p$.

PROOF. (a) By (7.3) and (7.5)

$$H(\mathcal{F})H(\mathcal{F}) = \omega_{k\mathcal{F}}\mu_u(H(\mathcal{F})) \subset \omega_{k\mathcal{F}}\mu_u[H(\mathcal{F})] \tag{7.48}$$

Thus $H(\mathcal{F})$ is closed under multiplication for $\mathcal{F}$ an open semiadditive filter by (7.47).

For a general semiadditive filter $\mathcal{F}$, the open filter $u\mathcal{F}$ is semiadditive by Proposition 7.6h, and it satisfies $H(u\mathcal{F}) = H(\mathcal{F})$. Finally $ku\mathcal{F} = \tilde{u}k\mathcal{F}$, so:

$$\omega_{ku\mathcal{F}}\varphi[A] = \omega_{k\mathcal{F}}\varphi[A]$$

by Proposition 3.6b. The preceding results then apply to $u\mathcal{F}$, showing that $H(u\mathcal{F}) = H(\mathcal{F})$ is a subsemigroup.

Since $H(\mathcal{F})$ is nonempty ($KH(\mathcal{F}) = u\mathcal{F}$), $\mathrm{Id}(H(\mathcal{F})) \neq \emptyset$ by Proposition 6.3b.

(b) If $\mathcal{F}_p$ is semiadditive, then by (a), $H(\mathcal{F}_p) = \{p\}$ is a subsemigroup. Therefore $p^2 = p$. Conversely if $p^2 = p$, then for $\mathcal{F} = \mathcal{F}_p$, $H(\mathcal{F})$ is the singleton set p. In that case the inclusion in (7.48) is an equality, so $p^2 = p$ implies that (7.47) holds with $\mathcal{F} = \mathcal{F}_p$. By Proposition 7.9, $\mathcal{F}_p$ is semiadditive. ■

In general the inclusion in (7.48) may be strict, so it does not follow that a closed subsemigroup has a semiadditive kernel. Of course a closed ideal always satisfies (7.47) but Proposition 7.9 is not needed in that case. The kernel of a closed ideal is an open translation $-$ invariant filter by Corollary 7.2 and so it is thick. Any thick family is semiadditive.

If $H(\mathcal{F})$ is an ideal then $\mathcal{F}$ is translation $-$ invariant, so $k\mathcal{F}$ is translation $+$ invariant. By Proposition 3.7, $\omega_{k\mathcal{F}}\varphi(x)$ is a $+$ invariant subset, so $y \in \omega_{k\mathcal{F}}\varphi(x)$ implies

$$\overline{f^T(y)} \subset \omega_{k\mathcal{F}}\varphi(x)$$

In the subsemigroup case we have the following result:

PROPOSITION 7.10. *Let $\mathcal{F}$ be a filter for T such that $H(\mathcal{F})$ is a subsemigroup of $\beta_u T$, e.g., $\mathcal{F}$ a semiadditive filter. Assume $\varphi : T \times X \to X$ is a uniform action. The relation $\omega_{k\mathcal{F}}\varphi$ is transitive. That is:*

$$y \in \omega_{k\mathcal{F}}\varphi(x) \Rightarrow \omega_{k\mathcal{F}}\varphi(y) \subset \omega_{k\mathcal{F}}\varphi(x) \tag{7.49}$$

PROOF. In the compact case, there exists for $z \in \omega_{k\mathcal{F}}\varphi(y)$, $p, q \in H(\mathcal{F})$ such that $y = px$ and $z = qy$. Hence $z = qpx$ and $qp \in H(\mathcal{F})$. For the general case, compactify and apply (3.29) ■

We now consider idempotent results.

PROPOSITION 7.11. *Let T be a uniform monoid. If $\mathcal{F}$ is a filter for T, then $\tau_0\mathcal{F} \subset k\tau_0 k\tau_0\mathcal{F}$ and both are filters. Furthermore:*

$$\begin{aligned} u\tau_0 k\tau_0\mathcal{F} &= \cup\{\mathcal{F}_e : e^2 = e \text{ and } \mathcal{F}_e \subset k\tau_0\mathcal{F}\} \\ &= \cup\{\mathcal{F}_e : e \in \mathrm{Id}(H(\tau_0\mathcal{F}))\} \qquad (7.50) \\ uk\tau_0 k\tau_0\mathcal{F} &= K(\mathrm{Id}(H(\tau_0\mathcal{F}))) \\ H(k\tau_0 k\tau_0\mathcal{F}) &= \overline{\mathrm{Id}(H(\tau_0\mathcal{F}))} \quad (\textit{closure in } \beta_u T) \qquad (7.51) \end{aligned}$$

PROOF. By Proposition 7.6g, $\tau_0\mathcal{F}$ is a filter. In general:

$$p \in H(\tau_0\mathcal{F}) \text{ iff } \tau_0\mathcal{F} \subset \tilde{u}\mathcal{F}_p = k\mathcal{F}_p \text{ iff } \mathcal{F}_p \subset k\tau_0\mathcal{F}$$

by definition of the hull and Theorem 5.2 ($\tilde{u}\mathcal{F}_p = k\mathcal{F}_p$). Thus the two descriptions of the union agree in (7.50). If $\mathcal{F}_e \subset k\tau_0\mathcal{F}$ and e is idempotent, then $\mathcal{F}_e$ is semiadditive by Corollary 7.4b, so $\mathcal{F}_e \subset \tau_0 k\tau_0\mathcal{F}$. Since $\mathcal{F}_e$ is also open:

$$\mathcal{F}_e = u\mathcal{F}_e \subset u\tau_0 k\tau_0\mathcal{F}$$

Thus the union is contained in $u\tau_0 k\tau_0\mathcal{F}$.

If $F \in u\tau_0 k\tau_0\mathcal{F}$, then there exists F_1 $k\tau_0\mathcal{F}$ semiadditive and $V \in \mathcal{U}_T$ such that $V(F_1) \subset F$. By semiadditivity, $\mathcal{S}(F_1)$ is a semiadditive filter contained in $k\tau_0\mathcal{F}$. On the other hand, $\tau_0\mathcal{F}$ is a semiadditive filter, and $\mathcal{S}(F_1) \cdot \tau_0\mathcal{F}$ is proper because $\mathcal{S}(F_1) \subset k\tau_0\mathcal{F}$. Hence by Proposition 7.6c, $\mathcal{S}(F_1) \cdot \tau_0\mathcal{F}$ is a semiadditive filter. By Corollary 7.4a, $H(\mathcal{S}(F_1) \cdot \tau_0\mathcal{F})$ contains an idempotent e. Thus $e \in H(\tau_0\mathcal{F})$; since $e \in H(\mathcal{S}(F_1))$, $u\mathcal{S}(F_1) \subset \mathcal{F}_e$. In particular, $F \supset V(F_1)$ is in $\mathcal{F}_e$. The reverse inclusion in (7.50) follows.

From Proposition 2.13:

$$\begin{aligned} uk\tau_0 k\tau_0\mathcal{F} &= u\tilde{u}k\tau_0 k\tau_0\mathcal{F} = uk(u\tau_0 k\tau_0\mathcal{F}) = uk(\cup\mathcal{F}_e) \\ &= u\cap k\mathcal{F}_e = u\cap\tilde{u}\mathcal{F}_e = u\cap u\tilde{u}\mathcal{F}_e = u\cap u\mathcal{F}_e = u\cap\mathcal{F}_e \end{aligned}$$

with e varying over the set of idempotents in $H(\tau_0\mathcal{F})$. The first equation in (7.51) follows from the definition of the kernel K, and the second follows from (5.52).

In particular, $uk\tau_0 k\tau_0\mathcal{F}$ is an open filter. Applying this result to discrete T for which every family is open, we see that $k\tau_0 k\tau_0\mathcal{F}$ is always a filter. ■

We can apply this result to $\mathcal{F} = k\mathcal{P}_+ = \{T\}$ and to $\mathcal{F} = k\mathcal{B}_T$. Each is a $-$ invariant filter, so $\tau_0\mathcal{F} = \mathcal{F}$ in each case. Recall that we denoted by $\beta_u^* T$ the closed subset $H(k\mathcal{B}_T)$.

COROLLARY 7.5. *For a uniform monoid T, the families $k\tau_0\mathcal{P}_+$ and $k\tau_0\mathcal{B}_T$ are filters, and:*

$$\begin{aligned} u\tau_0\mathcal{P}_+ &= \cup\{\mathcal{F}_e : e \in \mathrm{Id}(\beta_u T)\} \\ u\tau_0\mathcal{B}_T &= \cup\{\mathcal{F}_e : e \in \mathrm{Id}(\beta_u^* T)\} \end{aligned} \tag{7.52}$$

$$\begin{aligned} uk\tau_0\mathcal{P}_+ &= K(\mathrm{Id}(\beta_u T)) \\ uk\tau_0\mathcal{B}_T &= K(\mathrm{Id}(\beta_u^* T)) \end{aligned} \tag{7.53}$$

REMARK. (a) Since $t+t=t$ implies $t=0$ by cancellation, $t=0$ is the only idempotent in T. By Proposition 7.3, if each T_t is cocompact in T:

$$\mathrm{Id}(\beta_u T) = \{0\} \cup \mathrm{Id}(\beta_u^* T)$$

(b) When T is separable, we call the sets of $k\tau_0\mathcal{B}_T$ the IP dual sets (cf. Corollary 7.3). ■

For a proper family $\mathcal{F}$ for T and a uniform action $\varphi : T \times X \to X$, we defined $x \in X$ to be an $\mathcal{F}$ recurrent point if $x \in \omega_\mathcal{F}\varphi(x)$ and denoted by $|\omega_\mathcal{F}\varphi|$ the set of $\mathcal{F}$ recurrent points in X (see Theorem 3.1). By Proposition 3.6,

$$x \in |\omega_\mathcal{F}\varphi| \text{ iff } \mathcal{N}^\varphi(x, u[x]) \subset \mathcal{F}$$

By Proposition 7.7 $\mathcal{N}^\varphi(x, u[x])$ is a semiadditive filter.

For $y, x \in X$ we say that y *$\mathcal{F}$ prox to* x when:

$$(x,x) \in \omega_\mathcal{F}(\varphi \times \varphi)(y,x) \tag{7.54}$$

Then y $\mathcal{F}$ prox to x iff:

$$\mathcal{N}^{\varphi\times\varphi}((y,x), u[(x,x)]) \subset \mathcal{F} \tag{7.55}$$

Observe that for U a neighborhood of x:

$$N^{\varphi\times\varphi}((y,x), U \times U) = J^\varphi(\{y,x\}, U) \subset N^\varphi(x, U) \tag{7.56}$$

with equality when $y = x$. Hence:

$$\begin{aligned} \mathcal{N}^{\varphi\times\varphi}((y,x), u[(x,x)]) &\supset \mathcal{N}^\varphi(x, u[x]) \\ &= \mathcal{N}^{\varphi\times\varphi}((x,x), u[(x,x)]) \end{aligned} \tag{7.57}$$

Thus y $\mathcal{F}$ prox to x implies x is an $\mathcal{F}$ recurrent point and if x is $\mathcal{F}$ recurrent then x $\mathcal{F}$ prox to x.

PROPOSITION 7.12. *Let $\varphi : T \times X \to X$ be a uniform action and $\mathcal{F}$ a proper family for T. Let $x,y \in X$.*

a. The family $\mathcal{N}^{\varphi\times\varphi}((y,x),u[(x,x)])$ is proper iff $(x,x) \in \overline{(f\times f)^T(y,x)}$. In that case, $\mathcal{N}^{\varphi\times\varphi}((y,x),u[(x,x)])$ is a semiadditive filter for T. In particular if y $\mathcal{F}$ prox to x, then y $\tau_0\mathcal{F}$ prox to x.

b. Let $\varphi_1 : T \times X_1 \to X_1$ be a uniform action and $h : \varphi \to \varphi_1$ a continuous action map. If y $\mathcal{F}$ prox to x, then $h(y)$ $\mathcal{F}$ prox to $h(x)$. If h is an embedding, i.e., a homeomorphism of X onto $h(X) \subset X_1$, then $h(y)$ $\mathcal{F}$ prox to $h(x)$ iff y $\mathcal{F}$ prox to x.

c. Assume that X is compact and $\Phi : \beta_u T \times X \to X$ is the Ellis action extending φ. Define

$$\begin{aligned} \mathrm{Iso}_x &= \{p \in \beta_u T : px = x\} = H(\mathcal{N}^{\varphi}(x,u[x])) \\ \mathrm{Prox}_{(y,x)} &= \{p \in \beta_u T : py = px = x\} = H(\mathcal{N}^{\varphi\times\varphi}((y,x),u[(x,x)])) \end{aligned} \tag{7.58}$$

Iso_x *is a closed co-ideal of $\beta_u T$.* $\mathrm{Prox}_{(y,x)}$ *is nonempty iff $\mathcal{N}^{\varphi\times\varphi}((y,x),u[(x,x)])$ is proper. In that case,* $\mathrm{Prox}_{(y,x)}$ *is a closed subsemigroup of* Iso_x *in fact, an* Iso_x *ideal.*

d. Assume X is compact and $\mathcal{F}$ is a filter. The following three conditions are equivalent:

(1) y $k\mathcal{F}$ prox to x.

(2) y $\tau_0 k\mathcal{F}$ prox to x.

(3) $H(\mathcal{F}) \cap \mathrm{Prox}_{(y,x)} \neq \emptyset$.

On the other hand, y $\mathcal{F}$ prox to x iff $H(\mathcal{F}) \subset \mathrm{Prox}_{(y,x)}$, i.e., $py = px = x$ for all $p \in H(\mathcal{F})$. y $k\tau_0 k\tau_0\mathcal{F}$ prox to x iff $\mathrm{Id}(H(\tau_0\mathcal{F})) \subset \mathrm{Prox}_{(y,x)}$. Furthermore:

$$\omega_{\tau_0 k\tau_0\mathcal{F}}\varphi(x) = \overline{\{z \in X : x\ k\tau_0\mathcal{F} \text{ prox to } z\}} \tag{7.59}$$

For y $k\mathcal{F}$ prox to x, it is sufficient that

(4) $ey = x$ for some $e \in \mathrm{Id}(H(\mathcal{F}))$.

If $H(\mathcal{F})$ is a subsemigroup, e.g., if $\mathcal{F}$ is semiadditive, so $\mathcal{F} = \tau_0\mathcal{F}$, then condition (4) is necessary as well.

PROOF. (a) $\mathcal{N}^{\varphi\times\varphi}((y,x),u[(x,x)])$ is clearly proper iff the orbit closure of (y,x) contains (x,x). Semiadditivity follows as in Proposition 7.7 by using (7.56). It follows that $\mathcal{N}^{\varphi\times\varphi}((y,x),u[(x,x)])$ is contained in $\tau_0\mathcal{F}$ if it is contained in $\mathcal{F}$.

(b) This is obvious from Proposition 3.6c.

(c) Because Φ_x and Φ_y are continuous, the equations $px = x$ and $py = px = x$ determine closed subsets of $\beta_u T$. The hull definitions follow from (7.4) and (5.52).

Iso_x contains the identity $j_u(0)$, so it is nonempty; thus it is a co-ideal. If $p \in \text{Iso}_x$ and $q \in \text{Prox}_{(y,x)}$, then $pq \in \text{Prox}_{(y,x)}$, so $\text{Prox}_{(y,x)}$ is an Iso_x ideal.

(d) (1) $\Leftrightarrow$ (2) by (a).

(1) $\Leftrightarrow$ (3). Equation (7.2) implies $y\ k\mathcal{F}$ prox to x iff for some $p \in H(\mathcal{F})$:

$$p(y,x) = (py, px) = (x,x)$$

Similarly, Proposition 7.1 implies that $y\ \mathcal{F}$ prox to x iff $py = px = x$ for all $p \in H(\mathcal{F})$.

(4) $\Rightarrow$ (3). If e is idempotent and $ey = x$, then $ex = e^2y = ey = x$, as well. On the other hand if $H(\mathcal{F})$ is a closed semigroup, then (3) implies:

$$H(\mathcal{F}) \cap \text{Prox}_{(y,x)}$$

is a closed subsemigroup which therefore contains some idempotent. Hence (3) $\Rightarrow$ (4) in the subsemigroup case.

Applied to the filter $k\tau_0 k\tau_0\mathcal{F}$, we have $y\ k\tau_0 k\tau_0\mathcal{F}$ prox to x iff $H(k\tau_0 k\tau_0\mathcal{F}) \subset \text{Prox}_{(y,x)}$. Since the latter is closed, (7.51) implies this condition is equivalent to $\text{Id}(H(\tau_0\mathcal{F})) \subset \text{Prox}_{(y,x)}$. Also $\omega_{\tau_0 k\tau_0\mathcal{F}}\varphi(x)$ is the closure of $\{z : z = ex$ for some $e \in \text{Id}(H(\tau_0\mathcal{F}))\}$. Since $\tau_0\mathcal{F}$ is a semiadditive filter, (7.59) follows from (4) $\Leftrightarrow$ (1) applied to $\tau_0\mathcal{F}$. ∎

REMARK. The results in (d) are a bit simpler to state if $\mathcal{F}$ is a semiadditive filter. $H(\mathcal{F})$ is then a subsemigroup by Proposition 7.9 and $\tau_0\mathcal{F} = \mathcal{F}$ implies $\text{Id}(H(\tau_0\mathcal{F})) = \text{Id}(H(\mathcal{F}))$ and $\tau_0 k\tau_0\mathcal{F} = \tau_0 k\mathcal{F}$. ∎

By applying these results to the case $\mathcal{F} = k\mathcal{B}_T$ and $y = x$, we obtain a description of recurrent ($= \mathcal{B}_T$ recurrent) points:

$$|\omega\varphi| = |\omega_{\tau_0\mathcal{B}_T}\varphi| \tag{7.60}$$

The closure of $|\omega\varphi|$ in X is called the Birkhoff center or the *center* of X (with respect to φ). When X is compact and T is separable, Proposition 5.18 says that φ is central iff the set $|\omega\varphi|$ is dense in X; i.e., the center is the entire space.

PROPOSITION 7.13. *Let T be a uniform monoid and $\mu_u : T \times \beta_u T \to \beta_u T$ the extension of the translation action.*

$$\begin{aligned}
|\omega\mu_u| &= \{p \in \beta_u T : p \in \omega\mu_u(p)\} \\
&= \{p \in \beta_u^* T : qp = p \textit{ for some } q \in \beta_u^* T\} \\
&= \{p \in \beta_u^* T : ep = p \textit{ for some } e \in \text{Id}(\beta_u^* T)\} \\
&= \{p \in \beta_u^* T : p = eq \textit{ for some } e \in \text{Id}(\beta_u^* T) \textit{ and } q \in \beta_u T\} \qquad (7.61)
\end{aligned}$$

$$\overline{|\omega\mu_u|} = \omega_{\tau_0\mathcal{B}_T}\mu_u(\beta_u T) \qquad (\textit{closure in } \beta_u T) \tag{7.62}$$

Assume $\varphi : T \times X \to X$ is a uniform action with X compact:

$$|\omega\varphi| = \{x : qx = x \text{ for some } q \in \beta_u^* T\}$$
$$= \{x : ex = x \text{ for some } e \in \mathrm{Id}(\beta_u^* T)\} \tag{7.63}$$

If $\mathcal{F}$ is the open translation invariant filter $K(|\omega\mu_u|) = u \cap \{\mathcal{F}_p : p \in |\omega\mu_u|\}$, then:

$$\omega_{k\mathcal{F}}\varphi(x) = \overline{\omega\varphi(x) \cap |\omega\varphi|} \tag{7.64}$$

is the center of the restriction of φ to $\omega\varphi(x)$.

PROOF. The equations of (7.63) follow from Corollary 7.5d with $x = y$ and $\mathcal{F} = k\mathcal{B}_T$. Applied to μ_u the first three equations of (7.61) follow, because any recurrent point of $\beta_u T$ lies in $\beta_u^* T$ by Proposition 7.2. In particular, note that $p = eq$ and $e^2 = e$ implies $ep = e^2 q = eq = p$.

For each $q \in \beta_u T$, $\omega_{\tau_0 \mathcal{B}_T}\mu_u(q)$ is the closure of $\{eq : e \in \mathrm{Id}(\beta_u^* T)\}$ by (7.2) and (7.53). The equality of closed sets in (7.62) follows from (7.61).

The set $|\omega\mu_u|$ in $\beta_u^* T$ is invariant, so the open filter $\mathcal{F}$ is translation invariant by Corollary 7.2. By (7.2):

$$\omega_{k\mathcal{F}}\varphi(x) = \overline{\{px : p \in |\omega\mu_u|\}} = \Phi_x(\overline{|\omega\mu_u|})$$

If $y = px$ for some $p \in |\omega\mu|$, then $p \in \beta_u^* T$ implies $y \in \omega\varphi(x)$ and by (7.60) $ep = p$ for some $e \in \mathrm{Id}(\beta_u^* T)$.. Then

$$ey = epx = px = y$$

and $y \in |\omega\varphi|$ by (7.63). Conversely if $y \in \omega\varphi(x) \cap |\omega\varphi|$, then $y = p_1 x$ and $y = ey$ for some $p_1 \in \beta_u^* T$ and some $e \in \mathrm{Id}(\beta_u^* T)$. Let $p = ep_1 \in \beta_u^* T$:

$$px = ep_1 x = ey = y \qquad ep = e^2 p_1 = ep_1 = p$$

Then $y = px$ with $p \in |\omega\mu|$.

It is clear that $y \in \omega\varphi(x)$ is recurrent for φ iff it is recurrent for the restriction of φ to $\omega\varphi(x)$. Thus the closure of $\omega\varphi(x) \cap |\omega\varphi|$ is the center of $\omega\varphi(x)$. ■

REMARK. Using the notation of (6.35), (7.61) says that the set of recurrent points $|\omega\mu_u|$ is $|\beta_u^* T|$. ■

For a uniform action $\varphi : T \times X \to X$, a point $x \in X$ is called a minimal point if $x \in M$ where M is minimal with respect to inclusion among nonempty, closed + invariant subsets of X. For the action μ_u on $\beta_u T$, we saw that nonempty closed + invariant subsets are closed ideals and minimal subsets are minimal ideals. A point $p \in \beta_u T$ is minimal in the semigroup sense, i.e., $p \succ q \Rightarrow p \sim q$ iff the ideal $\beta_u T p$ contains no proper subideals (Note: Since $\beta_u T$ has an identity, $p \in \beta_u T p$ for all p, i.e., $\beta_u T = |\beta_u T|$). Thus $p \in \mathrm{Min}(\beta_u T)$ iff p is a minimal point for the action μ_u iff $\beta_u T p$ is a minimal ideal. We now consider minimal idempotents.

PROPOSITION 7.14. *Let $\varphi : T \times X \to X$ be a uniform action with X compact and let $\Phi : \beta_u T \times X \to X$ be the associated Ellis action. Let $x, y \in X$.*

a. Assume φ is a minimal action and J is a closed ideal in $\beta_u T$. J meets Iso_x. In fact, there exist minimal idempotents e of $\beta_u T$ such that $ex = x$ and $e \in J$. The following conditions on J are equivalent:

(1) J is a minimal ideal.

(2) The restriction of $\Phi_x : J \to X$ is a minimal action map from μ_u on J to φ.

(3) There exists $e \in \mathrm{Id}(\mathrm{Min}(\beta_u T))$ such that $J = \beta_u T e$.

(4) For all $p \in J$, $J = \beta_u T p$.

(5) For all $p \in J$:

$$J = H(\tilde{\gamma}\mathcal{F}_p) \tag{7.65}$$

In particular, p is a minimal element of $\beta_u T$ iff $u\tilde{\gamma}\mathcal{F}_p$ is a maximal open translation invariant filter for T.

b. Let $\mathcal{F}$ be a translation invariant filter. The following conditions on x and y are equivalent:

(1) $y\; k\mathcal{F}$ prox to x and x is a minimal point in X.

(2) There exists $e \in H(\mathcal{F}) \cap \mathrm{Id}(\mathrm{Min}(\beta_u T))$ such that $ey = x$.

(3) x is a minimal point and there exists $p \in H(\mathcal{F})$ such that $py = px$.

(4) x is a minimal point and $(y,x)\; k\mathcal{F}$ adheres to $1_{\omega\varphi(x)} = \{(z,z) : z \in \omega\varphi(x)\} \subset X \times X$.

c. Define the open filterdual $C = \cup\{\mathcal{F}_e : e \in \mathrm{Id}(\mathrm{Min}(\beta_u T))\}$

$$\gamma C = uk\tau k\tau \mathcal{B}_T$$

$$u\tilde{\gamma}kC = u\tau k\tau \mathcal{B}_T \tag{7.66}$$

$$\omega_C\varphi(x) = \overline{\{z : x\; \mathcal{B}_T\, \mathrm{Prox}\, z \text{ and } z \text{ is minimal}\}} \tag{7.67}$$

PROOF. (a) Since Φ_x is a continuous action map from μ_u to φ, $\Phi_x(J)$ is a closed + invariant subset of X, so by minimality, $\Phi_x(J) = X$.

(1) $\Rightarrow$ (2) is obvious and (2) $\Rightarrow$ (1) follows from Proposition 5.16.

(1) $\Leftrightarrow$ (3) by Theorem 6.1b.

(1) $\Leftrightarrow$ (4). This is obvious.

(4) $\Leftrightarrow$ (5). By (7.10).

In particular, $u\tilde{\gamma}\mathcal{F}_p$ is a maximal open translation invariant filter iff:

$$H(u\tilde{\gamma}\mathcal{F}_p) = H(\tilde{\gamma}\mathcal{F}_p) = \beta_u T p$$

is a minimal ideal. In general any ideal contains a minimal ideal, so it meets $\text{Iso}_x \cap \text{Id}(\text{Min}(\beta_u T))$ by (3).

(b) (1) $\Rightarrow$ (4). This is obvious.

(4) $\Rightarrow$ (3). By (4) and Proposition 3.11h, (y,x) $k\mathcal{F}$ adheres to some point (z,z) in $1_{\omega\varphi}(x)$. Then for some $p \in H(\mathcal{F})$, $py = px = z$.

(3) $\Rightarrow$ (2). By applying (a) to the restriction of φ to the minimal subset $\omega\varphi(x)$, we find a minimal idempotent e in the ideal $\beta_u T p$ such that $ex = x$. Since $e = qp$ for some $q \in \beta_u T$, $ey = qpy = qpx = ex = x$

(2) $\Rightarrow$ (1). It is clear that y $k\mathcal{F}$ prox to x. Because e is a minimal idempotent, $\beta_u Te$ is a minimal ideal, so its image under Φ_y is a minimal subset of X containing x.

(c) We use a succession of now familiar properties:

$$\begin{aligned} k\gamma C &= \tilde{\gamma} \cap \{k\mathcal{F}_e\} = \cap\{\tilde{\gamma}k\mathcal{F}_e\} = \cap\{\tilde{\gamma}\tilde{u}\mathcal{F}_e\} = \cap\{\tilde{u}\tilde{\gamma}\mathcal{F}_e\} \\ &= \tilde{u} \cap \{\tilde{\gamma}\mathcal{F}_e\} = \tilde{u}u \cap \{\tilde{\gamma}\mathcal{F}_e\} = \tilde{u}u \cap \{u\tilde{\gamma}\mathcal{F}_e\} \end{aligned}$$

As e varies over $\text{Id}(\text{Min}(\beta_u T))$, $u\tilde{\gamma}\mathcal{F}_e$ varies over all maximal open translation invariant filters. By (7.17):

$$\tilde{\gamma}kC = k\gamma C = \tilde{u}u\tau k\tau\mathcal{B}_T = \tilde{u}\tau k\tau\mathcal{B}_T$$

Equations (7.66) follow. Equation (7.67) follows from the definition of C, (7.2) and (b). ■

REMARK. For x minimal and y such that y $\mathcal{B}_T$ prox x, the set $\mathcal{N}^{\varphi\times\varphi}((y,x), U \times U)$ where U is a neighborhood of x, is what Furstenberg calls a *central set.* So C is the family generated by central sets. ■

The odd name $\mathcal{F}$ prox is intended to remind the reader of the ordinary proximal relation. For a family $\mathcal{F}$ we say that x and y in X are $\mathcal{F}$ *proximal* if (x,y) $\mathcal{F}$ adheres to the diagonal 1_X with respect to the product action $\varphi \times \varphi$. In the compact case, with $\mathcal{F}$ a filter, x and y are $k\mathcal{F}$ proximal iff $px = py$ for some $p \in H(\mathcal{F})$. So by Proposition 7.14b this is equivalent to y $k\mathcal{F}$ prox to x in the compact minimal case. Ordinary proximality occurs when $k\mathcal{F} = \mathcal{B}_T$, i.e., $\omega(\varphi \times \varphi)(x,y)$ meets 1_X. In the general compact case with $\mathcal{F}$ a filter, x and y are $\mathcal{F}$ proximal iff $px = py$ for all $p \in H(\mathcal{F})$. When $\mathcal{F} = k\mathcal{B}_T$, x and y are then *asymptotic*, i.e.:

$$\omega(\varphi \times \varphi)(x,y) \subset 1_X$$

We have seen that both the center and the mincenter of the restriction of φ to $\omega\varphi(x)$ can be described as $\omega_{k\mathcal{F}}\varphi(x)$ for an appropriately chosen translation invariant filter $\mathcal{F}$. The set of nonwandering points of $\omega\varphi(x)$ also takes the form $\omega_{k\mathcal{F}}\varphi(x)$ for a translation invariant family $\mathcal{F}$.

Given proper families $\mathcal{F}$ and $\mathcal{F}_1$, define

$$q_{\mathcal{F}_1}\mathcal{F} = \{F : \{t : F \cap g^{-t}(F) \in \mathcal{F}\} \in \mathcal{F}_1\} \tag{7.68}$$

PROPOSITION 7.15. *Let $\mathcal{F}$, $\mathcal{F}_1$, etc., be proper families for a uniform monoid T.*

a. $q_{\mathcal{F}_1}\mathcal{F}$ is a proper family with:

$$\tau\mathcal{F} \subset q_{\mathcal{F}_1}\mathcal{F} \subset \mathcal{F} \tag{7.69}$$

If $\mathcal{F}$ is thick, then $q_{\mathcal{F}_1}\mathcal{F} = \mathcal{F}$ for all $\mathcal{F}_1$. The operator preserves inclusions in each variable, i.e., if $\mathcal{F}_1 \subset \tilde{\mathcal{F}}_1$ and $\mathcal{F} \subset \tilde{\mathcal{F}}$, then $q_{\mathcal{F}_1}\mathcal{F} \subset q_{\tilde{\mathcal{F}}_1}\tilde{\mathcal{F}}$. In general for any collection $\{\mathcal{F}_\alpha\}$ of proper families:

$$q_{\cap_\alpha \mathcal{F}_\alpha}\mathcal{F} = \cap q_{\mathcal{F}_\alpha}\mathcal{F} \tag{7.70}$$

b. If $\mathcal{F}$ is translation invariant, then $q_{\mathcal{F}_1}\mathcal{F}$ is translation invariant.

c. If $\tau_0\mathcal{F} \subset \mathcal{F}_1$, then $q_{\mathcal{F}_1}(\tau_0\mathcal{F}) = \tau_0\mathcal{F}$. In general:

$$\tau\mathcal{F} \subset \tau_0\mathcal{F} = q_{\tau_0\mathcal{F}}(\tau_0\mathcal{F}) \subset q_{\tau_0\mathcal{F}}\mathcal{F} \subset q_{\mathcal{F}}\mathcal{F} \subset \mathcal{F} \tag{7.71}$$

If $\mathcal{F}$ is translation invariant, then $\gamma\tau_0\mathcal{F} \subset q_{\tau_0\mathcal{F}}\mathcal{F}$.

d. If $\mathcal{F}_1$ and $\mathcal{F}$ are filters, then $q_{\mathcal{F}_1}\mathcal{F}$ is a filter.

PROOF. (a) If $F \subset F_1$, then:

$$\{t : F \cap g^{-t}(F) \in \mathcal{F}\} \subset \{t : F_1 \cap g^{-t}(F_1) \in \mathcal{F}\}$$

$q_{\mathcal{F}_1}\mathcal{F}$ is hereditary because $\mathcal{F}_1$ is hereditary. If $\{t : F \cap g^{-t}(F) \in \mathcal{F}\} \neq \emptyset$, then $F \supset F \cap g^{-t}(F)$ is in $\mathcal{F}$. Therefore $q_{\mathcal{F}_1}\mathcal{F} \subset \mathcal{F}$. If $F \in \tau\mathcal{F}$, then $T = \{t : F \cap g^{-t}(F) \in \mathcal{F}\}$; since $T \in \mathcal{F}_1$, $F \in q_{\mathcal{F}_1}\mathcal{F}$. Monotonicity and the intersection condition (7.70) are easy exercises.

(b) For all $s \in T$:

$$g^{-s}(F \cap g^{-t}(F)) = g^{-s}(F) \cap g^{-t}(g^{-s}(F))$$

If $\mathcal{F}$ is translation invariant, it follows that the set:

$$\{t : g^{-s}(F) \cap g^{-t}(g^{-s}(F)) \in \mathcal{F}\}$$

is the same for all s. Hence $F \in q_{\mathcal{F}_1}\mathcal{F}$ iff $g^{-s}(F) \in q_{\mathcal{F}_1}\mathcal{F}$.

(c) If $F \in \tau_0\mathcal{F}$, then F contains some $\tau_0\mathcal{F}$ semiadditive set F_1. Therefore:

$$F_1 \subset \{t : F_1 \cap g^{-t}(F_1) \in \tau_0\mathcal{F}\} \subset \{t : F \cap g^{-t}(F) \in \tau_0\mathcal{F}\}$$

Thus F_1 and F, too, lie in $q_{\mathcal{F}_1}(\tau_0\mathcal{F})$ when $\tau_0\mathcal{F} \subset \mathcal{F}_1$. Hence $\tau_0\mathcal{F} \subset q_{\mathcal{F}_1}(\tau_0\mathcal{F})$. The reverse inclusion follows from (7.69). In particular we obtain the equation in (7.71); the inclusions follow from monotonicity. If $\mathcal{F}$ is translation invariant, then $q_{\tau_0\mathcal{F}}\mathcal{F}$ is also by (b). This family contains $\tau_0\mathcal{F}$, so it contains $\gamma\tau_0\mathcal{F}$.

(d) Notice that for F_1, F_2 subsets of T:

$$\{t : F_1 \cap g^{-t}(F_1) \in \mathcal{F}\} \cap \{t : F_2 \cap g^{-t}(F_2) \in \mathcal{F}\} \subset$$
$$\{t : (F_1 \cap F_2) \cap g^{-t}(F_1 \cap F_2) \in \mathcal{F}\}$$

provided $\mathcal{F}$ is a filter. If $F_1, F_2 \in q_{\mathcal{F}_1}\mathcal{F}$ and $\mathcal{F}_1$ as well as $\mathcal{F}$ is a filter, then $\mathcal{F}_1 \cap F_2 \in q_{\mathcal{F}_1}\mathcal{F}$. ■

PROPOSITION 7.16. *Let $\varphi : T \times X \to X$ be a uniform action with X compact. Assume that $\mathcal{F}_1$ and $\mathcal{F}$ are proper families for T with $\mathcal{F}$ a translation invariant filter. Let $\mathcal{F}_2 = q_{\mathcal{F}_1}(k\mathcal{F})$. For $x, y \in X$, $y \in \omega_{\mathcal{F}_2}\varphi(x)$ iff $y \in \omega_{k\mathcal{F}}\varphi(x)$ and $y \in \Omega_{\mathcal{F}_1}(\varphi|\omega_{k\mathcal{F}}\varphi(x))(y)$, where $\varphi|\omega_{k\mathcal{F}}\varphi(x)$ is the system obtained by restricting to the invariant set $\omega_{k\mathcal{F}}\varphi(x)$.*

PROOF. $\mathcal{F}_2 = q_{\mathcal{F}_1}(k\mathcal{F}) \subset k\mathcal{F}$, so $y \in \omega_{\mathcal{F}_2}\varphi(x)$ implies $y \in \omega_{k\mathcal{F}}\varphi(x)$. Since $k\mathcal{F}$ is a filterdual and X is compact, $\omega_{k\mathcal{F}}\varphi(x)$ is a nonempty invariant subset of X. We need only consider $y \in \omega_{k\mathcal{F}}\varphi(x)$.

Assume $y \notin \Omega_{\mathcal{F}_1}(\varphi|\omega_{k\mathcal{F}}\varphi(x))(y)$. There exists U_0 a closed neighborhood of y in $\omega_{k\mathcal{F}}\varphi(x)$ such that $N(U_0, U_0) \notin \mathcal{F}_1$. There exists U a closed neighborhood of y in X such that $U \cap \omega_{k\mathcal{F}}\varphi(x) \subset U_0$. Let $F = N(x, U)$. We prove that $F \notin \mathcal{F}_2$. It then follows that $y \notin \omega_{\mathcal{F}_2}\varphi(x)$.

Fix $t \in T$ such that:

$$F_t = F \cap g^{-t}(F) \in k\mathcal{F}$$

then $\mathcal{F} \cdot [F_t]$ is a proper filter, which we denote $\mathcal{F}_t$.

Because $\mathcal{F} \subset \mathcal{F}_t$ and X is compact, $\omega_{k\mathcal{F}_t}\varphi(x)$ is a nonempty subset of $\omega_{k\mathcal{F}}\varphi(x)$. Because $F_t \in \mathcal{F}_t$, $\overline{f^{F_t}(x)} \supset \omega_{k\mathcal{F}_t}\varphi(x)$. But $F_t \subset F$ implies

$$\overline{f^{F_t}(x)} \subset \overline{f^{F}(x)} \subset U$$

and $F_t \subset g^{-t}(F)$ implies

$$\overline{f^{F_t}(x)} \subset \overline{f^{-t}(U)} = f^{-t}(U)$$

because:

$$g^{-t}(F) = g^{-t}N(x, U) = N(x, f^{-t}(U))$$

Hence:

$$\omega_{k\mathcal{F}_t}\varphi(x) \subset U \cap f^{-t}(U) \cap \omega_{k\mathcal{F}}\varphi(x)$$

Thus $z \in \omega_{k\mathcal{F}_t}\varphi(x)$ implies

$$z \in U \cap \omega_{k\mathcal{F}}\varphi(x) \subset U_0 \qquad f^t(z) \in U \cap \omega_{k\mathcal{F}}\varphi(x) \subset U_0$$

by + invariance of $\omega_{k\mathcal{F}}\varphi(x)$.

We have shown that:

$$\{t : F \cap g^{-t}(F) \in k\mathcal{F}\} \subset \{t : U_0 \cap f^{-t}(U_0) \neq \emptyset\} = N(U_0, U_0)$$

The latter set is not in $\mathcal{F}_1$, so neither is the former. Thus $F \notin q_{\mathcal{F}_1}k\mathcal{F} = \mathcal{F}_2$.

Conversely suppose $y \in \Omega_{\mathcal{F}_1}(\varphi|\omega_{k\mathcal{F}}\varphi(x))(y)$. If U is an open neighborhood of y in X, let $U_0 = U \cap \omega_{k\mathcal{F}}\varphi(x)$, a neighborhood of y in $\omega_{k\mathcal{F}}\varphi(x)$. By assumption, $N(U_0, U_0) \in \mathcal{F}_1$.

For $t \in N(U_0, U_0)$, let $U_t = U \cap f^{-t}(U)$. $U_t \cap \omega_{k\mathcal{F}}\varphi(x) \supset U_0 \cap f^{-t}(U_0)$ is nonempty, so for $z \in U_0 \cap f^{-t}(U_0)$, U_t is open in X containing $z \in \omega_{k\mathcal{F}}\varphi(x)$. Hence, $N(x, U_t) \in k\mathcal{F}$. But:

$$N(x, U_t) = N(x, U) \cap N(x, f^{-t}(U)) =$$
$$N(x, U) \cap g^{-t}N(x, U)$$

Thus:

$$N(U_0, U_0) \subset \{t : N(x, U) \cap g^{-t}N(x, U) \in k\mathcal{F}\}$$

Since $N(U_0, U_0) \in \mathcal{F}_1$, it follows that $N(x, U) \in q_{\mathcal{F}_1}(k\mathcal{F}) \in \mathcal{F}_2$. Hence $y \in \omega_{\mathcal{F}_2}\varphi(x)$. ■

COROLLARY 7.6. *For a uniform monoid T, the family:*

$$q_{\mathcal{B}_T}\mathcal{B}_T = \{F : \{t : F \cap g^{-t}(F) \in \mathcal{B}_T\} \in \mathcal{B}_T\}$$

is a proper translation invariant family of subsets of T. If $\varphi : T \times X \to X$ is a uniform action with X compact and $x \in X$, then $\omega_{q_{\mathcal{B}_T}\mathcal{B}_T}\varphi(x)$ is the set of nonwandering points for the restriction of φ to $\omega\varphi(x)$.

We return to the idea of proximality and its opposite. Let $\varphi : T \times X \to X$ be a uniform action and let $\mathcal{F}$ be a filter for T such that $H(\mathcal{F})$ is a subsemigroup, e.g., $\mathcal{F}$ semiadditive suffices. Under these circumstances we call a point x of X $\mathcal{F}$ *distal* if x $k\mathcal{F}$ prox to z implies $x = z$. Call x *distal* if it is $k\mathcal{B}_T$ distal.

PROPOSITION 7.17. *Let $\varphi : T \times X \to X$ be a uniform action with X compact and let $\Phi : \beta_u T \times X \to X$ be its Ellis action extension, so that $\Phi^{\#} : \beta_u T \to S_\varphi$ is the continuous semigroup homomorphism onto the enveloping semigroup. Let $\mathcal{F}$ be a filter for T such that $H(\mathcal{F})$ is a subsemigroup.*

a. The following conditions on $x \in X$ are equivalent:

(1) x is $\mathcal{F}$ distal.

(2) For $p \in H(\mathcal{F})$ and $z \in X$, $px = pz = z$ implies $x = z$.

(3) $\mathrm{Id}(H(\mathcal{F})) \subset \mathrm{Iso}_x$, i.e., for every idempotent e in $H(\mathcal{F})$, $ex = x$.

(4) For every idempotent u in $\Phi^{\#}(H(\mathcal{F}))$ $ux = x$.

(5) For $p \in H(\mathcal{F})$ and $z \in \omega_{k\mathcal{F}}\varphi(x)$, $px = pz$ implies $x = z$.

When x is $\mathcal{F}$ distal, it is $k\mathcal{F}$ recurrent, i.e., $x \in \omega_{k\mathcal{F}}\varphi(x)$ and:

$$y \in \omega_{k\mathcal{F}}\varphi(x) \Rightarrow \omega_{k\mathcal{F}}\varphi(y) = \omega_{k\mathcal{F}}\varphi(x) \tag{7.72}$$

b. If $\mathcal{F}_1$ is a filter such that $H(\mathcal{F}_1)$ is a subsemigroup and $\mathcal{F}_1 \subset \mathcal{F}$, then x $\mathcal{F}_1$ distal implies x $\mathcal{F}$ distal. In particular if x is $\tau_0\mathcal{F}$ distal, then x is $\mathcal{F}$ distal. The point x is $\tau_0\mathcal{F}$ distal iff x is a $k\tau_0 k\tau_0\mathcal{F}$ recurrent point. Thus x is distal; i.e., $k\mathcal{B}_T$ distal iff x is a $k\tau_0\mathcal{B}_T$ recurrent point.

c. A point x is distal iff $ex = x$ for all idempotents e in $\beta_u^* T$. This implies x is $\mathcal{F}$ distal if $\mathcal{F}$ is full and a fortiori if $\mathcal{F}$ is translation invariant. The following conditions on $x \in X$ are equivalent:

(1) There exists a translation invariant filter $\mathcal{F}_1$ such that x is $\mathcal{F}_1$ distal.

(2) There exists an ideal J in $\beta_u T$ such that $\mathrm{Id}(J) \subset \mathrm{Iso}_x$.

(3) $\mathrm{Id}(\mathrm{Min}(\beta_u T)) \subset \mathrm{Iso}_x$, i.e., for every minimal idempotent e in $\beta_u T$ $ex = x$.

(4) For every minimal idempotent e in $\beta_u T$, x is $u\tilde{\gamma}\mathcal{F}_e$ distal.

When these conditions hold x is a minimal point; i.e., $\omega\varphi(x)$ is a minimal subset of X containing x.

d. If $\varphi_1 : T \times X_1 \to X_1$ is a uniform action with X_1 compact and $h : \varphi \to \varphi_1$ is a continuous action map, then x $\mathcal{F}$ distal implies $h(x)$ is $\mathcal{F}$ distal.

PROOF. (a) (1) $\Leftrightarrow$ (2) $\Leftrightarrow$ (3) is immediate from Proposition 7.12d. (3) $\Leftrightarrow$ (4) because (6.36) implies

$$\Phi^{\#}(\mathrm{Id}(H(\mathcal{F}))) = \mathrm{Id}(\Phi^{\#}H(\mathcal{F})))$$

(5) $\Rightarrow$ (2). If $px = pz = z$, then $px = z$ implies $z \in \omega_{k\mathcal{F}}\varphi(x)$, so $px = pz$ implies $x = z$ by (5).

Now assume (3). Since $H(\mathcal{F})$ contains some idempotent e, (3) implies that $x = ex \in \omega_{k\mathcal{F}}\varphi(x)$. Let $y \in \omega_{k\mathcal{F}}\varphi(x)$ so that $y = p_0x$ for some $p_0 \in H(\mathcal{F})$. $H(\mathcal{F})p_0$ is a closed subsemigroup of $H(\mathcal{F})$. Then there exists an idempotent e and $q_0 \in H(\mathcal{F})$ such that $e = q_0p_0$. By (3) $x = ex$. But $ex = q_0p_0x = q_0y \in \omega_{k\mathcal{F}}\varphi(y)$; thus $x \in \omega_{k\mathcal{F}}\varphi(y)$. (7.72) then follows from (7.49). If $px = pz$ with $p \in H(\mathcal{F})$ and $z \in \omega_{k\mathcal{F}}\varphi(x)$, then by applying (7.72) to $y = pz$, there exists $q \in H(\mathcal{F})$ such that $qpz = z$. Hence $qpx = qpz = z$. Since (3) $\Rightarrow$ (2), we obtain $x = z$; this proves (3) $\Rightarrow$ (5).

(b) Monotonicity results are obvious. Notice that for any filter $\mathcal{F}$, $\tau_0\mathcal{F}$ is a semiadditive filter, so $H(\tau_0\mathcal{F})$ is a subsemigroup. Applying Proposition 7.1 to the filter $k\tau_0k\tau_0\mathcal{F}$:

$$x \in \omega_{k\tau_0k\tau_0\mathcal{F}}\varphi(x) \text{ iff } H(k\tau_0k\tau_0\mathcal{F}) \subset \mathrm{Iso}_x$$

Now apply (7.51).

(c) By (a) x is $k\mathcal{B}_T$ distal iff $ex = x$ for all idempotents e in $H(k\mathcal{B}_T) = \beta_u^*T$. A filter $\mathcal{F}$ is full iff $H(\mathcal{F}) \subset \beta_u^*T$, so x is then $\mathcal{F}$ distal by (b).

(4) $\Rightarrow$ (1). This is obvious.

(1) $\Rightarrow$ (2). Use $J = H(\mathcal{F}_1)$ and apply (1) $\Rightarrow$ (3) of (a).

(2) $\Rightarrow$ (3). We can assume J is a minimal ideal in β_uT. If e is a minimal idempotent then β_uTe is a minimal ideal as well. By Corollary 6.1 of Theorem 6.1 there exists an idempotent e_1 in J such that $ee_1 = e_1$. By (2) $e_1x = x$, so $ex = ee_1x = e_1x = x$, as well. It follows as well that x is a minimal point.

(3) $\Rightarrow$ (4). By (7.10) $H(u\tilde{\gamma}\mathcal{F}_e) = H(\tilde{\gamma}\mathcal{F}_e)$ is the minimal ideal β_uTe. Then every idempotent e_1 in $H(u\tilde{\gamma}\mathcal{F}_e)$ is a minimal idempotent. Thus, (3) implies $e_1x = x$, and $u\tilde{\gamma}\mathcal{F}_e$ satisfies condition (3) of (a).

(d) For $e \in \mathrm{Id}(H(\mathcal{F}))$, $eh(x) = h(ex) = h(x)$. ∎

REMARK. If e is any idempotent in β_uT such that $ex = x$, then $H(\mathcal{F}_e) = \{e\}$ implies x is $\mathcal{F}_e$ distal. With $e = j_u(0)$, we see that any point is $\mathcal{F}_e$ distal for some idempotent e. If x is a recurrent point then x is $\mathcal{F}_e$ distal for some idempotent $e \in \beta_u^*T$ (for which $\mathcal{F}_e$ is full). ∎

The entire system φ is called $\mathcal{F}$ distal when every point is $\mathcal{F}$ distal. In particular, φ is distal when every point is $k\mathcal{B}_T$ distal. Ellis showed, among other results that we now present, that φ is distal iff the enveloping semigroup S_φ is a group of (not necessarily continuous) bijections of X.

THEOREM 7.1. *Let $\varphi : T \times X \to X$ be a uniform action with X compact and let $\Phi : \beta_uT \times X \to X$ be its Ellis action extension, so that $\Phi^\# : \beta_uT \to S_\varphi$ is the continuous semigroup homomorphism onto the enveloping semigroup.*

a. Let $\mathcal{F}$ be a filter for T such that $H(\mathcal{F})$ is a subsemigroup of $\beta_u T$. The following conditions are equivalent.

(1) φ is $\mathcal{F}$ distal.

(2) For every $e \in \mathrm{Id}(H(\mathcal{F}))$ and $x \in X$, $ex = x$.

(3) There exists $e \in \mathrm{Min}(\mathrm{Id}(H(\mathcal{F})))$ such that $ex = x$ for all $x \in X$.

(4) Every idempotent in $\Phi^{\#}(H(\mathcal{F}))$ is an injective mapping in X^X.

(5) Every idempotent in $\Phi^{\#}(H(\mathcal{F}))$ is a surjective mapping in X^X.

(6) $\Phi^{\#}(H(\mathcal{F}))$ is a group of bijective mappings in X^X.

(7) For every nonempty set I, the product action φ_* on X^I is $\mathcal{F}$ distal.

(8) 1_X is an $\mathcal{F}$ distal point for the product action φ_* on X^X.

(9) For the product action $\varphi \times \varphi$ $\omega_{k\mathcal{F}}(\varphi \times \varphi)$ is a symmetric relation on $X \times X$.

(10) $\omega_{k\mathcal{F}}(\varphi \times \varphi)$ is an equivalence relation on $X \times X$.

b. The following are equivalent:

(1) φ is distal, i.e., $k\mathcal{B}_T$ distal.

(2) There exists a translation invariant filter $\mathcal{F}$ such that φ is $\mathcal{F}$ distal.

(3) The enveloping semigroup S_φ is a group.

(4) There exists $e \in \mathrm{Min}(\mathrm{Id}(\beta_u T))$ such that $ex = x$ for all $x \in X$.

(5) 1_X is a $k\tau\mathcal{B}_T$ recurrent point for the action φ_* on X^X.

(6) The action φ_* on X^X restricts to a minimal action on S_φ.

c. The following conditions are equivalent; when they hold, φ is called weakly rigid*:*

(1) There exists a full filter $\mathcal{F}$ with $H(\mathcal{F})$ a subsemigroup such that φ is $\mathcal{F}$ distal.

(2) $1_X \in \Phi^{\#}(\beta_u^* T)$

(3) 1_X is a recurrent point for the product action φ_* on X^X.

(4) For every finite set I, every point of X^I is recurrent for the product action φ_*.

(5) The family $\cup_{x\in X}\mathcal{N}^{\varphi}(x,u[x])$ is contained in some full filter $\mathcal{F}$ for T.

PROOF. (a) (1) $\Leftrightarrow$ (2). By Proposition 7.17a.

(2) $\Rightarrow$ (3). By Theorem 6.1 minimal idempotents exist.

(3) $\Rightarrow$ (4) and (5). Since $\Phi^{\#}$ is a homomorphism and $\Phi^{\#}(e) = 1_X$, $\Phi^{\#}(H(\mathcal{F})) = \Phi^{\#}(H(\mathcal{F})e)$. If u is an idempotent in $\Phi^{\#}(H(\mathcal{F}))$ then by (6.36) there exists an idempotent e_1 in the closed subsemigroup $H(\mathcal{F})e$ such that $\Phi^{\#}(e_1) = u$. By minimality of e, $e_1 \sim e$, i.e., $ee_1 = e$. Therefore:

$$u = \Phi^{\#}(e_1) = \Phi^{\#}(e)\Phi^{\#}(e_1) = \Phi^{\#}(ee_1) = \Phi^{\#}(e) = 1_X$$

(4) $\Rightarrow$ (2) and (5) $\Rightarrow$ (2). If $u \in X^X$ is an idempotent, then $u(u(x)) = u(x)$. If u is injective then $u(x) = x$ for all x. If u is surjective, then every z in X equals $u(x)$ for some x, so $u(z) = z$ for all z. Thus u is the identity map. Applied to $u = \Phi^{\#}(e)$, this implies (2).

(2) $\Leftrightarrow$ (6). If $\Phi^{\#}(H(\mathcal{F}))$ is a group of bijections, then the identity element u is the identity map; this implies (2). On the other hand, (2) implies that 1_X is the only idempotent in $\Phi^{\#}(H(\mathcal{F}))$. Now let g be any element of the closed subsemigroup $\Phi^{\#}(H(\mathcal{F}))$ of X^X. The $\Phi^{\#}(H(\mathcal{F}))g$ is a closed subsemigroup and so it contains an idempotent. But the only idempotent is 1_X; thus there exists $\bar{g} \in \Phi^{\#}(H(\mathcal{F}))$ such that $\bar{g}g = 1_X$. Applied to $\bar{g}$, there exists $\bar{\bar{g}}$ such that $\bar{\bar{g}}\,\bar{g} = 1_X$. Then $\bar{\bar{g}} = \bar{\bar{g}}\,\bar{g}g = g$. Hence $\bar{g}$ and g are inverse elements of X^X, and $\Phi^{\#}(H(\mathcal{F}))$ is a group of bijections.

(2) $\Rightarrow$ (7). If $\alpha \in X^I$ and $e \in \mathrm{Id}(H(\mathcal{F}))$, then for all $i \in I$:

$$(e\alpha)(i) = e(\alpha(i)) = \alpha(i)$$

Then $e\alpha = \alpha$. Thus, φ_* satisfies (2) and hence φ_* is $\mathcal{F}$ distal because (2) $\Leftrightarrow$ (1).

(7) $\Rightarrow$ (8). This is obvious.

(8) $\Rightarrow$ (1). $\mathrm{Ev}_x : X^X \to X$ is a continuous action map from φ_* to φ. Then 1_X $\mathcal{F}$ distal for φ_* implies x is $\mathcal{F}$ distal for φ by Proposition 7.17d.

(7) $\Rightarrow$ (10). Since $\varphi \times \varphi$ is $\mathcal{F}$ distal, the relation $\omega_{k\mathcal{F}}(\varphi \times \varphi)$ is an equivalence relation by (7.72).

(10) $\Rightarrow$ (9). This is obvious.

(9) $\Rightarrow$ (1). If y $k\mathcal{F}$ prox to x, then $(x,x) \in \omega_{k\mathcal{F}}(\varphi \times \varphi)(y,x)$. By symmetry $(y,x) \in \omega_{k\mathcal{F}}(\varphi \times \varphi)(x,x)$. But 1_X is a closed + invariant subset of $X \times X$. So $(y,x) \in 1_X$ and $y = x$. Hence y is $\mathcal{F}$ distal.

(b) (1) $\Rightarrow$ (2). This is obvious.

(2) $\Rightarrow$ (3). By (a) $\Phi^{\#}(H(\mathcal{F}))$ is a group of bijections, and it is a closed subsemigroup of S_φ. If e is an idempotent of $H(\mathcal{F})$, then by translation invariance $j_u(t)e \in H(\mathcal{F})$ for all $t \in T$. But $j_u(t)e \in H(\mathcal{F})$ and

$$\Phi^{\#}(j_u(t)e) = \Phi^{\#}(j_u(t))\Phi^{\#}(e) = f^t 1_X = f^t \tag{7.73}$$

imply $\Phi^{\#}(H(\mathcal{F}))$ contains the closure in X^X of $\{f^t : t \in T\}$. Thus $\Phi^{\#}$ maps $H(\mathcal{F})$ onto S_φ.

(3) $\Rightarrow$ (1). S_φ is a group that contains 1_X, so the identity element is 1_X. For every idempotent e in $\beta_u T$, $\Phi^{\#}(e) = 1_X$, the only idempotent in the group, so φ is $k\mathcal{B}_T$ distal by (a).

(1) $\Leftrightarrow$ (4). This is obvious from (a).

(4) $\Leftrightarrow$ (5) $\Leftrightarrow$ (6). By Theorem 3.1, 1_X is $k\tau\mathcal{B}_T$ recurrent in X^X iff it is a minimal point iff its orbit closure S_φ is a minimal subset. By Proposition 7.14b this occurs iff 1_X is the image under $\Phi^{\#}$ of some minimal idempotent.

(c) (1) $\Rightarrow$ (2).

$$1_X \in \Phi^{\#}(H(\mathcal{F})) \subset \Phi^{\#}(\beta_u^* T)$$

because $\mathcal{F}$ is full.

(2) $\Rightarrow$ (3). $\Phi^{\#}(\beta_u^*(T)) = \omega\varphi_*(1_X)$ by (7.11).

(3) $\Rightarrow$ (4). We can assume $I = \{1, \dots, n\}$. For $(x_1, \dots, x_n) \in X^I$, $\mathrm{Ev}_{x_1} \times \dots \times \mathrm{Ev}_{x_n}$ is a continuous action map from X^X to X^I. Then 1_X recurrent implies $(x_1, \dots, x_n)$ is recurrent.

(4) $\Rightarrow$ (5). If $U_1, \dots, U_n$ open in X contain $x_1, \dots, x_n$ respectively, (4) implies that $N((x_1, \dots, x_n), U_1, \times \cdots \times U_n) = \cap_{i=1}^n N(x_i, U_i)$ meets every T_t.

(5) $\Rightarrow$ (1). Since each $\mathcal{N}(x, u[x])$ is semiadditive we can replace $\mathcal{F}$ by $\tau_0 \mathcal{F}$ if necessary and assume that the full filter $\mathcal{F}$ is semiadditive. Since $\mathcal{N}(x, u[x]) \subset \mathcal{F}$, $p \in H(\mathcal{F})$ implies $px = x$. Then $\Phi^{\#}(H(\mathcal{F})) = \{1_X\}$; thus φ is $\mathcal{F}$ distal. ■

Thus, φ is weakly rigid when 1_X is a recurrent point for φ_* on X^X; i.e., $1_X \in \omega\varphi_*(1_X)$ and then every point x of X is a recurrent point for φ. Recall that φ is central when every point x is nonwandering. In that case for every nonempty open subset U of X the hitting time set $N(U, U)$ is in $\mathcal{B}_T$, so the generated full family $\overline{\mathcal{F}}_\varphi$ is proper. Proposition 4.8c says that 1_X is nonwandering for φ_*; i.e., $1_X \in \Omega\varphi_*(1_X)$, iff all products of copies of φ are central or equivalently if $\overline{\mathcal{F}}_\varphi$ is contained in some full filter. By Proposition 4.2 $\overline{\mathcal{F}}_\varphi$ is a full filter when φ is transitive. Recall that φ is transitive when for every pair U_1, U_2 of nonempty open subsets of X $N(U_1, U_2)$ is in $\mathcal{B}_T$. In that case the generated full family $\overline{\mathcal{T}}_\varphi$ [cf. (4.3)] is proper. By Theorem 4.1, φ is weak mixing when $\overline{\mathcal{T}}_\varphi$ is a full filter or equivalently when $\overline{\mathcal{T}}_\varphi \subset \tau\mathcal{B}_T$.

We can generate $\mathcal{F}_\varphi$ and $\mathcal{T}_\varphi$ by letting V vary in $\mathcal{U}_X$, x, x_1 and x_2 vary in X, setting $U = V(x)$, $U_1 = V(x_1)$, and $U_2 = V(x_2)$. We make the definitions uniform by defining for $V \in \mathcal{U}_X$:

$$N^{\varphi}(V) = \{t : V^{-1} \circ f^t \circ V \supset 1_X\} = \bigcap_{x \in X} N^{\varphi}(V(x), V(x))$$

$$\begin{aligned} N^{\varphi}(V,V) &= \{t : V^{-1} \circ f^t \circ V = X \times X\} \\ &= \bigcap_{x_1, x_2 \in X} N^{\varphi}(V(x_1), V(x_2)) \end{aligned} \tag{7.74}$$

Let $\tilde{\mathcal{F}}_{\varphi}$ and $\tilde{\mathcal{T}}_{\varphi}$ denote the families generated by the $N^{\varphi}(V)$s and the $N^{\varphi}(V,V)$s and $k\mathcal{B}_T$:

$$\begin{aligned} \tilde{\mathcal{F}}_{\varphi} &= [\{N^{\varphi}(V) : V \in \mathcal{U}_X\}] \cdot k\mathcal{B}_T \\ \tilde{\mathcal{T}}_{\varphi} &= [\{N^{\varphi}(V,V) : V \in \mathcal{U}_X\}] \cdot k\mathcal{B}_T \end{aligned} \tag{7.75}$$

We call the uniform action φ *uniformly central* if $\tilde{\mathcal{F}}_{\varphi}$ is a proper family and *uniformly weak mixing* if $\tilde{\mathcal{T}}_{\varphi}$ is a proper family.

PROPOSITION 7.18. *Let $\varphi : T \times X \to X$ be a uniform action.*

a. $\overline{\mathcal{F}}_{\varphi} \subset \tilde{\mathcal{F}}_{\varphi}$. If $\tilde{\mathcal{F}}_{\varphi}$ is a proper family, then it is an open full filter, so $\tilde{\mathcal{F}}_{\varphi} \subset \mathcal{B}_T$. In that case, φ and all products of copies of φ are central.

b. $\overline{\mathcal{T}}_{\varphi} \subset \tilde{\mathcal{T}}_{\varphi}$. If $\tilde{\mathcal{T}}_{\varphi}$ is a proper family, then it is an open translation invariant filter, so $\tilde{\mathcal{T}}_{\varphi} \subset \tau\mathcal{B}_T$. In that case, φ is weak mixing.

PROOF. The first inclusions are clear in each case.

$$V_1 \subset V_2 \Rightarrow N(V_1) \subset N(V_2) \quad \text{and} \quad N(V_1, V_1) \subset N(V_2, V_2) \tag{7.76}$$

So $N(V_1 \cap V_2) \subset N(V_1) \cap N(V_2)$ and similarly for $N(V,V)$. Each family is a filter when it is proper.

To show $\tilde{\mathcal{F}}_{\varphi}$ and $\tilde{\mathcal{T}}_{\varphi}$ are open, we begin with $V \in \mathcal{U}_X$ choose $W \in \mathcal{U}_X$ such that $W = W^{-1}$, $W \circ W \subset V$, then we use uniformity of the action to choose $\tilde{W} \in \mathcal{U}_T$ so that $(t_1, t) \in \tilde{W}$ implies $(f^{t_1}(x), f^t(x)) \in W$ for all x, i.e., $f^t \circ f^{-t_1} \subset W$. Hence:

$$\begin{aligned} W^{-1} \circ f^t \circ W &\subset W \circ f^t \circ 1_X \circ W \subset W \circ f^t \circ f^{-t_1} \circ f^{t_1} \circ W \\ &\subset W \circ W \circ f^{t_1} \circ W \subset V^{-1} \circ f^{t_1} \circ V \end{aligned}$$

It follows that:

$$\tilde{W}^{-1}(N(W)) \subset N(V) \qquad \tilde{W}^{-1}(N(W,W)) \subset N(V,V)$$

The families $\tilde{\mathcal{F}}_{\varphi}$ and $\tilde{\mathcal{T}}_{\varphi}$ are open by Proposition 2.16a.

If $\tilde{\mathcal{T}}_{\varphi}$ is proper then φ is clearly transitive, so φ is a dense action by Proposition 4.2a. We use this to show $\tilde{\mathcal{T}}_{\varphi}$ is translation invariant.

Fix $s \in T$ and $V \in \mathcal{U}_X$. As usual choose $W \in \mathcal{U}_X$ such that $W = W^{-1}$ and $W \circ W \subset V$, then use uniform continuity of f^s to choose $\tilde{V} \in \mathcal{U}_X$ so that $\tilde{V}^{-1} = \tilde{V} \subset W$ and $f^s \circ \tilde{V} \subset W \circ f^s$. Then:

$$W \circ f^s \circ (\tilde{V} \circ f^t \circ \tilde{V}) \subset W \circ W \circ f^{t+s} \circ \tilde{V} \subset V^{-1} \circ f^{t+s} \circ V \tag{7.77}$$

$$\begin{aligned}(\tilde{V} \circ f^{t+s} \circ \tilde{V}) \circ f^{-s} \circ W &\subset \tilde{V} \circ f^t \circ W \circ f^s \circ f^{-s} \circ W \\ &\subset \tilde{V} \circ f^t \circ W \circ 1_X \circ W \subset V \circ f^t \circ V \end{aligned} \tag{7.78}$$

If $t \in N(\tilde{V},\tilde{V})$ and f^s has a dense image then by (7.74) and (7.77), $V^{-1} \circ f^{t+s} \circ V$ contains $W \circ f^s \circ [X \times X] = X \times X$, so $t+s \in N(V,V)$. Inverting we have $[X \times X] \circ f^{-s} \circ W = X \times X$. From (7.78), $t+s \in N(\tilde{V},\tilde{V})$ implies $t \in N(V,V)$. Hence,

$$g^{-s}N(\tilde{V},\tilde{V}) \subset N(V,V) \qquad N(\tilde{V},\tilde{V}) \subset g^{-s}N(V,V) \tag{7.79}$$

The last inclusion implies that the filter $\tilde{\mathcal{T}}_\varphi$ is translation – invariant. If $g^{-s}(F)$ contains $N(V,V)$, then the first equation of (7.79) implies that:

$$N(\tilde{V}\tilde{V}) \cap T_s \subset F$$

Thus the filter $\tilde{\mathcal{T}}_\varphi$ is invariant. An invariant filter is full and thick, so it is contained in $\tau\mathcal{B}_T$. This implies φ is weak mixing by Theorem 4.1a. ∎

PROPOSITION 7.19. *Let $\varphi : T \times X \to X$ be a uniform action with X compact.*

a. The family $\tilde{\mathcal{F}}_\varphi$ is proper iff $\mathcal{F}_\varphi$ is contained in some full filter in which case $\tilde{\mathcal{F}}_\varphi$ is the smallest full filter containing $\mathcal{F}_\varphi$.

b. The family $\tilde{\mathcal{T}}_\varphi$ is proper, i.e., φ is uniformly weak mixing, iff φ is weak mixing. In that case:

$$\overline{\mathcal{T}}_\varphi = \tilde{\mathcal{T}}_\varphi \tag{7.80}$$

For $p \in \beta_u T$ and A a closed subset of X we let $p[A]$ denote $\omega_{\mathcal{F}_p}\varphi[A]$. Then:

$$\begin{aligned} H(\tilde{\mathcal{T}}_\varphi) &= \{p \in \beta_u T : p[A] = X \\ &\qquad \textit{for every closed subset } A \textit{ of } X \textit{ with } \operatorname{Int} A \neq \emptyset\} \\ &= \{p \in \beta_u T : \Omega_{\mathcal{F}_p}\varphi = X \times X\} \end{aligned} \tag{7.81}$$

In particular the closed invariant subset $H(\tilde{\mathcal{T}}_\varphi)$ is nonempty iff φ is weak mixing.

If T is separable and X is metrizable, then for φ weak mixing, the subset $\{x \in X : \omega_{k\tilde{\mathcal{T}}_\varphi}\varphi(x) = X\}$ is residual.

PROOF. If W is a symmetric element of $\mathcal{U}_X$ such that $W^2 \subset V$ and $\{W(x_1),\dots,W(x_n)\}$ covers X then it is easy to check that:

$$N^{\varphi}(W) \subset \cap_{i=1}^{n} N^{\varphi}(W(x_i),W(x_i)) \subset N^{\varphi}(V)$$
$$N^{\varphi}(W,W) \subset \cap_{i,j=1}^{n} N^{\varphi}(W(x_i),W(x_j)) \subset N^{\varphi}(V,V) \quad (7.82)$$

Hence closing $\mathcal{F}_{\varphi}$ under intersection and intersecting with $k\mathcal{B}_T$ we obtain $\tilde{\mathcal{F}}_{\varphi}$, proving (a).

Similarly $\tilde{\mathcal{T}}_{\varphi}$ is proper iff $\mathcal{T}_{\varphi}$ is contained in some full filter. That is, by Theorem 4.1 uniform weak mixing is equivalent to weak mixing in the compact case. Furthermore in the weak mixing case, $\overline{\mathcal{T}}_{\varphi}$ is a full filter. Thus (7.80) follows. Furthermore we then clearly have

$$\Omega_{\overline{\mathcal{T}}_{\varphi}} \varphi = X \times X \quad (7.83)$$

To prove (7.81), observe that for any family $\mathcal{F}$, (3.47) implies that $\Omega_{\mathcal{F}}\varphi(x) = X \Leftrightarrow \omega_{\mathcal{F}}\varphi[A] = X$ for all closed sets A such that $x \in \text{Int}A$. This proves the equality of the two sets in (7.81) and by (7.83), $\tilde{\mathcal{T}}_{\varphi} \subset \mathcal{F}_p$ implies that p is in this set. The family $\tilde{\mathcal{T}}_{\varphi}$ is open, so $p \in H(\tilde{\mathcal{T}}_{\varphi})$ iff $\tilde{\mathcal{T}}_{\varphi} \subset \mathcal{F}_p$.

Conversely suppose $p[A] = X$ for all A with $\text{Int}A \neq \emptyset$. For $V \in \mathcal{U}_X$, choose $W \in \mathcal{U}_X$ such that W is open, $W = W^{-1}$ and $W \circ W \subset V$. Now choose points $x_1,\dots,x_n \in X$ so that $\{W(x_i) : i = 1,\dots,n\}$ covers X. Since $\omega_{\mathcal{F}_p}\varphi[\overline{W(x_i)}] = X$ for all i, we can choose for $i,j = 1,\dots,n$ subsets $F_{ij} \in \mathcal{F}_p$ so that $t \in F_{ij}$ implies

$$\overline{W(x_i)} \cap f^{-t}(W(x_j)) \neq \emptyset$$

and so

$$W(x_i) \cap f^{-t}(W(x_j)) \neq \emptyset$$

Let $F = \cap_{i,j=1}^{n} F_{ij}$. By (7.82), $F \subset N(V,V)$. Since $\mathcal{F}_p$ is a filter, both F and $N(V,V)$ lie in $\mathcal{F}_p$. Hence $\tilde{\mathcal{T}}_{\varphi} \subset \mathcal{F}_p$, so p is in the hull.

Because $\tilde{\mathcal{T}}_{\varphi}$ is a translation invariant filter when it is proper, its hull, when nonempty, is a closed invariant subset of $\beta_u^* T$.

When T is separable and X is metrizable, the filter $\tilde{\mathcal{T}}_{\varphi}$ is countably generated. The last result follows from Proposition 4.6b. ∎

We conclude by applying a deft argument of Furstenberg.

LEMMA 7.2. *Let $\mathcal{F}$ and $\mathcal{F}_1$ be proper families for T such that $\mathcal{F}_1$ has a countably generated dual and contains $\mathcal{F} \cdot k\mathcal{F}$. Assume that $\varphi : T \times X \to X$ is a uniform action that is $\mathcal{F}$ transitive and $x \in X$ such that:*

$$\omega_{k\mathcal{F}}\varphi(x) = X \quad (7.84)$$

i.e., x is a $k\mathcal{F}$ transitive point. If X is separable and completely metrizable, then there exists a residual subset R_x such that for all $y \in R_x$:

$$\omega_{\mathcal{F}_1}(\varphi \times \varphi)(x,y) = X \times X \tag{7.85}$$

PROOF. Let U_1, U_2, O be nonempty open subsets of X and $F \in k\mathcal{F}_1$; then:

$$N(x,U_1) \cap N(O,U_2) \cap F \neq \emptyset \tag{7.86}$$

because the three sets are in $k\mathcal{F}$, $\mathcal{F}$ and $k\mathcal{F}_1$, respectively and by assumption $k\mathcal{F} \cdot \mathcal{F} \cdot k\mathcal{F}_1$ is proper. So for each such U_1, U_2, and F, the set:

$$G(U_1,U_2,F) = \{y : N(x,U_1) \cap N(y,U_2) \cap F \neq \emptyset\} \tag{7.87}$$

is open and dense in X. Let U_1 and U_2 vary over a countable base for the topology and F vary over a countable family of generators for $k\mathcal{F}_1$. The intersection is the required residual set R_x. ■

PROPOSITION 7.20. *Let $\varphi : T \times X \to X$ be a uniform action with X a compact metric space and T separable.*

a. Assume φ is weak mixing. The set of $k\tilde{\mathcal{T}}_\varphi$ transitive points for $\varphi \times \varphi$ is residual in $X \times X$, and for each such point (x,y), both x and y are $k\tilde{\mathcal{T}}_\varphi$ transitive points for φ. On the other hand, if x is a $k\tilde{\mathcal{T}}_\varphi$ transitive point for φ, then there exists a residual subset R_x of X such that $y \in R_x$ implies (x,y) is a $k\tilde{\mathcal{T}}_\varphi$ transitive point for $\varphi \times \varphi$.

b. Assume φ is weak mixing and minimal points are dense in X; i.e., the mincenter for φ is all of X. The set of transitive points (that is, $\mathcal{B}_T$ transitive points) for $\varphi \times \varphi$ is residual in $X \times X$, and for each such point (x,y), both x and y are transitive points for φ. On the other hand, if x is a transitive point for φ, then there exists a residual subset R_x of X such that $y \in R_x$ implies (x,y) is a transitive point for $\varphi \times \varphi$.

PROOF. For any proper family $\mathcal{F}$, $\omega_{\mathcal{F}}(\varphi \times \varphi)(x,y) = X \times X$ implies $\omega_{\mathcal{F}}\varphi(x) = X$ by (3.28). If φ is weak mixing then so is $\varphi \times \varphi$, so the set of $k\tilde{\mathcal{T}}_\varphi$ transitive points is dense in $X \times X$ by applying Proposition 7.18b to $\varphi \times \varphi$; a fortiori the transitive points are residual.

For x a $k\tilde{\mathcal{T}}_\varphi$ transitive point, we apply Lemma 7.2 with $\mathcal{F} = k\mathcal{F}_1 = \tilde{\mathcal{T}}_\varphi$. Because $\tilde{\mathcal{T}}_\varphi$ is a filter, $\mathcal{F} \cdot k\mathcal{F} = k\mathcal{F} = \mathcal{F}_1$. Since X is metric and T is separable, $\tilde{\mathcal{T}}_\varphi$ is countably generated. Lemma 7.2 implies the existence of the set R_x described in (a).

For x a transitive point, $\omega_{k\tau k\tau\mathcal{B}_T}\varphi(x)$ is the mincenter of $\omega\varphi(x) = X$ by Theorem 3.1. Thus every transitive point is $k\tau k\tau\mathcal{B}_T$ recurrent. On the other hand, φ is $k\tau\mathcal{B}_T$ central because the minimal points, which are $k\tau\mathcal{B}_T$ recurrent points by

Theorem 3.1, are dense. Because φ is topologically transitive, it is $k\tau\mathcal{B}_T$ transitive by Proposition 4.3c. Since φ is also weak mixing, i.e., $\tau\mathcal{B}_T$ transitive, it is in fact $\tau k\tau\mathcal{B}_T$ transitive by Proposition 4.11a. Thus we can apply Lemma 7.2 with $\mathcal{F} = \tau k\tau\mathcal{B}_T$ and $\mathcal{F}_1 = \mathcal{B}_T$. Again $k\mathcal{B}_T$ is countably generated because T is separable. ■

REMARK. If φ is weak mixing and the minimal points are dense in X, then the same is true for every product action φ_* on X^I. This is because there is a dense subset in the product which is the union of subspaces on which φ_* restricts to an isomorphism of φ (see the long proof of Proposition 4.8d). It follows that in (b) for every $y \in R_x$, (x,y) is in fact a $k\tau k\tau\mathcal{B}_T$ transitive point for $\varphi \times \varphi$. ■

8

Equicontinuity

Lyapunov stability of a point x with respect to an action $\varphi : T \times X \to X$ is equicontinuity of the family of functions $\{f^t : t \in T\} \subset C(X;X)$ at the point x. As with all equicontinuity notions, it is really a uniform concept.

For a uniform action $\varphi : T \times X \to X$, $V \in \mathcal{U}_X$ and $F \subset T$, we define

$$\begin{aligned} V_\varphi^F &= \cap_{t\in F}(f^t \times f^t)^{-1}V \\ &= \{(x_1,x_2) : (f^t x_1, f^t x_2) \in V \text{ for all } t \in F\} \\ Eq_{V,\varphi}^F &= \{x : (x,x) \in \operatorname{Int} V_\varphi^F\} \end{aligned} \tag{8.1}$$

For a family $\mathcal{F}$ of subsets of T, we define

$$\begin{aligned} Eq_{V,\varphi}^{\mathcal{F}} &= \cup_{F\in\mathcal{F}} Eq_{V,\varphi}^F \\ Eq_\varphi^{\mathcal{F}} &= \cap_{V\in\mathcal{U}_X} Eq_{V,\varphi}^{\mathcal{F}} \end{aligned} \tag{8.2}$$

We call x an $\mathcal{F}, V$ *equicontinuity point* or an $\mathcal{F}$ *equicontinuity point* if $x \in Eq_V^{\mathcal{F}}$ or $Eq^{\mathcal{F}}$, respectively. Notice that as usual we drop the subscript φ when the action is understood.

Observe that $x \in Eq_{V,\varphi}^F$ iff:

$$F \subset J^{\varphi\times\varphi}(U \times U, V) \tag{8.3}$$

for some neighborhood U of x, so, $x \in Eq_{V,\varphi}^{\mathcal{F}}$ iff:

$$J^{\varphi\times\varphi}(U \times U, V) \in \mathcal{F} \tag{8.4}$$

for some neighborhood U of x. Notice that we can always shrink U to obtain $U \times U \subset V$.

When the family $\mathcal{F}$ is not mentioned, the case $\mathcal{F} = k\mathcal{B}_T$ is assumed. So $Eq_V = Eq_V^{k\mathcal{B}_T}$ and $Eq = Eq^{k\mathcal{B}_T}$ are the set of V *equicontinuity points* and *equicontinuity points* for φ, respectively.

LEMMA 8.1. *For a uniform action $\varphi : T \times X \to X$:*

$$(V_\varphi^F)^{-1} = (V^{-1})_\varphi^F \qquad Eq_V^F = Eq_{V^{-1}}^F$$
$$V^{F_1} \cap V^{F_2} = V^{F_1 \cup F_2} \qquad Eq_V^{F_1} \cap Eq_V^{F_2} = Eq_V^{F_1 \cup F_2}$$
$$V^F \circ W^F \subset (V \circ W)^F \tag{8.5}$$

If $F_1 \subset F_2$, $\mathcal{F}_2 \subset \mathcal{F}_1$, and $V_2 \subset V_1$, then:

$$V_2^{F_2} \subset V_1^{F_1} \qquad Eq_{V_2}^{F_2} \subset Eq_{V_1}^{F_1}$$
$$Eq_{V_2}^{\mathcal{F}_2} \subset Eq_{V_1}^{\mathcal{F}_1} \qquad Eq^{\mathcal{F}_2} \subset Eq^{\mathcal{F}_1} \tag{8.6}$$

Each Eq_V^F and $Eq_V^{\mathcal{F}}$ is open. If X is metrizable, then each $Eq^{\mathcal{F}}$ is a G_δ. If V is closed, then each V_φ^F is closed.

For each $s \in T$:

$$V^{g^s(F)} = (f^s \times f^s)^{-1}(V^F) = f^{-s} \circ V^F \circ f^s$$
$$f^{-s}(Eq_V^F) \subset Eq_V^{g^s(F)} \tag{8.7}$$

Let $\varphi_1 : T \times X_1 \to X_1$ be a uniform action and $h : \varphi \to \varphi_1$ a uniformly continuous action map. If $h \circ V \subset W \circ h$, or equivalently $h \times h(V) \subset W$, for $V \in \mathcal{U}_X$ and $W \in \mathcal{U}_{X_1}$, then $h \circ V_\varphi^F \subset W_{\varphi_1}^F \circ h$, so $h \times h(V_\varphi^F) \subset W_{\varphi_1}^F$. If in addition h is an open map, then:

$$Eq_{V,\varphi}^F \subset h^{-1}(Eq_{W,\varphi_1}^F) \qquad Eq_{V,\varphi}^{\mathcal{F}} \subset h^{-1}(Eq_{W,\varphi_1}^{\mathcal{F}})$$
$$Eq_\varphi^{\mathcal{F}} \subset h^{-1}(Eq_{\varphi_1}^{\mathcal{F}}) \tag{8.8}$$

If $h, \tilde{h} : \varphi \to \varphi_1$ are continuous action maps, then $h \circ \tilde{h}^{-1} = \{(\tilde{h}(x), h(x)) : x \in X\} \subset X_1 \times X_1$, and for $W \in \mathcal{U}_{X_1}$:

$$h \circ \tilde{h}^{-1} \subset W \;\Rightarrow\; h \circ \tilde{h}^{-1} \subset W_{\varphi_1}^T \tag{8.9}$$

If W is closed and $x \in$ Trans$_\varphi$ [i.e., $\omega\varphi(x) = X$], then:

$$(\tilde{h}(x), h(x)) \in W_{\varphi_1}^T \;\Rightarrow\; h \circ \tilde{h}^{-1} \subset W_{\varphi_1}^T \tag{8.10}$$

If $W \circ \tilde{V} \circ W \subset V$ for $\tilde{V}, W, V \in \mathcal{U}_X$, then:

$$Eq_{\tilde{V}}^{\mathcal{F}} \subset Eq_V^{u\mathcal{F}} \subset Eq_V^{\mathcal{F}} \tag{8.11}$$

PROOF. The results through (8.8) are easy exercises. For example, a metrizable uniformity is countably generated, so $\mathcal{U}_X$ metrizable implies $Eq^{\mathcal{F}}$ is a G_δ.

For (8.9), $h \circ \tilde{h}^{-1} \subset W$ implies $(\tilde{h}(f^t(x)), h(f^t(x))) \in W$ for all t. But $\tilde{h}$ and h are action maps, so $(f_1^t(\tilde{h}(x)), f_1^t(h(x))) \in W$ for all t, i.e., $(\tilde{h}(x), h(x)) \in W_{\varphi_1}^T$. Conversely $(\tilde{h}(x), h(x)) \in W_{\varphi_1}^T$ implies $(\tilde{h}(f^t(x)), h(f^t(x))) \in W$ for all t. If $x \in \mathrm{Trans}_\varphi$ and W closed, $(\tilde{h}(y), h(y)) \in W$ for all $y \in X$. Thus $h \circ \tilde{h}^{-1} \subset W$, so (8.10) follows from (8.9).

For (8.11) we choose $\tilde{W} \in \mathcal{U}_T$ such that $\tilde{W} = \tilde{W}^{-1}$ and $(t, t_1) \in \tilde{W}$ implies $(f^t(x), f^{t_1}(x)) \in W$ for all $x \in X$. If $F \subset J(U \times U, \tilde{V})$, then:

$$\tilde{W}(F) \subset J(U \times U, W \circ \tilde{V} \circ W) \subset J(U \times U, V)$$

■

COROLLARY 8.1. *If $\mathcal{F}$ is a translation + invariant family [$F \in \mathcal{F}, s \in T$ imply $g^s(F) \in \mathcal{F}$], then each $Eq_V^{\mathcal{F}}$ and $Eq^{\mathcal{F}}$ is – invariant, i.e.:*

$$f^{-s}(Eq_V^{\mathcal{F}}) \subset Eq_V^{\mathcal{F}} \qquad f^{-s}(Eq^{\mathcal{F}}) \subset Eq^{\mathcal{F}} \tag{8.12}$$

If φ is an open action, then for any proper family $\mathcal{F}$, $Eq^{\mathcal{F}}$ is + invariant.

PROOF. The first result is clear from (8.7). The second follows from (8.8) applied to $h = f^s$. ■

Now define $\mathcal{U}_\varphi$ to be the family of subsets generated by $\{V_\varphi^T : V \in \mathcal{U}_X\}$. By (8.5) $\mathcal{U}_\varphi$ is a uniformity on the set X. Since $V_\varphi^T \subset V$:

$$\mathcal{U}_X \subset \mathcal{U}_\varphi \tag{8.13}$$

Thus the identity map from X equipped with $\mathcal{U}_\varphi$ to X equipped with $\mathcal{U}_X$ is uniformly continuous. However the topology associated with $\mathcal{U}_\varphi$ is usually strictly finer than the original $\mathcal{U}_X$ topology. Hence the map from $(X, \mathcal{U}_X)$ to $(X, \mathcal{U}_\varphi)$ is usually not continuous. A fortiori, the uniformity $\mathcal{U}_\varphi$ is usually strictly larger than $\mathcal{U}_X$.

PROPOSITION 8.1. *If $\mathcal{U}_X$ is complete, then $\mathcal{U}_\varphi$ is complete. If $\mathcal{U}_X$ is metrizable, then $\mathcal{U}_\varphi$ is metrizable.* $\mathrm{Trans}_\varphi = \{x : \omega\varphi(x) = X\}$ *is a closed subset of X with respect to the $\mathcal{U}_\varphi$ topology.*

PROOF. If $\mathcal{U}_X$ is metrizable and hence countably generated, then clearly $\mathcal{U}_\varphi$ is countably generated and hence metrizable. Furthermore for d any bounded pseudometric on X, define

$$d_\varphi(x_1, x_2) = \sup\{d(f^t(x_1), f^t(x_2)) : t \in T\} \tag{8.14}$$

If d is a metric generating $\mathcal{U}_X$, then d_φ is a metric generating $\mathcal{U}_\varphi$. In general letting d vary over a generating set in the gage of $\mathcal{U}_X$, we obtain a generating set in the gage of $\mathcal{U}_\varphi$.

If $\{x_\alpha\}$ is a $\mathcal{U}_\varphi$ Cauchy net, then by (8.13), $\{x_\alpha\}$ is $\mathcal{U}_X$ Cauchy. If $\mathcal{U}_X$ is complete, there exists $x \in X$ such that $\{x_\alpha\}$ converges to x rel $\mathcal{U}_X$. For $V \in \mathcal{U}_X$ with V closed, $V^T_\varphi \in \mathcal{U}_\varphi$, and for α, α' large enough in the index set, $(x_{\alpha'}, x_\alpha) \in V^T_\alpha$. But V^T_φ is $\mathcal{U}_X$ closed in $X \times X$. Letting $x_{\alpha'}$ tend to x, $(x, x_\alpha) \in V^T_\varphi$ for α large enough. As such V^T_φs generate $\mathcal{U}_\varphi$ we see that $\{x_\alpha\}$ converges to x rel $\mathcal{U}_\varphi$.

For each fixed $t \in T$, I claim that the map from X to $C(X)$ associating $x \mapsto \overline{f^{T_t}(x)}$ is uniformly continuous when $\mathcal{U}_\varphi$ is used on X and the original uniformity on $C(X)$. It then follows that:

$$\text{Trans}_\varphi = \cap_{t \in T}\{x : \overline{f^{T_t}(x)} = X\}$$

is a $\mathcal{U}_\varphi$ closed subset of X.

To prove the claim, choose for $V \in \mathcal{U}_X$, $W \in \mathcal{U}_X$ such that $W = W^{-1}$ and $W^2 \subset V$. If $(x_1, x_2) \in W^T_\varphi$ and $z_1 \in \overline{f^{T_t}(x_1)}$, then $W(z_1)$ contains $f^s(x_1)$ for some $s \in T_t$. Since:

$$(x_1, x_2) \in W^T_\varphi \qquad f^s(x_2) \in W(f^s(x_1)) \subset V(z_1)$$

By symmetry, $(x_1, x_2) \in W^T_\varphi$ implies that $\overline{f^{T_t}(x_1)}$ and $\overline{f^{T_t}(x_2)}$ are V related elements of $C(X)$. ■

PROPOSITION 8.2. *Let $\varphi : T \times X \to X$ be a uniform action. Assume that for each $s \in T$ the set T_s is cobounded; i.e., there exists a compact subset A_s of T such that $T = A_s \cup T_s$.*

If $F_1 \equiv F_2$ (mod $k\mathcal{B}_T$) then $Eq_V^{F_1} = Eq_V^{F_2}$ for all $V \in \mathcal{U}_X$. In particular for $F \in k\mathcal{B}_T$:

$$Eq_V^F = Eq_V^T = Eq_V$$

For any family $\mathcal{F}$:

$$Eq_V^{\mathcal{F}} = Eq_V^{\mathcal{F} \cdot k\mathcal{B}_T} \qquad Eq^{\mathcal{F}} = Eq^{\mathcal{F} \cdot k\mathcal{B}_T} \tag{8.15}$$

PROOF. For any compact subset A of T, $\varphi : A \times X \to X$ is uniformly continuous, so $V^A \in \mathcal{U}_X$ for any $V \in \mathcal{U}_X$. In particular:

$$(x,x) \in \text{Int}\, V^T \Leftrightarrow (x,x) \in \text{Int}(V^F \cap V^A) = \text{Int}\, V^{F \cup A}$$

If $F_1 \equiv F_2$ mod $k\mathcal{B}_T$, then by hypothesis, there exists a compact subset A of T such that $F_1 \cup A = F_2 \cup A$. ■

REMARK. It follows that if T satisfies the hypothesis of Proposition 8.2, then $x \in Eq_\varphi$ iff $\{f^t : t \in T\}$ is a family of functions equicontinuous at the point x. ■

PROPOSITION 8.3. *Let $\varphi : T \times X \to X$ be a uniform action and $\mathcal{F}, \mathcal{F}_1, \mathcal{F}_2$ proper families for T satisfying $\mathcal{F} \cdot \mathcal{F}_1 \subset \mathcal{F}_2$. For $V \in \mathcal{U}_X$, assume $x \in Eq_V^{\mathcal{F}}$. If A is a closed subset of X such that for every $W \in \mathcal{U}_X$, $N^{\varphi}(W(x),A) \in \mathcal{F}_1$, i.e., $\mathcal{N}^{\varphi}(u[x],A) \subset \mathcal{F}_1$, then there exists $W \in \mathcal{U}_X$ such that $J^{\varphi}(W(x),V(A)) \in \mathcal{F}_2$. Furthermore:*

$$\Omega_{k\mathcal{F}_2}\varphi(x) \subset \overline{V(A)} \tag{8.16}$$

In particular:

$$\Omega_{\mathcal{F}_1}\varphi(x) \times \Omega_{k\mathcal{F}_2}\varphi(x) \subset \overline{V} \tag{8.17}$$

including of course the possibility that either $\Omega_{\mathcal{F}_1}\varphi(x)$ or $\Omega_{k\mathcal{F}_2}\varphi(x)$ is empty.

Assume $V = V^{-1}$. If in addition X is compact and $\mathcal{F}_2$ is a filterdual, then:

$$\Omega_{\mathcal{F}_1}\varphi(x) \subset \overline{V}(\omega_{\mathcal{F}_2}\varphi(x)) \tag{8.18}$$

PROOF. Because $x \in Eq_V^{\mathcal{F}}$, there exists $W \in \mathcal{U}_X$ such that $J^{\varphi\times\varphi}(W(x) \times W(x), V) = F \in \mathcal{F}$. Let $F_1 = N^{\varphi}(W(x),A)$ and $F_2 = F_1 \cap F$. By assumption $F_1 \in \mathcal{F}_1$, so $F_2 \in \mathcal{F}_2$. If $t \in F_2 \subset F_1$ then for some $x_t \in W(x)$, $f^t(x_t) \in A$, so $t \in F_2 \subset F$ implies $f^t(W(x)) \subset V(f^t(x_t)) \subset V(A)$.

If $y \in \Omega_{k\mathcal{F}_2}\varphi(x)$ and $W_0 \subset W$, then $\tilde{F}_2 = N(W_0(x), W_0(y)) \in k\mathcal{F}_2$, so it meets $F_2 \subset J^{\varphi}(W_0(x), V(A))$. If $t \in \tilde{F}_2 \cap F_2$ then $f^t(W_0(x))$ meets $W_0(y)$, and it is contained in $V(A)$. Since W_0 can be chosen arbitrarily small, $y \in \overline{V(A)}$.

In particular if $y \in \Omega_{k\mathcal{F}_2}\varphi(x)$ and $z \in \Omega_{\mathcal{F}_1}\varphi(x)$, then for an arbitrary closed $W_1 \in \mathcal{U}_X$, we can apply the previous result to $A = W_1(z)$, to obtain $y \in \overline{V \circ W_1(z)}$, so $W_1(z) \times W_1(y) \cap V \neq \emptyset$. Hence, $(z,y) \in \overline{V}$, proving (8.17).

In addition, $F_2 = J(W(x), V(A)) \in \mathcal{F}_2$ for some $W \in \mathcal{U}_X$. Then for all $\tilde{F}_2 \in k\mathcal{F}_2$, $\tilde{F}_2$ meets F_2, and $f^{\tilde{F}_2}(x)$ meets $V(A)$. When X is compact and $k\mathcal{F}_2$ is a filter:

$$\omega_{\mathcal{F}_2}\varphi(x) = \cap_{\tilde{F}_2 \in k\mathcal{F}_2} \overline{f^{\tilde{F}_2}(x)}$$

meets $\overline{V(A)}$. For all closed $W_1 \in \mathcal{U}_X$ with $W_1 = W_1^{-1}$, $z \in W_1 \circ V \circ W_1(\omega_{\mathcal{F}_2}\varphi(x))$. Intersecting over W_1 we obtain (8.18). ∎

PROPOSITION 8.4. *Let $\varphi : T \times X \to X$ be a uniform action and $\mathcal{F}, \mathcal{F}_1, \mathcal{F}_2$ proper families for T satisfying $\mathcal{F} \cdot \mathcal{F}_1 \subset \mathcal{F}_2$. Assume $x \in Eq^{\mathcal{F}}$. If A is a closed subset of X such that for every $W \in \mathcal{U}_X$, $N^{\varphi}(W(x),W(A)) \in \mathcal{F}_1$, i.e., $\mathcal{N}^{\varphi}(u[x],u[A]) \subset \mathcal{F}_1$, then for every $V \in \mathcal{U}_X$ there exists $W \in \mathcal{U}_X$ such that $J^{\varphi}(W(x),V(A)) \in \mathcal{F}_2$. Furthermore:*

$$\Omega_{k\mathcal{F}_2}\varphi(x) \subset A \tag{8.19}$$

In particular, either $\Omega_{\mathcal{F}_1}\varphi(x) = \Omega_{k\mathcal{F}_2}\varphi(x)$ is a singleton set or at least one of the sets $\Omega_{\mathcal{F}_1}\varphi(x)$, $\Omega_{k\mathcal{F}_2}\varphi(x)$ is empty. Furthermore:

$$\Omega_{\mathcal{F}_1}\varphi(x) \subset \omega_{\mathcal{F}_2}\varphi(x) \qquad \Omega_{k\mathcal{F}_2}\varphi(x) \subset \omega_{k\mathcal{F}_1}\varphi(x) \tag{8.20}$$

PROOF. For all $W_0 \in \mathcal{U}_X$, $x \in Eq^{\mathcal{F}}_{W_0}$, Proposition 8.3 applies to $\tilde{A} = \overline{W_0(A)}$. Given V, choose W_0 so that $W_0^4 \subset V$, and apply Proposition 8.3 to obtain $W \subset W_0$ so that $J^{\varphi}(W(x), W_0(\tilde{A})) \in \mathcal{F}_2$. $W_0(\tilde{A}) \subset V(A)$ implies $J^{\varphi}(W(x), V(A)) \in \mathcal{F}_2$. By (8.16):

$$\Omega_{k\mathcal{F}_2}\varphi(x) \subset \overline{W_0(\tilde{A})} \subset V(A)$$

Intersecting on V we obtain (8.19). Similarly, (8.17) implies $\Omega_{\mathcal{F}_1}\varphi(x) \times \Omega_{k\mathcal{F}_2}\varphi(x) \subset 1_X$.

We obtain (8.20) from the proof of (8.18) which yields that for every $z \in \Omega_{\mathcal{F}_1}\varphi(x)$ and $V \in \mathcal{U}_X$ there exists $W \in \mathcal{U}_X$ such that $J^{\varphi}(W(x), V(z)) \in \mathcal{F}_2$.

$$J^{\varphi}(W(x), V(z)) \subset N^{\varphi}(x, V(z))$$

Therefore $z \in \omega_{\mathcal{F}_2}\varphi(x)$. The second part of (8.20) follows from the first because $\mathcal{F} \cdot \mathcal{F}_1 \subset \mathcal{F}_2$ implies $\mathcal{F} \cdot k\mathcal{F}_2 \subset k\mathcal{F}_1$. ■

COROLLARY 8.2. *If $x \in Eq^{\mathcal{F}}_{\varphi}$ for a uniform action φ and $\mathcal{F}$ is a filter then $\Omega_{\mathcal{F}}\varphi(x) = \omega_{\mathcal{F}}\varphi(x)$ and $\Omega_{k\mathcal{F}}\varphi(x) = \omega_{k\mathcal{F}}\varphi(x)$.*

PROOF. Apply (8.20) with $\mathcal{F}_1 = \mathcal{F}_2 = \mathcal{F}$. The reverse inclusions are obvious. ■

COROLLARY 8.3. *If $x \in Eq_{\varphi}$ then for any full family $\mathcal{F}$:*

$$\Omega_{\mathcal{F}}\varphi(x) = \omega_{\mathcal{F}}\varphi(x) \qquad \Omega_{k\mathcal{F}}\varphi(x) = \omega_{k\mathcal{F}}\varphi(x)$$

In particular, $\Omega\varphi(x) = \omega\varphi(x)$, so if x is nonwandering, it is recurrent. Furthermore if X is compact and x is a limit of minimal points, it is a minimal point.

PROOF. Apply (8.20) with $\mathcal{F}_1 = \mathcal{F}_2 = \mathcal{F}$. Since $\mathcal{F}$ is full $k\mathcal{B}_T \cdot \mathcal{F} = \mathcal{F}$. With $\mathcal{F} = \mathcal{B}_T$ we obtain $\Omega\varphi(x) = \omega\varphi(x)$. With $\mathcal{F} = k\tau\mathcal{B}_T$, $\Omega_{k\tau\mathcal{B}_T}\varphi(x) = \omega_{k\tau\mathcal{B}_T}\varphi(x)$. In the compact case, x is minimal iff $x \in \omega_{k\tau\mathcal{B}_T}\varphi(x)$ by Theorem 3.1. If x is a limit of points in $|\omega_{k\tau\mathcal{B}_T}\varphi|$, $x \in \Omega_{k\tau\mathcal{B}_T}\varphi(x)$. Therefore $x \in \omega_{k\tau\mathcal{B}_T}\varphi(x)$, hence x is minimal. ■

PROPOSITION 8.5. *Let $\varphi : T \times X \to X$ be a uniform action with X compact, $\mathcal{F}$ a filter for T and $x \in X$.*

a. For $V \in \mathcal{U}_X$, $x \in Eq^{\mathcal{F}}_V$ implies $\Omega_{k\mathcal{F}}(\varphi \times \varphi)(x,x) \subset \overline{V}$ and $\Omega_{k\mathcal{F}}(\varphi \times \varphi)(x,x) \subset Int\, V$ implies $x \in Eq^{\mathcal{F}}_V$.

b. The following are equivalent:

(1) $x \in Eq^{\mathcal{F}}$

(2) $\Omega_{k\mathcal{F}}(\varphi \times \varphi)(x,x) = \omega_{k\mathcal{F}}(\varphi \times \varphi)(x,x)$

(3) $\Omega_{k\mathcal{F}}(\varphi \times \varphi)(x,x) \subset 1_X$

PROOF. (a) If $x \in Eq_V^{\mathcal{F}}$, then for some neighborhood U of x, $J^{\varphi\times\varphi}(U\times U,V)\in\mathcal{F}$. Then the complement $N^{\varphi\times\varphi}(U\times U,(X\times X)\setminus V)$ is not in $k\mathcal{F}$. Hence no point of $(X\times X)\setminus\overline{V}$ is in $\Omega_{k\mathcal{F}}(\varphi\times\varphi)(x,x)$.

On the other hand, if $\Omega_{k\mathcal{F}}(\varphi\times\varphi)(x,x)\subset \operatorname{Int}V$ then by Proposition 3.10:

$$\omega_{k\mathcal{F}}(\varphi\times\varphi)[U\times U]\subset \operatorname{Int}V$$

for some neighborhood $U\times U$ of (x,x), since $X\times X$ is compact and $k\mathcal{F}$ is a filterdual. Furthermore these three properties then imply

$$\overline{(f\times f)^F(U\times U)}\subset \operatorname{Int}V$$

for some $F\in\mathcal{F}$. Hence:

$$F\subset J^{\varphi\times\varphi}(U\times U,V)$$

therefore $x\in Eq_V^{\mathcal{F}}$.

(b) (1) $\Rightarrow$ (2). Clearly $x\in Eq_\varphi^{\mathcal{F}}$ implies $(x,x)\in Eq_{\varphi\times\varphi}^{\mathcal{F}}$, so by Corollary 8.2, we then have (2).

(2) $\Rightarrow$ (3). 1_X is a closed $\varphi\times\varphi+$ invariant set containing (x,x), so it contains $\omega_{k\mathcal{F}}(\varphi\times\varphi)(x,x)$.

(3) $\Rightarrow$ (1). By (3), $\Omega_{k\mathcal{F}}(\varphi\times\varphi)(x,x)\subset\operatorname{Int}V$ for all $V\in\mathcal{U}_X$, so by (a) $x\in Eq_V^{\mathcal{F}}$ for all $V\in\mathcal{U}_X$. ■

COROLLARY 8.4. *For $\varphi: T\times X\to X$ a uniform action with X compact and $x\in X$ the following are equivalent:*

(1) $x\in Eq_\varphi$

(2) $\Omega(\varphi\times\varphi)(x,x)=\omega(\varphi\times\varphi)(x,x)$

(3) $\Omega(\varphi\times\varphi)(x,x)\subset 1_X$

PROOF. Apply Proposition 8.5 with $\mathcal{F}=k\mathcal{B}_T$. ■

For a proper family $\mathcal{F}$, a uniform action $\varphi: T\times X\to X$ is called $\mathcal{F}$ *equicontinuous* if $X=Eq_\varphi^{\mathcal{F}}$, or equivalently by (8.2), if for every $V\in\mathcal{U}_X$, $X=Eq_{V,\varphi}^{\mathcal{F}}$. As usual φ is called *equicontinuous* if it is $k\mathcal{B}_T$ equicontinuous.

On the other hand, φ is called $k\mathcal{F}$ *sensitive* if for some $V\in\mathcal{U}_X$ $Eq_{V,\varphi}^{\mathcal{F}}=\emptyset$, or equivalently by (8.2), if for some $V\in\mathcal{U}_X$, $Eq_{V,\varphi}^{F}=\emptyset$ for all $F\in\mathcal{F}$. By (8.4) this says that for every nonempty open subset U of X, $J^{\varphi\times\varphi}(U\times U,V)\notin\mathcal{F}$. Thus φ is $k\mathcal{F}$ sensitive iff there exists $V\in\mathcal{U}_X$ such that for every nonempty open subset U of X:

$$N^{\varphi\times\varphi}(U\times U,(X\times X)\setminus V)\in k\mathcal{F} \tag{8.21}$$

We use equicontinuity to refer to the strongest — among invariant families — notion of $\mathcal{F}$ equicontinuity, namely, $k\mathcal{B}_T$ equicontinuity. *Sensitivity* corresponds to the weakest version of $\mathcal{F}$ sensitivity, namely, $\mathcal{B}_T$ sensitivity; thus sensitivity follows the pattern established by centrality, transitivity, mixing, etc.

PROPOSITION 8.6. *Let $\varphi : T \times X \to X$ be a uniform action with X compact and $\mathcal{F}$ a filter for T. The following are equivalent:*

(1) φ is $\mathcal{F}$ equicontinuous.

(2) For every $V \in \mathcal{U}_X$, there exists $F \in \mathcal{F}$ such that $X = Eq^F_{V,\varphi}$.

(3) For every $V \in \mathcal{U}_X$, there exists $F \in \mathcal{F}$ such that $V^F_\varphi \in \mathcal{U}_X$.

(4) For every $V \in \mathcal{U}_X$, there exists $W \in \mathcal{U}_X$ such that $J^{\varphi\times\varphi}(W,V) \in \mathcal{F}$.

(5) $\Omega_{k\mathcal{F}}(\varphi \times \varphi)(1_X) \subset 1_X$; i.e., 1_X is $\Omega_{k\mathcal{F}}(\varphi \times \varphi)+$ invariant.

On the other hand, φ is $k\mathcal{F}$ sensitive iff there exists $V \in \mathcal{U}_X$ such that:

$$\Omega_{k\mathcal{F}}(\varphi \times \varphi)(x,x) \setminus V \neq \emptyset$$

for all $x \in X$.

PROOF. (1) $\Rightarrow$ (2). $X = \bigcup_{F\in\mathcal{F}} Eq^F_{V,\varphi}$, so by compactness there exist $F_1, \dots, F_k \in \mathcal{F}$ such that X is covered by the open sets $\{Eq^{F_i}_{V,\varphi} : i = 1, \dots, k\}$. The union is contained in $Eq^F_{V,\varphi}$ with $F = \cap_{i=1}^k F_i$. Because $\mathcal{F}$ is a filter $F \in \mathcal{F}$.

(2) $\Rightarrow$ (3). $Eq^F_{V,\varphi} = X$ iff V^F_φ is a neighborhood of the diagonal, so $V^F_\varphi \in \mathcal{U}_X$ by compactness.

(3) $\Rightarrow$ (4). With $W = V^F_\varphi$, $F \subset J^{\varphi\times\varphi}(W,V)$, so $J^{\varphi\times\varphi}(W,V) \in \mathcal{F}$.

(4) $\Rightarrow$ (5). As in the proof of Proposition 8.5, $J(W,V) \in \mathcal{F}$ and $(x,x) \in \operatorname{Int} W$ imply $\Omega_{k\mathcal{F}}(\varphi \times \varphi)(x,x) \subset \overline{V}$. Intersect over $V \in \mathcal{U}_X$.

(5) $\Rightarrow$ (1). Apply Proposition 8.5b to each $x \in X$.

If $Eq^{\mathcal{F}}_V = \emptyset$, then by Proposition 8.5a each $\Omega_{k\mathcal{F}}(\varphi \times \varphi)(x,x)$ meets $(X \times X) \setminus (\operatorname{Int} V)$. Therefore with $V_0 = \operatorname{Int} V$, $\Omega_{k\mathcal{F}}(\varphi \times \varphi)(x,x) \setminus V_0 \neq \emptyset$ for all x. On the other hand, if $V \in \mathcal{U}_X$ is closed and $\Omega_{k\mathcal{F}}(\varphi \times \varphi)(x,x) \setminus V \neq \emptyset$, then by Proposition 8.5a, $x \notin Eq^{\mathcal{F}}_V$. ∎

For a proper family $\mathcal{F}$, a uniform action $\varphi : T \times X \to X$ is called $\mathcal{F}$ *almost equicontinuous* if for every $V \in \mathcal{U}_X$, $Eq^{\mathcal{F}}_{V,\varphi}$ is a dense open subset of X. φ is called *almost equicontinuous* if it is $k\mathcal{B}_T$ almost equicontinuous. If $Eq^{\mathcal{F}}_\varphi$ is dense, then clearly φ is $\mathcal{F}$ almost equicontinuous. If X is a completely metrizable space, then φ $\mathcal{F}$ almost equicontinuous implies $Eq^{\mathcal{F}}_\varphi$ is residual, the intersection of a countable family of dense open sets.

Almost equicontinuity is especially important when combined with transitivity because of the *Auslander–Yorke Dichotomy Theorem* as follows:

THEOREM 8.1. *Let $\varphi : T \times X \to X$ be a uniform action which is transitive, i.e., $\Omega\varphi = X \times X$. Let $\mathcal{F}$ be a translation invariant family for T.* Either *φ is $k\mathcal{F}$ sensitive, i.e., for some $V \in \mathcal{U}_X$ $Eq_V^{\mathcal{F}} = \emptyset$ or φ is $\mathcal{F}$ almost equicontinuous, i.e., for all $V \in \mathcal{U}_X$ $Eq_V^{\mathcal{F}}$ is an open dense subset of X. If φ is $\mathcal{F}$ almost equicontinuous then every transitive point for φ is an $\mathcal{F}$ equicontinuity point, i.e.:*

$$\mathrm{Trans}_\varphi \subset Eq_\varphi^{\mathcal{F}} \tag{8.22}$$

PROOF. Each $Eq_V^{\mathcal{F}}$ is a $-$ invariant open set because $\mathcal{F}$ is translation invariant (cf. Corollary 8.1). Because φ is transitive $Eq_V^{\mathcal{F}} \neq \emptyset$ then implies $Eq_V^{\mathcal{F}}$ is dense. Therefore either all the $Eq_V^{\mathcal{F}}$s are dense or some are empty. If $Eq_V^{\mathcal{F}} \neq \emptyset$ and $x \in \mathrm{Trans}_\varphi$, then since $\overline{f^T(x)} = X$, $f^t(x) \in Eq_V^{\mathcal{F}}$ for some $t \in T$. Since $Eq_V^{\mathcal{F}}$ is $-$ invariant, $x \in Eq_V^{\mathcal{F}}$. If φ is almost equicontinuous, this is true for all $V \in \mathcal{U}_X$, so by intersecting, we obtain (8.22). ∎

In the transitive case, special results are obtained from the following result, the *Hinge Lemma*.

LEMMA 8.2. *With $\mathcal{F}$ a proper family for T and $\varphi : T \times X \to X$ a uniform action, assume φ is $k\mathcal{F}$ transitive, i.e., $\Omega_{k\mathcal{F}}\varphi = X \times X$. Suppose U is a nonempty open subset of X and V is a closed element of $\mathcal{U}_X$. The following conditions are equivalent:*

(1) $J^{\varphi\times\varphi}(U \times U, V) \in \mathcal{F}$

(2) $J^{\varphi\times\varphi}(U \times U, V) = T$, *i.e.,* $U \times U \subset V_\varphi^T$

(3) There exists $F \in \mathcal{F}$ such that $(x, f^s(x)) \in U \times U$ implies $(f^t(x), f^{s+t}(x)) \in V$ for all $t \in F$.

(4) $(x, f^s(x)) \in U \times U$ implies $(f^t(x), f^{s+t}(x)) \in V$ for all $t \in T$.

(5) $s \in N^\varphi(U, U)$ implies $f^s \subset V$; i.e., $(x_1, f^s(x_1)) \in U \times U$ for some x_1, implies $(x, f^s(x)) \in V$ for all $x \in X$.

In particular if $\Omega_{k\mathcal{F}}\varphi = X \times X$ then for V closed in $\mathcal{U}_X$:

$$Eq_V^{\mathcal{F}} = Eq_V^T \tag{8.23}$$

PROOF. We show

$$\begin{array}{ccccc} (1) & \Rightarrow & (3) & \Rightarrow (5) \Rightarrow (2) \\ \Uparrow & & \Uparrow & \\ (2) & \Rightarrow & (4) & \end{array}$$

(1) $\Rightarrow$ (3) and (2) $\Rightarrow$ (4). With $F = J^{\varphi\times\varphi}(U\times U, V)$, $(x, f^s(x), t) \in U\times U\times F$ implies $(f^t(x), f^{s+t}(x)) \in V$.

(2) $\Rightarrow$ (1) and (4) $\Rightarrow$ (3). These are obvious.

(3) $\Rightarrow$ (5). Assume for $x, x_1 \in X$ that $(x_1, f^s(x_1)) \in U\times U$. We prove $(x, f^s(x)) \in V$. Fix $W \in \mathcal{U}_X$ with $W = W^{-1}$ and small enough that $W(x_1) \subset U \cap f^{-s}(U)$. Choose W_0 such that $W_0 \subset W$ and $f^s \circ W_0 \subset W \circ f^s$. By assumption $x \in \Omega_{k\mathcal{F}}\varphi(x_1)$, so $N(W_0(x_1), W_0(x)) \in k\mathcal{F}$. If $F \in \mathcal{F}$ satisfies (3) we can choose $t \in F \cap N(W_0(x_1), W_0(x))$ and $x_2 \in W_0(x_1)$ with $f^t(x_2) \in W_0(x)$.

$$f^s(x_2) \in W(f^s(x_1)) \qquad f^{t+s}(x_2) \in W(f^s(x))$$

by choice of W_0. Hence:

$$(x_2, f^s(x_2)) \in W(x_1) \times W(f^s(x_1)) \subset U \times U$$

By (3), $t \in F$ implies $(f^t(x_2), f^{s+t}(x_2)) \in V$. Thus $(x, f^s(x)) \in W \circ V \circ W$. Since this is true for all $W \in \mathcal{U}_X$ sufficiently small, we can intersect over such Ws to obtain $(x, f^s(x))$ an element of the closed set V.

(5) $\Rightarrow$ (2). Assume $(x, y, t) \in U\times U\times T$. We prove that $(f^t(x), f^t(y)) \in V$. Fix $W \in \mathcal{U}_X$ with $W = W^{-1}$ small enough that $W(x) \subset U$ and $W(y) \subset U$. Choose $W_0 \in \mathcal{U}_X$ with $W_0 \subset W$ and such that:

$$f^t(W_0(x)) \subset W(f^t(x)) \qquad f^t(W_0(y)) \subset W(f^t(y))$$

By assumption $y \in \Omega_{k\mathcal{F}}\varphi(x)$, so with $s \in N(W_0(x), W_0(y))$ there exists $x_1 \in W_0(x)$ such that $f^s(x_1) \in W_0(y)$. In particular, $(x_1, f^s(x_1)) \in U \times U$. By (5):

$$(f^t(x_1), f^{s+t}(x_1)) = (f^t(x_1), f^s(f^t(x_1))) \in V$$

But:

$$f^t(x_1) \in W(f^t(x)) \qquad f^{s+t}(x_1) \in W(f^t(y))$$

Hence, $(f^t(x), f^t(y)) \in W \circ V \circ W$. Intersecting over $W \in \mathcal{U}_X$ $(f^t(x), f^t(y)) \in V$ because V is closed.

In particular, (8.23) follows from the equivalence of (1) and (2). ■

Recall that the uniform action φ on X induces a uniform action φ_* on $C^u(X;X)$ by $(f^t)_*(h) = f^t \circ h$. We call φ $\mathcal{F}$ *uniformly rigid* if 1_X is an $\mathcal{F}$ recurrent point for φ_* on $C^u(X;X)$. This is equivalent to:

$$\{t : f^t \subset V\} \in \mathcal{F} \tag{8.24}$$

for all $V \in \mathcal{U}_X$. φ is called *uniformly rigid* when it is $\mathcal{B}_T$ uniformly rigid. As with sensitivity, uniform rigidity is the weakest notion of $\mathcal{F}$ uniform rigidity among full families $\mathcal{F}$.

PROPOSITION 8.7. *Let $\varphi : T \times X \to X$ be a uniform action and let $\mathcal{F}$, $\mathcal{F}_1$, etc., be proper families for T.*

a. If φ is $\mathcal{F}$ uniformly rigid, then induced actions on X^Y (pointwise convergence), $C(Y;X)$ (uniform convergence), and $C(X)$ are $\mathcal{F}$ uniformly rigid.

b. Let $\hat{\varphi}$ denote the same set action map but with X equipped with the uniformity $\mathcal{U}_\varphi$ (see (8.13)). The action $\hat{\varphi}$ is equicontinuous. If φ is $\mathcal{F}$ uniformly rigid, then $\hat{\varphi}$ is $\mathcal{F}$ uniformly rigid. In that case every point of X is $\mathcal{F}$ recurrent with respect to $\hat{\varphi}$ and a fortiori with respect to φ.

c. If $\mathcal{F}_1 \subset \mathcal{F}_2$ and φ is $\mathcal{F}_1$ uniformly rigid, then φ is $\mathcal{F}_2$ uniformly rigid. In particular if $\mathcal{F}$ is a full family and φ is $\mathcal{F}$ uniformly rigid, then φ is uniformly rigid, i.e., $\mathcal{B}_T$ uniformly rigid.

d. Let $h : \varphi \to \varphi_1$ be a uniformly continuous action map. If h is a dense map and φ is $\mathcal{F}$ uniformly rigid, then φ_1 is $\mathcal{F}$ uniformly rigid.

If h is a uniform embedding, i.e., a uniform isomorphism onto its image, and φ_1 is $\mathcal{F}$ uniformly rigid, then φ is $\mathcal{F}$ uniformly rigid.

e. If φ is uniformly rigid and A is a closed + invariant subset, then for $t \in T$:

$$(f^{-t})(A) = A \tag{8.25}$$

If φ is a surjective action, then A is an invariant subset.

If $V \in \mathcal{U}_X$ is closed then

$$(f^t \times f^t)^{-1}(V_\varphi^T) = V_\varphi^T \tag{8.26}$$

If φ is surjective, then V_φ^T is invariant.

f. If φ is a uniformly rigid action, then φ is injective and dense. If in addition $\mathcal{U}_\varphi$ is complete, then φ is bijective, and $\hat{\varphi}$ is uniformly reversible.

g. If φ is uniformly reversible and $\mathcal{F}$ uniformly rigid, then the reverse system $\overline{\varphi}$ is $\mathcal{F}$ uniformly rigid. If φ is uniformly rigid and X is compact, then φ is uniformly reversible.

PROOF. (a) If $f^t \subset V$ for $V \in \mathcal{U}_X$, then clearly, f^t_* is contained in the V associated elements of the uniformities for X^Y, $C(Y;X)$ and $C(X)$.

(b) If $f^t \subset V$ for $V \in \mathcal{U}_X$ then by (8.9), $f^t \subset V_\varphi^T$ (use $h = f^t$ and $\tilde{h} = 1_X$). It follows that $\hat{\varphi}$ is a uniform action on X equipped with $\mathcal{U}_\varphi$ and $\hat{\varphi}$ is $\mathcal{F}$ uniformly rigid. Furthermore $t \in N(x, V_\varphi^T(x))$ for all $x \in X$, so x is $\mathcal{F}$ recurrent with respect to $\hat{\varphi}$. Since $V_\varphi^T \subset V$, $t \in N(x, V(x))$.

(c) This is obvious.

(d) If $V \in \mathcal{U}_{X_1}$ is closed, then $(h \times h)^{-1}(V) \in \mathcal{U}_X$ and

$$\{t : f^t \subset (h \times h)^{-1}(V)\} = \{t : (h \times h)(f^t) \subset V\}$$

is in $\mathcal{F}$ when φ is $\mathcal{F}$ uniformly rigid. If h is dense, then

$$(h \times h)(f^t) = h \circ f^t \circ h^{-1} = f_1^t \circ h \circ h^{-1}$$

is dense in f_1^t. Thus when V is closed:

$$\{t : f^t \subset (h \times h)^{-1}(V)\} = \{t : f_1^t \subset V\}$$

Hence φ_1 is $\mathcal{F}$ uniformly rigid.

On the other hand, $f_1^t \subset V$ implies

$$f^t \subset (h \times h)^{-1}(f_1^t) \subset (h \times h)^{-1}V$$

If h is a uniform embedding, then $\{(h \times h)^{-1}V : V \in \mathcal{U}_{X_1}\}$ generates $\mathcal{U}_X$. If φ_1 is $\mathcal{F}$ uniformly rigid, φ also is.

(e) If $x \in A$, then $f^t(x) \in A$ by $+$ invariance. If $f^t(x) \in A$, then:

$$\omega\varphi(x) = \omega\varphi(f^t(x)) \subset A$$

because A is closed and $+$ invariant. If φ is uniformly rigid, then by (b) $x \in \omega\varphi(x)$; (8.25) follows. By (a) $\varphi \times \varphi$ is uniformly rigid, so (8.26) follows from (8.25).

(f) Applying by (8.25) to 1_X, $(f^t \times f^t)^{-1}(1_X) = 1_X$, so f^t is injective. If $x \in X$, then:

$$x \in \omega\varphi(x) \subset \overline{f^{T_t}(X)} \subset \overline{f^t(X)}$$

implies $f^t(X)$ is dense.

Assume $\mathcal{U}_\varphi$ is complete. Applying the preceding argument to $\hat{\varphi}$, we can find a net $\{x_\alpha\}$ in X such that $\{f^t(x_\alpha)\}$ converges rel $\mathcal{U}_\varphi$ to x. In particular, $\{f^t(x_\alpha)\}$ is $\mathcal{U}_\varphi$ Cauchy. By (8.26), $\{x_\alpha\}$ is $\mathcal{U}_\varphi$ Cauchy, so by $\mathcal{U}_\varphi$ completeness $\{x_\alpha\}$ converges to some $\tilde{x}$, rel $\mathcal{U}_\varphi$. Then $\{f^t(x_\alpha)\}$ converges to $f^t(\tilde{x})$; hence $f^t(\tilde{x}) = x$ and f^t is surjective.

When φ is bijective, each V_φ^T with V closed is invariant by (8.26). Then each f^{-t} is $\mathcal{U}_\varphi$ uniformly continuous, and by Proposition 1.3 $\hat{\varphi}$ is reversible.

(g) If $V = V^{-1}$ in $\mathcal{U}_X$ and $f^t \subset V$, then $f^{-t} \subset V$. Therefore

$$N^{\varphi_*}(1_X, V^X) = N^{\bar{\varphi}_*}(1_X, V^X)$$

where φ_* and $\overline{\varphi}_*$ are the induced actions on $C^u(X;X)$.

If φ is uniformly rigid, then each f^t is injective and dense. Then if X is compact, f^t is surjective, and the inverse map f^{-t} is uniformly continuous. So by Proposition 1.3 again, φ is a reversible action. ■

In the category of uniform spaces with continuous maps, the dense maps are the epimorphisms. So for Y,X uniform spaces we define the subspaces of $C(Y;X)$

$$\begin{aligned} C\,\mathrm{epi}(Y;X) &= \{g \in C(Y;X) : \overline{g(Y)} = X\} \\ C^u\,\mathrm{epi}(Y;X) &= C\,\mathrm{epi}(Y;X) \cap C^u(Y;X) \end{aligned} \tag{8.27}$$

On these subspaces we obtain a strong converse to Proposition 8.7a.

PROPOSITION 8.8. *Let $\varphi : T \times X \to X$ be a uniform action, Y a uniform space, and $\mathcal{F}$ a full family for T. Let φ_* denote the induced action on $C(Y;X)$.*

$C\,\mathrm{epi}(Y;X)$ and $C^u\,\mathrm{epi}(Y;X)$ are closed subspaces of $C(Y;X)$. If φ is a dense action, then they are both φ_* + invariant. In that case, let $\varphi_*|C\,\mathrm{epi}$ and $\varphi_*|C^u\,\mathrm{epi}$ denote the restrictions of φ_* to the corresponding subspaces.

If φ is $\mathcal{F}$ uniformly rigid, then it is a dense action, and both $\varphi_*|C\,\mathrm{epi}$ and $\varphi_*|C^u\,\mathrm{epi}$ are $\mathcal{F}$ uniformly rigid. In particular every h in $C\,\mathrm{epi}(Y;X)$ is $\mathcal{F}$ recurrent, i.e.:

$$|\omega_{\mathcal{F}}(\varphi_*|C\,\mathrm{epi})| = C\,\mathrm{epi}(Y;X)$$

If φ is a dense action and there exists an $\mathcal{F}$ nonwandering point for $\varphi_*|C\,\mathrm{epi}$, i.e., if $|\Omega_{\mathcal{F}}(\varphi_*|C\,\mathrm{epi})| \neq \emptyset$, then φ is $\mathcal{F}$ uniformly rigid.

PROOF. For $V \in \mathcal{U}_X$ recall that $V^Y \in \mathcal{U}_C$ is $\{(g_1,g_2) : (g_1(y),g_2(y)) \in V$ for all $y \in Y\}$. If $g(Y)$ is not dense, then for some $x \in X$ and some $V = V^{-1}$ in $\mathcal{U}_X$, $V^2(x)$ is disjoint from $g(Y)$. Then $V(x)$ is disjoint from $g_1(Y)$ for all $g_1 \in V^Y(g)$. Thus $C \backslash C\,\mathrm{epi}$ is open. $C\,\mathrm{epi}$ is closed and so is its intersection with the closed set C^u. If f^t is dense and g is dense, then so is $f^t \circ g$. Hence, $C\,\mathrm{epi}$ and $C^u\,\mathrm{epi}$ are φ_* + invariant when φ is a dense action.

If φ is $\mathcal{F}$ uniformly rigid, then it is uniformly rigid, i.e., $\mathcal{B}_T$ uniformly rigid because $\mathcal{F}$ is full. Then φ is dense by Proposition 8.7f. By (d) the restrictions are $\mathcal{F}$ uniformly rigid.

Now assume $W, V \in \mathcal{U}_X$ with V closed, $W = W^{-1}$ and $W \circ W \subset V$. For any $t \in T$ and $g \in C(Y;X)$, we prove that:

$$f_*^{-t}(W^Y(g)) \cap W^Y(g) \cap C\,\mathrm{epi}(Y;X) \neq \emptyset \Rightarrow f^t \subset V \tag{8.28}$$

If g_1 is in the intersection, then for all $y \in Y$, $(g(y), f^t(g_1(y)))$ and $(g(y), g_1(y))$ are in W. Then for all y:

$$(g_1(y), f^t(g_1(y))) \in W \circ W \subset V$$

Because $g_1(Y)$ is dense in X and V is closed, $(x, f^t(x)) \in V$ for all $x \in X$, proving (8.28).

If φ is dense but not $\mathcal{F}$ uniformly rigid, then for some $V \in \mathcal{U}_X$, which we may assume closed, $\{t : f^t \subset V\} \notin \mathcal{F}$. Choosing $W \in \mathcal{U}_X$ so that $W = W^{-1}$ and $W \circ W \subset V$, (8.28) implies that $N^{\varphi_*|C\mathrm{epi}}(W^Y(g), W^Y(g)) \notin \mathcal{F}$ for all $g \in C\,\mathrm{epi}$. ■

Dense maps play a crucial role in this result. If Y is compact and zero dimensional and φ is $\mathcal{F}$ mixing, then Proposition 4.10c implies that φ_* is $\mathcal{F}$ mixing on $C(Y;X)$. Then $\Omega_{\mathcal{F}}\varphi_* = C \times C$. On the other hand, if φ is not $\mathcal{F}$ uniformly rigid then $|\Omega_{\mathcal{F}}\varphi_*|C\text{epi}| = \emptyset$. However we do have the following result.

COROLLARY 8.5. *Let $\varphi : T \times X \to X$ be a uniform action with X a compact manifold without boundary. Let $\mathcal{F}$ be a full family for T. If 1_X is an $\mathcal{F}$ nonwandering point for the action φ_* on $C(X;X)$, then φ is $\mathcal{F}$ uniformly rigid.*

PROOF. Because X is a manifold, $C(X;X)$ is locally connected. The components, which are therefore clopen, are the homotopy classes of maps on X. If $g : X \to X$ is not surjective, then the induced map on mod 2 homology in the dimension of X is not injective. Every g homotopic to 1_X induces the identity on homology. Thus the component of 1_X consists of surjective maps. Then there exists $W_0 \in \mathcal{U}_X$ such that:

$$W_0^X(1_X) \subset C\text{epi}(X;X) \tag{8.29}$$

The rest of the argument follows that of Proposition 8.8 using (8.28); just choose $W \subset W_0$. ■

Contrast these uniform results with our earlier pointwise results: For φ_* on X^X, 1_X is recurrent when φ is weakly rigid (cf. Theorem 7.1c). 1_X is nonwandering when φ and all products of copies of φ are central (cf. Proposition 4.8c).

Extending the language of Chapter 1, we call the uniformity $\mathcal{U}_X$ + *invariant* with respect to an action $\varphi : T \times X \to X$ if it is generated by elements $V \in \mathcal{U}_X$ such that $(f^t \times f^t)(V) \subset V$ for all $t \in T$, or equivalently, such that $V = V_\varphi^T$. Thus $\mathcal{U}_X$ is + invariant when $\mathcal{U}_X = \mathcal{U}_\varphi$. We call the uniformity *invariant* if it is generated by elements $V \in \mathcal{U}_X$ such that $V = (f^t \times f^t)^{-1}(V)$ for all $t \in T$. Thus (8.26) says that if φ is uniformly rigid, then the uniformity $\mathcal{U}_\varphi$ is invariant.

PROPOSITION 8.9. *For a $\mathcal{F}$ full family for T, assume that φ is a $k\mathcal{F}$ transitive and $\mathcal{F}$ almost equicontinuous action. For every $V \in \mathcal{U}_X$, $Eq_V^{\mathcal{T}}$ is open and dense in X, so φ is almost equicontinuous (i.e., $k\mathcal{B}_T$ almost equicontinuous). Furthermore, φ is $k\mathcal{F}$ uniformity rigid, and:*

$$\text{Trans}_\varphi = Eq_\varphi^{\mathcal{F}} = Eq_\varphi = \{x : \omega_{k\mathcal{F}}\varphi(x) = X\} \tag{8.30}$$

If, in addition φ is uniformly reversible, then the reverse action $\overline{\varphi}$ is $k\mathcal{F}$ transitive and almost equicontinuous. If X is compact, then φ is uniformly reversible.

PROOF. Since $k\mathcal{B}_T \subset \mathcal{F}$, we have

$$Eq_V^T \subset Eq_V^{k\mathcal{B}_T} \subset Eq_V^{\mathcal{F}}$$

If V is closed, then by (8.23) they are equal. Thus $\mathcal{F}$ almost equicontinuity implies almost equicontinuity when φ is $k\mathcal{F}$ transitive (the reverse implication is always true). $k\mathcal{F}$ uniform rigidity follows from (5) of the Hinge Lemma 8.2.

In particular from (8.23) we have $Eq_\varphi^{\mathcal{F}} = Eq_\varphi$. Because $k\mathcal{B}_T$ is translation invariant, $\text{Trans}_\varphi \subset Eq_\varphi$ by (8.22). If $x \in Eq_\varphi$, then by Corollary 8.3, $\omega_{k\mathcal{F}}\varphi(x) = \Omega_{k\mathcal{F}}\varphi(x)$, and this set is X by $k\mathcal{F}$ transitivity. Finally $\omega_{k\mathcal{F}}\varphi(x) = X$ implies $\omega\varphi(x) = X$ because $\mathcal{F}$ is full, so $x \in \text{Trans}_\varphi$. Equation (8.30) follows.

Suppose φ is reversible. For any pair of open sets U_1, U_2, $N^\varphi(U_1,U_2) = N^{\overline{\varphi}}(U_2,U_1)$, so $\overline{\varphi}$ is $k\mathcal{F}$ transitive. When $V = V^{-1}$ is closed in $\mathcal{U}_X$, (5) of the Hinge Lemma is the same for φ and $\overline{\varphi}$, i.e., $N^\varphi(U,U) = N^{\overline{\varphi}}(U,U)$ and $f^s \subset V$ iff $f^{-s} \subset V$. Then $\overline{\varphi}$ is almost equicontinuous because φ is. In fact for $V = V^{-1}$ closed in $\mathcal{U}_X$:

$$Eq_{V,\varphi}^T = Eq_{V,\overline{\varphi}}^T \tag{8.31}$$

If X is compact, then φ is reversible by Proposition 8.7f. ■

Recall that φ is minimal when $\text{Trans}_\varphi = X$, and φ is equicontinuous when $Eq_\varphi = X$.

THEOREM 8.2. *Assume $\varphi : T \times X \to X$ is a transitive uniform action.*

a. The following conditions are equivalent; When they hold, φ is minimal and equicontinuous:

(1) φ is minimal and almost equicontinuous.

(2) φ is equicontinuous.

(3) $\mathcal{U}_\varphi$ is topologically equivalent to $\mathcal{U}_X$, i.e., while the inclusion $\mathcal{U}_\varphi \subset \mathcal{U}_X$ may be proper. the induced topologies on X agree.

When φ is minimal and equicontinuous, it is uniformly rigid and $\mathcal{U}_\varphi$ is the unique φ invariant uniformity topologically equivalent to $\mathcal{U}_X$.

If φ is uniformly reversible as well as minimal and equicontinuous, then the reverse action $\overline{\varphi}$ is minimal and equicontinuous.

b. Assume in addition that X is compact. The following conditions are equivalent, and these imply that φ is minimal and equicontinuous:

(1) φ is minimal and $\tau\mathcal{B}_T$ almost equicontinuous.

(2) φ is equicontinuous.

(3) $\mathcal{U}_X = \mathcal{U}_\varphi$

(4) $\Omega(\varphi \times \varphi)(1_X) = 1_X$

When X is compact, a minimal equicontinuous action is $k\tau\mathcal{B}_T$ uniformly rigid and uniformly reversible.

PROOF. (a) Apply Proposition 8.9 with $\mathcal{F} = \mathcal{B}_T$. By (8.30) (1) and (2) are equivalent; (2) is equivalent to $Eq_V^T = X$ for all $V \in \mathcal{U}_X$ by (8.23). Then (2) is equivalent to the condition that each V_φ^T is a neighborhood of the diagonal 1_X for all $V \in \mathcal{U}_X$; this is (3). If φ is reversible, then by (8.31) (2) for φ, i.e., $Eq_{V,\varphi}^T = X$ for all $V \in \mathcal{U}_X$ implies the same for $\overline{\varphi}$.

By Proposition 8.9 a minimal equicontinuous action is uniformly rigid, so by Proposition 8.7 the uniformity $\mathcal{U}_\varphi$ is invariant.

Suppose $\mathcal{U}_1$ and $\mathcal{U}_2$ are uniformities, each generated by invariant elements and each inducing the original $\mathcal{U}_X$ topology. We prove $\mathcal{U}_1 \subset \mathcal{U}_2$. Hence $\mathcal{U}_1 \subset \mathcal{U}_\varphi$ and $\mathcal{U}_\varphi \subset \mathcal{U}_1$.

Let W be an invariant element of $\mathcal{U}_1$ and $x \in X$. Then W is a neighborhood of the diagonal. There exists $V_1 \in \mathcal{U}_2$ such that $V_1(x) \times V_1(x) \subset W$. Choose V to be an invariant element of $\mathcal{U}_2$ so that $V^2 \subset V_1$. If $(x_1,x_2) \in V$, then by minimality there exists $t \in T$ such that $f^t(x_1) \in V(x) \subset V_1(x)$. By invariance, $x_2 \in V(x_1)$ implies

$$f^t(x_2) \in V(f^t(x_1)) \subset V^2(x) \subset V_1(x)$$

Thus, $(f^t(x_1), f^t(x_2)) \in W$. By invariance $(x_1,x_2) \in W$. Because $V \subset W$, $W \in \mathcal{U}_2$.

(b) (2) $\Rightarrow$ (1) by (a).

(1) $\Rightarrow$ (3). By Theorem 3.1 a compact minimal action φ satisfies $\omega_{k\tau\mathcal{B}_T}\varphi(x) = X$ for all x, so φ is $k\tau\mathcal{B}_T$ transitive. By Proposition 8.9 with $\mathcal{F} = \tau\mathcal{B}_T$, φ is almost equicontinuous, so by (a), $\mathcal{U}_\varphi$ and $\mathcal{U}_X$ are topologically equivalent. Since a compact space has a unique uniformity, $\mathcal{U}_\varphi = \mathcal{U}_X$.

(3) $\Rightarrow$ (2). This is obvious.

(2) $\Leftrightarrow$ (4). Apply Proposition 8.6 to $\mathcal{F} = k\mathcal{B}_T$.

Since φ is $k\tau\mathcal{B}_T$ transitive, it is $k\tau\mathcal{B}_T$ uniformly rigid and reversible by Proposition 8.9. ∎

PROPOSITION 8.10. *Let $\varphi : T \times X \to X$ be a transitive uniform action with X compact. Assume that either T is separable or* $\text{Trans}_\varphi \neq \emptyset$.

a. If every point of X is minimal, i.e., X is a union of minimal subsets, then φ is minimal.

b. The following conditions on φ are equivalent:

(1) φ *is $k\tau\mathcal{B}_T$ transitive and $\tau\mathcal{B}_T$ almost equicontinuous.*

(2) φ is distal and $\tau\mathcal{B}_T$ almost equicontinuous.

(3) φ is minimal and equicontinuous.

PROOF. (a) Observe first that if $x \in \text{Trans}_\varphi$ and so $X = \omega\varphi(x)$ then x a minimal point implies φ is minimal. In particular if X is metrizable, then by Proposition 4.6, $\text{Trans}_\varphi \neq \emptyset$, and so φ is minimal.

For the general result we assume M_0 and M_1 are distinct and hence disjoint minimal subsets of X. We derive a contradiction by reducing to the metric case. Choose a continuous function $u : X \to [0,1]$ such that $M_0 \subset u^{-1}(0)$ and $M_1 \subset u^{-1}(1)$. Let E be the subalgebra of $\mathcal{B}^u(X) = \mathcal{B}(X)$ generated by $\{u \circ f^t : t \in T\}$. Since T is separable, E is a separable, $\varphi +$ invariant subalgebra of $\mathcal{B}^u(X)$. The compactification $j_E : X \to X_E$ induces a uniform action φ_E on X_E so that j_E is a surjective action map. Since each point $x \in X$ is minimal, $j_E(x)$ is minimal in X_E. Thus φ_E satisfied the hypotheses of (a), and in addition separability of E implies X_E is metrizable; hence φ_E is minimal (see paragraph 1). But the induced function $u_E : X_E \to [0,1]$ maps $j_E(M_0)$ and $j_E(M_1)$ to distinct points. Hence $j_E(M_0)$ and $j_E(M_1)$ are disjoint minimal subsets of X_E. This contradiction establishes that X was minimal to begin with.

(b) By Proposition 8.9 (1) implies φ is $k\tau\mathcal{B}_T$ uniformly rigid and so is distal by Theorem 7.1, i.e., (1) $\Rightarrow$ (2). By Theorem 3.1, φ is minimal iff $\omega_{k\tau\mathcal{B}_T}\varphi(x) = X$ for all $x \in X$. (3) $\Rightarrow$ (1) follows. Finally (2) implies each point is minimal. By (a), φ is minimal, so equicontinuity, and hence (3), follow from Theorem 8.2b. ∎

THEOREM 8.3. *Let $\varphi : T \times X \to X$ be a uniform action with X compact and* $\text{Trans}_\varphi \neq \emptyset$. *Assume φ is not an equicontinuous, minimal action.*

a. If φ is $\tau\mathcal{B}_T$ almost equicontinuous, then for all $x \in \text{Trans}_\varphi$, $x \notin \Omega_{k\tau\mathcal{B}_T}\varphi(x)$, so φ has as an almost homeomorphic factor a nontrivial $\tau\mathcal{B}_T$ eversion.

b. If φ is $\tau k\tau\mathcal{B}_T$ almost equicontinuous, then for all $x \in \text{Trans}_\varphi$, $x \notin \Omega_{k\tau k\tau\mathcal{B}_T}\varphi(x)$, so φ has as an almost homeomorphic factor a nontrivial $\tau k\tau\mathcal{B}_T$ eversion.

PROOF. If for some $x \in \text{Trans}_\varphi$, $x \in \Omega_{k\tau\mathcal{B}_T}\varphi(x)$ in (a) or $x \in \Omega_{k\tau k\tau\mathcal{B}_T}\varphi(x)$ in (b), then by Proposition 8.9 φ is almost equicontinuous, and $x \in \text{Trans}_\varphi$ satisfies $\omega_{k\tau\mathcal{B}_T}\varphi(x) = X$ in (a) and $\omega_{k\tau k\tau\mathcal{B}_T}\varphi(x) = X$ in (b). By Theorem 3.1 φ is minimal in (a), and in (b) X is the mincenter of φ. Therefore in (a) we see that φ is minimal and equicontinuous by Theorem 8.2b. In (b) $x \in \text{Trans}_\varphi = Eq_\varphi$ is a limit of minimal points, so is minimal by Corollary 8.3. Thus again φ is minimal and equicontinuous by Theorem 8.2b.

The eversion results follow from Theorem 4.3. ∎

PROPOSITION 8.11. *Let $\tilde{\varphi} : T \times \tilde{X} \to \tilde{X}$ be a nontrivial $\mathcal{F}$ eversion with fixed point e. The action $\tilde{\varphi}$ is $\mathcal{F}$ almost equicontinuous. In fact $Eq^{\mathcal{F}}_{\tilde{\varphi}} \supset \tilde{X}\setminus\{e\}$, with equality if $\mathcal{F}$ is a filter.*

PROOF. Given $x \in \tilde{X}\backslash\{e\}$ and $V \in \mathcal{U}_{\tilde{X}}$, let U_1 be a closed neighborhood of e with $x \notin U_1$ and $U_1 \times U_1 \subset V$. By definition of an $\mathcal{F}$ eversion, $J^{\tilde{\varphi}}(\tilde{X}\backslash U_1, U_1) \in \mathcal{F}$. Clearly $J^{\tilde{\varphi}}(\tilde{X}\backslash U_1, U_1)$ is contained in $J^{\tilde{\varphi}\times\tilde{\varphi}}((\tilde{X}\backslash U_1)\times(\tilde{X}\backslash U_1), V)$. Hence $x \in Eq_V^{\mathcal{F}}$.

On the other hand if $t \in J^{\tilde{\varphi}}(\tilde{X}\backslash U_1, U_1)$, then:

$$e = f^t(e) \in f^t(U_1) \qquad f^t(U_1) \supset \tilde{X}\backslash U_1$$

because the eversion is surjective. Hence:

$$(f^t \times f^t)(U_1 \times U_1) \supset e \times (\tilde{X}\backslash U_1)$$

If $V(e) \neq \tilde{X}$ $t \notin J^{\tilde{\varphi}\times\tilde{\varphi}}(U_1 \times U_1, V)$. If $\mathcal{F}$ is a filter, $J(\tilde{X}\backslash U_1, U_1) \in \mathcal{F}$ implies that the disjoint set $J(U_1 \times U_1, V)$ is not in $\mathcal{F}$. ■

THEOREM 8.4. *Let $\varphi : T \times X \to X$ be a uniform action such that* $\mathrm{Trans}_\varphi \neq \emptyset$. *The following conditions are equivalent:*

(1) φ is almost equicontinuous.

(2) The restriction of φ to the invariant subset Trans_φ *is equicontinuous and minimal.*

(3) On the subset Trans_φ, *the uniformity $\mathcal{U}_\varphi$ is topologically equivalent to $\mathcal{U}_X$; i.e., $\mathcal{U}_\varphi$ induces on* Trans_φ *the original $\mathcal{U}_X$ topology.*

(4) For every $V \in \mathcal{U}_X$, there exists $x \in$ Trans_φ *and U a neighborhood of x in X such that for $s \in T$, $f^s(x) \in U$ implies $(f^t(x), f^{s+t}(x)) \in V$ for all $t \in T$.*

(5) For every $V \in \mathcal{U}_X$, there exists $x \in$ Trans_φ *and U a neighborhood of x in X such that for $s \in T$ $f^s(x) \in U$ implies $(y, f^s(y)) \in V$ for all $y \in X$, i.e., $f^s \subset V$.*

If $\mathcal{U}_X$ is complete on X, then $\mathcal{U}_\varphi$ is complete on Trans_φ.

PROOF. (1) $\Rightarrow$ (2). φ always restricts to a minimal action on Trans_φ when the latter set is nonempty. Every point of Trans_φ is an equicontinuity point for φ and hence for the restriction to Trans_φ by (8.22).

(2) $\Rightarrow$ (3) by Theorem 8.2a.

(3) $\Rightarrow$ (4). V_φ^T is a neighborhood of (x,x) for any $x \in \mathrm{Trans}_\varphi$. Let $U = V_\varphi^T(x)$. $y \in U$ implies $(f^t(x), f^t(y)) \in V$ for all $t \in T$, from which (4) follows.

(4) $\Rightarrow$ (5). Shrink V to assume V is closed. Then $\{y : (y, f^s(y)) \in V\}$ is a closed subset. If it includes all $f^t(x)$s for some $x \in \mathrm{Trans}_\varphi$, then it equals X.

(5) $\Rightarrow$ (4). Let $y = f^t(x)$.

(4) $\Rightarrow$ (1). Given $W \in \mathcal{U}_X$ choose $V \in \mathcal{U}_X$ closed, symmetric, and such that $V^2 \subset W$. Choose $x \in \mathrm{Trans}_\varphi$ and U open to satisfy (4). For $y \in U$ choose a

net $\{f^{s_\alpha}(x)\}$ in U converging to y. By (4) $(f^t(x), f^t(f^{s_\alpha}(x)) \in V$ for all α, so $(f^t(x), f^t(y)) \in V$. Hence $(y_1, y_2) \in U \times U$ implies $(f^t(y_1), f^t(y_2)) \in V^2 \subset W$. Thus $Eq_W^{k\mathcal{B}_T} \neq \emptyset$ for all $W \in \mathcal{U}_X$. φ is almost equicontinuous by the Auslander–Yorke Theorem 8.1.

If $\mathcal{U}_X$ is complete then $\mathcal{U}_\varphi$ is complete on X and hence on the $\mathcal{U}_\varphi$ closed subset Trans_φ by Proposition 8.1. ∎

Let $h : X \to X_1$ be a continuous map. We call h *almost open* if U open and nonempty implies $\mathrm{Int}\, h(U) \neq \emptyset$ (see Appendix). We call h *almost quasi-open* if U open and nonempty implies $\mathrm{Int}\, \overline{h(U)} \neq \emptyset$. If $h(X)$ is dense in X_1 and h is almost open from X to $h(X)$, e.g., if h is a dense embedding, then h is almost quasi-open by Proposition 5.5b.

PROPOSITION 8.12. *Let $h : \varphi \to \varphi_1$ be a uniformly continuous action map. Assume that $h : X \to X_1$ is almost quasi-open and $h(X)$ is dense in X_1. If for some proper family $\mathcal{F}$, φ is $\mathcal{F}$ almost equicontinuous, then φ_1 is $\mathcal{F}$ almost equicontinuous.*

PROOF. Let $V_1 \in \mathcal{U}_{X_1}$ be closed and let U_1 be a nonempty open set in X_1. Then $V \equiv (h \times h)^{-1} V_1$ is in $\mathcal{U}_X$ because h is uniformly continuous. Because $h(X)$ is dense in X_1, $U \equiv h^{-1}(U_1)$ is a nonempty open set in X. Since $Eq_V^{\mathcal{F}}$ is open and dense in X, there exists U_0 a nonempty open subset of U such that $J^{\varphi \times \varphi}(U_0 \times U_0, V) \in \mathcal{F}$. Since h is almost quasi-open, $\overline{h(U_0)}$ has a nonempty interior W_0. Since h is an action map and V_1 is closed:

$$J^{\varphi \times \varphi}(U_0 \times U_0, V) \subset J^{\varphi_1 \times \varphi_1}(\overline{h(U_0)} \times \overline{h(U_0)}, V_1) \subset J^{\varphi_1 \times \varphi_1}(W_0 \times W_0, V_1)$$

Consequently $J^{\varphi_1 \times \varphi_1}(W_0 \times W_0, V_1)$ is in $\mathcal{F}$. Thus $W_0 \subset Eq_{\varphi_1, V_1}^{\mathcal{F}}$, so $Eq_{\varphi_1, V_1}^{\mathcal{F}}$ meets U_1. Since this is true for every nonempty open set U_1, $Eq_{\varphi_1, V_1}^{\mathcal{F}}$ is dense in X_1. As the closed V_1s generate $\mathcal{U}_{X_1}$, φ_1 is $\mathcal{F}$ almost equicontinuous. ∎

PROPOSITION 8.13. *Let $h : \varphi \to \varphi_1$ be a continuous action map of uniform actions. Assume that $h : X \to X_1$ is a dense embedding, i.e., $h : X \to h(X)$ is a homeomorphism and $h(X)$ is dense in X_1. If* $\mathrm{Trans}_\varphi \neq \emptyset$, *then* $\mathrm{Trans}_{\varphi_1} \neq \emptyset$. *In fact:*

$$\mathrm{Trans}_\varphi = h^{-1}(\mathrm{Trans}_{\varphi_1}) \tag{8.32}$$

Assume now that $\mathrm{Trans}_\varphi \neq \emptyset$ *and φ is almost equicontinuous so that the uniformity $\mathcal{U}_\varphi$ on* Trans_φ *is topologically equivalent to $\mathcal{U}_X$.*

The following conditions are then equivalent:

(1) φ_1 is almost equicontinuous.

(2) $h : (\mathrm{Trans}_\varphi, \mathcal{U}_\varphi) \to (X_1, \mathcal{U}_{X_1})$ is uniformly continuous.

(3) $h : (\mathrm{Trans}_\varphi, \mathcal{U}_\varphi) \to (\mathrm{Trans}_{\varphi_1}, \mathcal{U}_{\varphi_1})$ *is a uniform embedding; i.e.,* $\mathcal{U}_\varphi$ *is the same as the uniformity induced on* Trans_φ *by pulling back* $\mathcal{U}_{\varphi_1}$.

If these conditions hold and $(\mathrm{Trans}_\varphi, \mathcal{U}_\varphi)$ *is complete, then:*

$$h(\mathrm{Trans}_\varphi) = \mathrm{Trans}_{\varphi_1} \tag{8.33}$$

In particular if $h : (X, \mathcal{U}_X) \to (X_1, \mathcal{U}_{X_1})$ *is uniformly continuous, then* φ_1 *is almost equicontinuous. Furthermore if* $(X, \mathcal{U}_X)$ *is complete, then (8.33) holds.*

PROOF. For each $t \in T$ and $x \in X$, $h(f^{T_t}(x)) = f_1^{T_t}(h(x))$. If $x \in \mathrm{Trans}_\varphi$, then $f^{T_t}(x)$ is dense in X for all t, so $f_1^{T_t}(h(x))$ is dense in $h(X)$ and in X_1 for all t. Thus $h(x) \in \mathrm{Trans}_{\varphi_1}$. If $h(x) \in \mathrm{Trans}_{\varphi_1}$, then $f_1^{T_t}(h(x))$ is dense in X_1 and a fortiori in $h(X)$. Since $h : X \to h(X)$ is a homeomorphism, $f^{T_t}(x)$ is dense in X; hence $x \in \mathrm{Trans}_\varphi$.

If φ is almost equicontinuous, then $\mathcal{U}_\varphi$ is topologically equivalent to $\mathcal{U}_X$ on Trans_φ by Theorem 8.4. For the equivalences:

(3) $\Rightarrow$ (2). $(\mathrm{Trans}_{\varphi_1}, \mathcal{U}_{\varphi_1}) \to (X_1, \mathcal{U}_{X_1})$ is always uniformly continuous since $\mathcal{U}_{X_1} \subset \mathcal{U}_{\varphi_1}$ on X_1.

(2) $\Rightarrow$ (1). Because φ is almost equicontinuous, the action $\hat{\varphi}$ on $(\mathrm{Trans}_\varphi, \mathcal{U}_\varphi)$ is an equicontinuous minimal action by Theorem 8.4, and h restricts to a dense embedding of Trans_φ into X_1. Because $h : \hat{\varphi}_{\mathrm{Trans}_\varphi} \to \varphi_1$ is a uniformly continuous action map by (2), Proposition 8.12 implies that φ_1 is almost equicontinuous.

(1) $\Rightarrow$ (3). Since φ and φ_1 are both equicontinuous, $\mathcal{U}_\varphi$ on Trans_φ and $\mathcal{U}_{\varphi_1}$ on $\mathrm{Trans}_{\varphi_1}$ induce the original topologies there by Theorem 8.4. Since $h : \mathrm{Trans}_\varphi \to \mathrm{Trans}_{\varphi_1}$ is a topological embedding, it follows that the uniformities $\mathcal{U}_\varphi$ and $h^* \mathcal{U}_{\varphi_1}$ on Trans_φ yield the same original topology and these are both generated by invariant elements [cf. (8.26)]. By Theorem 8.2a these two uniformities agree.

If (1)–(3) hold, then $h(\mathrm{Trans}_\varphi)$ is a dense subset of $(\mathrm{Trans}_{\varphi_1}, \mathcal{U}_{\varphi_1})$. If $\mathcal{U}_\varphi$ is complete on Trans_φ, then (3) implies $h(\mathrm{Trans}_\varphi)$ is a complete subset of $(\mathrm{Trans}_{\varphi_1}, \mathcal{U}_{\varphi_1})$, so it is a closed subset. Since it is a dense, closed subset of $\mathrm{Trans}_{\varphi_1}$, $h(\mathrm{Trans}_\varphi) = \mathrm{Trans}_{\varphi_1}$.

If $h : (X, \mathcal{U}_X) \to (X_1, \mathcal{U}_{X_1})$ is uniformly continuous then (2) holds because $(\mathrm{Trans}_\varphi, \mathcal{U}_\varphi) \to (X, \mathcal{U}_X)$ is uniformly continuous. If $\mathcal{U}_X$ is complete, then by Proposition 8.1 $\mathcal{U}_\varphi$ is complete on X and Trans_φ is $\mathcal{U}_\varphi$ closed and hence $\mathcal{U}_\varphi$ complete. ■

COROLLARY 8.6. *Let* $\varphi : T \times X \to X$ *be an almost equicontinuous uniform action. If* E *is a closed* $\varphi+$ *invariant subalgebra of* $\mathcal{B}^u(X)$ *that distinguishes points and closed sets, then* $j_E : X \to X_E$ *is a uniformly continuous dense embedding inducing an almost equicontinuous action* φ_E *on* X_E.

$$j_E(\mathrm{Trans}_\varphi) \subset \mathrm{Trans}_{\varphi_E}$$

with equality if $\mathcal{U}_\varphi$ is complete on Trans_φ *(and so a fortiori if $\mathcal{U}_X$ is complete on X).*

PROOF. The induced action φ_E is described in Proposition 5.12. Since $j_E : \varphi \to \varphi_E$, then satisfies the hypotheses of Proposition 8.12 it follows that φ_E is almost equicontinuous. The results on the set of transitive points follow from Proposition 8.13. ∎

PROPOSITION 8.14. *Let $h : \varphi \to \varphi_1$ be a surjective continuous action map, with φ an almost equicontinuous uniform action on a compact space X.*

If h is an almost open map, then φ_1 is almost equicontinuous.

If there exists $x_1 \in \text{Trans}_{\varphi_1}$ such that $h^{-1}(x_1) \subset \text{Trans}_\varphi$, then φ_1 is almost equicontinuous. In particular if there exists $x \in \text{Trans}_\varphi$ such that $x = h^{-1}(h(x))$, then φ_1 is almost equicontinuous.

If h is a minimal action map and $\text{Trans}_\varphi \neq \emptyset$, then φ_1 is almost equicontinuous.

If φ is minimal, then φ_1 is minimal and equicontinuous.

PROOF. If h is almost open, then φ_1 is almost equicontinuous by Proposition 8.12.

Assume now that $h^{-1}(x_1) \subset \text{Trans}_\varphi$, so $h^{-1}(x_1)$ is a compact subset of Eq_φ by Proposition 8.9. For $V_1 \in \mathcal{U}_{X_1}$, let $V_2 = V_2^{-1} \in \mathcal{U}_{X_1}$, with $V_2^2 \subset V_1$; let $V = (h \times h)^{-1}(V_2) \in \mathcal{U}_X$. By (8.23) $h^{-1}(x_1) \subset Eq^T_{V,\varphi}$, so there is a collection $\{U_1, \dots, U_n\}$ of open subsets of X, covering $h^{-1}(x_1)$ and such that $J^{\varphi\times\varphi}(U_i \times U_i, V) = T$ for $i = 1, \dots, n$. We can assume as well that each U_i meets $h^{-1}(x_1)$. By compactness we can choose an open set U of X_1 such that $h^{-1}(U) \subset \cup_{i=1}^n U_i$. For $x_2 \in U$ we can choose some U_i and $z_2 \in U_i$ such that $h(z_2) = x_2$. Choose $z_1 \in U_i \cap h^{-1}(x_1)$. Since $(z_1, z_2) \in U_i \times U_i$:

$$(f^t(z_1), f^t(z_2)) \in V = (h \times h)^{-1} V_2$$

for all $t \in T$. Hence, $(f_1^t(x_1), f_1^t(x_2)) \in V_2$ for all $t \in T$. If $x_2, x_3 \in U$:

$$(f_1^t(x_1), f_1^t(x_3)) \qquad (f_1^t(x_1), f_1^t(x_2)) \in V_2$$

Since $V_2^{-1} V_2 \subset V_1$, $(f_1^t(x_2), f_1^t(x_3)) \in V_1$ for all t, i.e., $J^{\varphi_1\times\varphi_1}(U \times U, V_1) = T$. It follows that $x_1 \in Eq_{\varphi_1}$ and so φ_1 is almost equicontinuous by the Auslander–Yorke Dichotomy Theorem (Theorem 8.1).

If $x \in \text{Trans}_\varphi$, then since h is surjective, $x_1 = h(x) \in \text{Trans}_{\varphi_1}$. If $x = h^{-1}(h(x))$, $h^{-1}(x_1) \subset \text{Trans}_\varphi$, so φ_1 is almost equicontinuous.

If h is minimal, then $h^{-1}(x_1) \subset \text{Trans}_\varphi$ for all $x_1 \in \text{Trans}_{\varphi_1}$ by (5.67). So φ_1 is almost equicontinuous.

If φ is minimal then h surjective implies φ_1 is minimal. Furthermore h is minimal, so φ_1 is almost equicontinuous; φ_1 is equicontinuous by Theorem 8.2. ∎

PROPOSITION 8.15. *Any factor or subsystem of an almost equicontinuous transitive action is uniformly rigid. Conversely if φ is a uniformly rigid action with X compact and* $\text{Trans}_\varphi \neq \emptyset$ *then there exists* φ_0*, an almost equicontinuous action with X_0 compact and* $\text{Trans}_{\varphi_0} \neq \emptyset$ *and uniformly continuous action maps $r : \varphi_0 \to \varphi$, $j : \varphi \to \varphi_0$ satisfying $r \circ j = 1_\varphi$.*

PROOF. An almost equicontinuous transitive system is uniformly rigid by Proposition 8.9 (with $\mathcal{F} = k\mathcal{B}_T$). By Proposition 8.7d any subsystem or factor is then uniformly rigid.

Now suppose X is compact, φ is uniformly rigid, and $x \in \text{Trans}_\varphi$. Assume φ is not almost equicontinuous (otherwise let $\varphi_0 = \varphi$). By Proposition 8.7a, the action $\hat{\varphi}$ on X with uniformity $\mathcal{U}_\varphi$ is uniformly rigid. Let $\hat{X}_x = \omega\hat{\varphi}(x)$, a $\mathcal{U}_\varphi$ closed subset of X with $x \in \hat{X}_x$ by uniform rigidity of $\hat{\varphi}$. Let $\hat{\mathcal{U}}$ be the restriction of $\mathcal{U}_\varphi$ to $\hat{X}_x$ and denote by $\hat{\varphi}_x$ the restriction of $\hat{\varphi}$. Clearly $\hat{\mathcal{U}} = \hat{\mathcal{U}}_{\hat{\varphi}_x}$ on $\hat{X}_x$, so by Proposition 8.1 $\text{Trans}_{\hat{\varphi}_x}$ is a closed subset of $\hat{X}_x$. Since it contains the dense subset $f^T(x)$, $\text{Trans}_{\hat{\varphi}_x} = \hat{X}_x$. That is, $\hat{\varphi}_x$ is a minimal equicontinuous action. Let E be a $\hat{\varphi}_x+$ invariant, closed subalgebra of $\mathcal{B}^u(\hat{X}_x)$ that contains $\hat{r}^*\mathcal{B}^u(X)$ where $\hat{r} : \hat{\varphi}_x \to \varphi$ is induced by the obvious uniformly continuous inclusion map from $(\hat{X}_x, \hat{\mathcal{U}})$ to $(X, \mathcal{U}_X)$. Choose E so that it separates points and closed sets. Then $j_E : \hat{X}_x \to X_E$ is a uniformly continuous dense embedding, inducing the almost equicontinuous action φ_E with:

$$\text{Trans}_{\varphi_E} = j_E(\hat{X}_x) = j_E(\text{Trans}_{\hat{\varphi}_x})$$

since $\hat{\mathcal{U}}$ is complete (cf. Corollary 8.6).

Because $\hat{r}^*\mathcal{B}^u(X) \subset E$, the map $\hat{r} : \hat{X}_x \to X$ factors through j_E to define $r_E : X_E \to X$, mapping φ_E to φ. Since $r_E(X_E)$ contains $x \in \text{Trans}_\varphi$, r_E is surjective by compactness.

Now because φ_E is almost equicontinuous and φ is not, Proposition 8.14 implies $r_E^{-1}(x)$ is not contained in Trans_{φ_E}. Let $z \in r_E^{-1}(x) \setminus \text{Trans}_{\varphi_E}$ and $\tilde{X} = \omega\varphi_E(z)$, a compact subset of $\tilde{X}$ containing z because $\tilde{X}_E$ is uniformly rigid. Thus $r_E(\tilde{X}) = X$.

Define the closed invariant equivalence relation:

$$[(r_E^{-1} \circ r_E) \cap (\tilde{X} \times \tilde{X})] \cup 1_{X_E}$$

on X_E. Denote the quotient map by $q : X_E \to X_0$. Since the equivalence relation is invariant, we obtain an action φ_0 on X_0 with q mapping φ_E to φ_0. Clearly r_E factors through q to obtain r_0: $\varphi_0 \to \varphi$ with $r_E = r_0 \circ q$.

Because q is surjective, and injective on the complement of $\tilde{X}$, which includes all of Trans_{φ_E}, φ_0 is almost equicontinuous by Proposition 8.14 applied to $h = q$.

The restriction of r_0 to the closed invariant subset $q(\tilde{X})$ is a homeomorphism by compactness and the definition of the equivalence relation. Let j_0 denote the inverse homeomorphism. ∎

PROPOSITION 8.16. *Let $\varphi : T \times X \to X$ be a uniform action with $\mathcal{U}_X = \mathcal{U}_\varphi$; i.e., $\mathcal{U}_X$ is $\varphi+$ invariant. Let $(\tilde{X}, \mathcal{U}_{\tilde{X}})$ be the completion of $(X, \mathcal{U}_X)$ and let $\tilde{\varphi} : T \times \tilde{X} \to \tilde{X}$ be the extension of the action φ. Then $\mathcal{U}_{\tilde{X}} = \mathcal{U}_{\tilde{\varphi}}$, i.e., $\mathcal{U}_{\tilde{X}}$ is $\tilde{\varphi}+$ invariant.*

Assume now that φ is minimal, so that φ and $\tilde{\varphi}$ are equicontinuous, minimal actions. The following conditions are equivalent:

(1) $\mathcal{U}_X = \mathcal{U}_\varphi$ is a totally bounded uniformity.

(2) For every nonempty open set U in X, there is a finite set $\{t_1, \dots, t_n\}$ in T such that $X = \cup_{i=1}^n f^{-t_i}(U)$.

(3) The completion $\tilde{X}$ is compact.

(4) The completion provides the unique compactification of φ. That is, if φ_1 is an action with X_1 compact and $j : \varphi \to \varphi_1$ is an action map with $j : X \to X_1$, a uniformly continuous, dense embedding, then j induces a homeomorphism of $\tilde{X}$ onto X_1 and thereby an action isomorphism of $\tilde{\varphi}$ onto φ_1.

(5) There exists a compactification $j : \varphi \to \varphi_1$ with $j : X \to X_1$ a uniformly continuous, dense embedding and φ_1 minimal.

(6) φ is $k\tau\mathcal{B}_T$ transitive.

PROOF. The set of closed $V \in \mathcal{U}_X$ such that $V = V_\varphi^T$ generates $\mathcal{U}_X$. Their closures in $\tilde{X} \times \tilde{X}$ generate $\mathcal{U}_{\tilde{X}}$ and satisfy $(\tilde{f}^t \times \tilde{f}^t)(\overline{V}) \subset \overline{V}$ for all $t \in T$. Then $\mathcal{U}_{\tilde{X}}$ is $\tilde{\varphi}$ $+$ invariant. When φ is minimal, $\tilde{\varphi}$ is at least transitive because X is dense in $\tilde{X}$. By Theorem 8.2a $\tilde{\varphi}$ is equicontinuous and minimal and by Proposition 8.9 (8.26) holds; i.e., $\mathcal{U}_X$ and $\mathcal{U}_{\tilde{X}}$ are invariant uniformities.

(2) $\Rightarrow$ (1). Given $V \in \mathcal{U}_X$ choose $V_1 = V_1^{-1}$ closed with $V_1^2 \subset V^T$ and choose $x \in X$. With $U = \operatorname{Int} V_1(x)$, choose $t_1, \dots, t_n$ so that $\{f^{-t_i}(U)\}$ covers X and choose $z_i \in f^{-t_i}(U)$. If $z \in f^{-t_i}(U)$, then:

$$(f^{t_i}(z_i), f^{t_i}(z)) \in V_1^2 \subset V^T$$

then $z \in V(z_i)$ by (8.26). Thus $\{V(z_i) : i = 1, \dots, n\}$ covers X.

(1) $\Rightarrow$ (4). Because j is uniformly continuous it extends to a map $\tilde{j}$ of $\tilde{X}$ to the compact space X_1. By (1) X is totally bounded, so $\tilde{X}$ is compact; hence $\tilde{j}$ is surjective. Since $\tilde{j}$ maps $\tilde{\varphi}$ onto φ_1, the latter is minimal and equicontinuous. By Theorem 8.2b $\mathcal{U}_{X_1} = \mathcal{U}_{\varphi_1}$, so by Theorem 8.2a $\mathcal{U}_X = \mathcal{U}_\varphi$ is the same uniformity as the pull back $j^* \mathcal{U}_{\varphi_1}$; i.e., $j : X \to X_1$ is a uniform embedding. Thus the restriction of j to a map of X to $j(X)$ is a uniform space isomorphism. Then $\tilde{j}$ is the induced isomorphism of the completions.

(4) $\Rightarrow$ (3). Start with any compactification $j : \varphi \to \varphi_1$ with j a dense embedding and apply (4) to get $\tilde{X}$ compact.

(3) $\Rightarrow$ (6). Since $\tilde{\varphi}$ is a compact minimal action, it is $k\tau\mathcal{B}_T$ transitive and so φ is as well, because X is dense in $\tilde{X}$.

(6) $\Rightarrow$ (5). There exists a compactification $j : \varphi \to \varphi_1$ with j a uniformly continuous dense embedding. Any such φ_1 is almost equicontinuous by Proposition 8.12. Since φ is $k\tau\mathcal{B}_T$ transitive, φ_1 is also with $\text{Trans}_{\varphi_1} \supset j(X)$. Then by Proposition 8.10b, φ_1 is minimal equicontinuous.

(5) $\Rightarrow$ (2). For U open and nonempty in X, there exists U_1 open and nonempty in X_1 such that $j^{-1}(U_1) = U$. Because φ_1 is minimal and X_1 is compact, there exists $\{t_1, \dots, t_n\}$ such that $\{f_1^{-t_i}(U_1)\}$ covers X_1. Since $j^{-1}(f_1^{-t_i}(U_1)) = f^{-t_i}(U)$, $\{f^{-t_i}(U)\}$ covers X. ■

We now describe Ellis's enveloping semigroup characterization of equicontinuity in the compact case.

PROPOSITION 8.17. *Let $\varphi : T \times X \to X$ be a uniform action with X compact. Let $\Phi : \beta_u T \times X \to X$ be the Ellis action extension so that $\Phi^\# : \beta_u T \to S_\varphi$ is the continuous semigroup homomorphism onto the enveloping semigroup. Assume $\mathcal{F}$ is a filter for T.*

a. A point $x \in X$ lies in $Eq_\varphi^{\mathcal{F}}$ iff Φ is continuous at (p,x) for every $p \in H(\mathcal{F})$.

b. φ is $\mathcal{F}$ equicontinuous iff Φ is continuous at every point (p,x) of $H(\mathcal{F}) \times X$. In that case, each $g \in \Phi^\#(H(\mathcal{F}))$ commutes with every element of S_φ, so it is a continuous action map from φ to itself. In addition the evaluation map Ev : *$\Phi^\#(H(\mathcal{F})) \times X \to X$ is continuous, and on the compact set $\Phi^\#(H(\mathcal{F})) \subset C(X;X)$, the pointwise topology, inherited from X^X, agrees with the uniform topology, inherited from $C(X;X)$.*

c. Assume the action φ is surjective, $\mathcal{F}$ equicontinuous, and $H(\mathcal{F})$ is a subsemigroup of $\beta_u T$. The action φ is $\mathcal{F}$ distal and $\Phi^\#(H(\mathcal{F}))$ is a compact, abelian, uniform group of homeomorphisms on X. Evaluation Ev : *$\Phi^\#(H(\mathcal{F})) \times X \to X$ and composition* Comp : *$\Phi^\#(H(\mathcal{F})) \times S_\varphi \to S_\varphi$ are uniform actions. If in addition $\mathcal{F}$ is a full filter, then φ is uniformly rigid and uniformly reversible.*

PROOF. (a) (8.11) implies that $Eq_\varphi^{\mathcal{F}} = Eq_\varphi^{u\mathcal{F}}$. Since $H(\mathcal{F}) = H(u\mathcal{F})$ as well, we can assume that $\mathcal{F}$ is an open filter.

Given $V_1 \in \mathcal{U}_X$, choose $V \in \mathcal{U}_X$ such that $V = V^{-1}$ and $V \circ V \subset V_1$.

Assume Φ is continuous at every point of $H(\mathcal{F}) \times \{x\}$ and define $G \equiv \{(p,y) \in \beta_u T \times X : (px, py) \in V\}$. G is a neighborhood of $H(\mathcal{F}) \times \{x\}$. Since $\mathcal{F}$ is a filter, there exist $F \in \mathcal{F}$ and U a compact neighborhood of x such that $H(F) \times U \subset G$. By (5.40) $H(F) = \overline{j_u(F)}$, so for all $t \in F$ and $y_1, y_2 \in U$ $(f^t(x), f^t(y_\alpha)) \in V$ for $\alpha = 1,2$. Then $(f^t(y_1), f^t(y_2)) \in V_1$; that is, $J^{\varphi \times \varphi}(U \times U, V_1)$ contains F. It follows that $x \in Eq_\varphi^{\mathcal{F}}$.

Now assume $x \in Eq_\varphi^{\mathcal{F}}$ and $p \in H(\mathcal{F})$. Choose V (as before) closed. There exists $F \in \mathcal{F}$ and an open neighborhood U of x such that $U \times U \subset V_\varphi^F$. Because

$\mathcal{F}$ is an open family, there exists $F_1 \in \mathcal{F}$ and $W \in \mathcal{U}_T$ such that $W(F_1) \subset F$. If $q \in H(F_1)$, then $F_1 \in \tilde{u}\mathcal{F}_q$, so $F \in \mathcal{F}_q$, i.e., $H(F_1) \subset \tilde{H}(F)$. It follows that

$$p \in H(\mathcal{F}) \subset H(F_1) \subset \tilde{H}(F) \subset H(F) \tag{8.34}$$

For $x_1, x_2 \in U$ we have $(f^t(x_1), f^t(x_2)) \in V$ for all $t \in F$. That is, with x_1, x_2 fixed $(\Phi_{x_1}(q), \Phi_{x_2}(q)) \in V$ for all $q \in j_u(F)$. Because V is closed, this implies $(qx_1, qx_2) \in V$ for all $q \in \overline{j_u(F)} = H(F)$. Hence, $(q, x_1) \in \tilde{H}(F) \times U$ implies $(qx, qx_1) \in V$. By continuity of Φ_x, we can choose an open neighborhood $\tilde{U}$ of p contained in $\tilde{H}(F)$ such that $q \in \tilde{U}$ implies $(px, qx) \in V$. Thus $(q, x_1) \in \tilde{U} \times U$ implies $qx_1 \in V_1(px)$, since $V \circ V \subset V_1$. Thus Φ is continuous at (p, x).

(b) $\mathcal{F}$ equicontinuity is equivalent to continuity at every point of $H(\mathcal{F}) \times X$ by (a). Then every $g \in \Phi^{\#}(H(\mathcal{F}))$ lies in $S_\varphi \cap C(X;X)$. Therefore g commutes with every $q \in S_\varphi$ by the remark following Lemma 6.2. In particular, $g \circ f^t = f^t \circ g$, so g is an action map.

By compactness the surjection $\Phi^{\#} \times 1_X$ from $H(\mathcal{F}) \times X$ to its image is a quotient map, so continuity of Ev on $\Phi^{\#}(H(\mathcal{F})) \times X$ follows from continuity of φ on $H(\mathcal{F}) \times X$.

For $g \in \Phi^{\#}(H(\mathcal{F}))$ and V open in $\mathcal{U}_X$ the set $\{(g_1, x) : (g(x), g_1(x)) \in V\}$ is open in $\Phi^{\#}(H(\mathcal{F})) \times X$ and contains the compact set $\{g\} \times X$. Then it contains $\tilde{U} \times X$ for some open neighborhood $\tilde{U}$ of g in $\Phi^{\#}(H(\mathcal{F}))$. For $g_1 \in \tilde{U}$, $(g, g_1) \in V^X$. Thus pointwise close in $\Phi^{\#}(H(\mathcal{F}))$ implies uniform close.

(c) For $p \in H(\mathcal{F})$ and $y \in X$, choose a net $\{t_\alpha\}$ such that $\{j_u(t_\alpha)\}$ converges to p in $\beta_u T$. Because the action φ is surjective, there exists $x_\alpha \in X$ such that $f^{t_\alpha}(x_\alpha) = y$. By going to a subnet we can assume $\{x_\alpha\}$ converges to some $x \in X$. Since Φ is continuous at (p, x), it follows that $\Phi^{\#}(p)(x) = px = y$. Thus, each element of $\Phi^{\#}(H(\mathcal{F}))$ is surjective. It now follows from Theorem 7.1a that φ is $\mathcal{F}$ distal and $\Phi^{\#}(H(\mathcal{F}))$ is a group of bijections on X. By continuity of Ev on $\Phi^{\#}(H(\mathcal{F})) \times X$, it follows that each of these maps is continuous, so the group consists of homeomorphisms. By (b) the topology is the uniform topology, so $\Phi^{\#}(H(\mathcal{F}))$ is a compact abelian subgroup of $C^u is(X;X)$, a topological group by Example 5 in Chapter 1. Thus $\Phi^{\#}(H(\mathcal{F}))$ is a topological group, so it is uniform. The action Ev is a uniform action, the restriction of the uniform action by $C^u is$ Comp is the restriction of the induced action on the product space X^X.

$(\Phi^{\#})^{-1}(1_X) \cap H(\mathcal{F})$ is a closed subsemigroup of $\beta_u T$, so it contains an idempotent e. If $\mathcal{F}$ is full, then $\mathcal{F}_e$, which contains $\mathcal{F}$ and hence $k\mathcal{B}_T$, is full as well [cf. (5.36)]. As p approaches e in $\beta_u T$, $\Phi^{\#}(p)$ approaches 1_X uniformly because Φ is continuous at every point of $\{e\} \times X$. Thus for any $V \in \mathcal{U}_X$ there exists $F \in \mathcal{F}_e$ such that $f^t \subset V$ for all $t \in F$. Just make sure that $(x, px) \in V$ for all $(p, x) \in H(F) \times X$. Since $\mathcal{F}_e \subset \mathcal{B}_T$, it follows that φ is uniformly rigid, i.e., $\mathcal{B}_T$ uniformly rigid. Since X is compact, φ is uniformly reversible by Proposition 8.7g. ∎

If $T = \mathbf{Z}_+$, X is the one-point compactification of $\mathbf{Z}$ and $\varphi : \mathbf{Z}_+ \times X \to X$ is the compactification of the translation action, with fixed point ∞, then with $\mathcal{F} = k\mathcal{B}_{\mathbf{Z}_+}$, $\Phi^\#(\beta_u^* \mathbf{Z}_+)$ consists of a single element of S_φ, the idempotent e such that $e(x) = \infty$ for all $x \in X$. Clearly Φ restricts to a continuous map on $\beta_u^* \mathbf{Z}_+ \times X$, and $\Phi^\#(\beta_u^* \mathbf{Z}_+)$ is a compact group (though not a subgroup of X^X). However φ is not distal, so it is certainly not equicontinuous.

Now suppose φ is a weakly rigid action, but not uniformly rigid. By weak rigidity there exists an idempotent e in $\beta_u^* T$ such that $\Phi^\#(e) = 1_X$. With $\mathcal{F}$ the full, semiadditive filter $\mathcal{F}_e$, $\Phi^\#(H(\mathcal{F})) = \{1_X\}$ is a compact abelian group of homomorphisms on X. Furthermore Φ is continuous when restricted to $H(\mathcal{F}) \times X = \{e\} \times X$. However φ is not $\mathcal{F}$ equicontinuous for any full filter $\mathcal{F}$ such that $H(\mathcal{F})$ is a subsemigroup because φ is not uniformly rigid. However in the translation invariant case, we do obtain the Ellis Theorem.

THEOREM 8.5. *Let $\varphi : T \times X \to X$ be a surjective uniform action with X compact. Let $\Phi : \beta_u T \times X \to X$ be the Ellis extension so that $\Phi^\# : \beta_u T \to S_\varphi$ is the continuous semigroup homomorphism onto the enveloping semigroup. The following conditions on φ are equivalent:*

(1) φ is equicontinuous (i.e., $k\mathcal{B}_T$ equicontinuous).

(2) φ is $\mathcal{F}$ equicontinuous for some translation invariant filter $\mathcal{F}$ for T.

(3) S_φ is a group of homeomorphisms on X.

(4) $Eq^T_{V,\varphi} = X$ for every $V \in \mathcal{U}_X$.

(5) φ is $k\tau\mathcal{B}_T$ uniformly rigid.

When these conditions hold, φ is distal and uniformly reversible. Furthermore the reverse action $\overline{\varphi}$ is equicontinuous as well.

PROOF. (4) $\Rightarrow$ (1) and (1) $\Rightarrow$ (2) are obvious.

(2) $\Rightarrow$ (3). By Proposition 8.17 $\Phi^\#(H(\mathcal{F}))$ is a group of homeomorphisms on X. In particular it contains 1_X. Because $\mathcal{F}$ is translation invariant, $H(\mathcal{F})$ is an ideal in $\beta_u T$. If $\Phi^\#(e) = 1_X$, then $\Phi^\#(j_u(t)e) = f^t$ for all $t \in T$. By compactness $\Phi^\#(H(\mathcal{F}))$ is all of S_φ.

(3) $\Rightarrow$ (4). This is an application of the Ellis Joint Continuity Theorem for which we refer to Ellis (1957) [see also Auslander (1988), Chapt. 4]. It says that if $\Phi : S \times X \to X$ is an Ellis action and for each $p \in S$, Φ^p is a homeomorphism of X, then Φ is continuous, i.e., jointly continuous. In particular applied to (3), $\mathrm{Ev} : S_\varphi \times X \to X$ is continuous, so $\Phi : \beta_u T \times X \to X$ is continuous. Consider the filter $k\mathcal{P}_+ = \{T\}$ with $H(k\mathcal{P}_+) = \beta_u T$. Proposition 8.17a implies that every point x of X is in $Eq_\varphi^{k\mathcal{P}_+}$. This is a rephrasing of (4).

(1) $\Rightarrow$ (5). By Proposition 8.17c φ is distal and so by Theorem 7.1, 1_X is $k\tau\mathcal{B}_T$ recurrent for φ_* on X^X. By Proposition 8.17b the topology on S_φ inherited from X^X agrees with the topology from $C(X;X)$. Hence, 1_X is $k\tau\mathcal{B}_T$ recurrent for φ_* on $C(X;X)$, i.e., φ is $k\tau\mathcal{B}_T$ uniformly rigid.

(5) $\Rightarrow$ (3). On the orbit closure $H = \overline{\{f^t : t \in T\}}$ in $C(X;X)$ the elements commute with each f^t. It follows that the induced uniformity is + invariant on H. That is, $V \in \mathcal{U}_X$ and $h_1, h_2 \in H$:

$$(h_1, h_2) \in V^X \Rightarrow (f^t h_1, f^t h_2) = (h_1 f^t, h_2 f^t) \in V^X$$

for all $t \in T$. If φ is $k\tau\mathcal{B}_T$ uniformly rigid then 1_X is a $k\tau\mathcal{B}_T$ recurrent point for the restriction φ_{*H} and so condition (6) of Proposition 8.16 holds for φ_{*H}. Because H is closed in the complete space $C(X;X)$ Proposition 8.16 implies that H is compact. Therefore the inclusion of $C(X;X)$ in X^X restricts to a homeomorphism of H onto S_φ. Thus the elements of S_φ are continuous. By Theorem 7.1 φ is distal and S_φ is a group of bijections which are, by continuity, homeomorphisms.

When these conditions hold, φ is uniformly rigid and uniformly reversible by Proposition 8.17c. $\overline{\varphi}$ is equicontinuous by (8.26). ■

Recall that an ambit (φ, x) is a uniform action on a compact space and a base point whose orbit is dense. An ambit map is a continuous action map preserving base points. With $\mu_u : T \times \beta_u T \to \beta_u T$ the compactification of the translation action, $(\mu_u, j_u(0))$ is the universal ambit. For any uniform action φ on a compact space, we denote by $\mu_\varphi : T \times S_\varphi \to S_\varphi$ the restriction of the product action φ_* to the enveloping semigroup, which is a closed + invariant subspace of X^X. Clearly $(\mu_\varphi, 1_X)$ is an ambit with the homomorphism $\Phi^\#$ an ambit map from $(\mu_u, j_u(0))$ to $(\mu_\varphi, 1_X)$.

PROPOSITION 8.18. *Let (φ, x) be an ambit with $\varphi : T \times X \to X$ a surjective, equicontinuous action. The enveloping semigroup S_φ is then a compact abelian group of homeomorphisms of X, and* Ev$_x : S_\varphi \to X$ *is a homeomorphism. The action φ is minimal and uniformly rigid, and* Ev$_x : (\mu_\varphi, 1_X) \to (\varphi, x)$ *is an ambit isomorphism.*

PROOF. By Proposition 4.5c, φ is a transitive action, so, by Theorem 8.2b, it is minimal. By Theorem 8.5, S_φ is a compact abelian group of homeomorphisms and φ is uniformly rigid. Ev$_x$ is surjective because it is an ambit map. By compactness it suffices to prove Ev$_x$ is injective.

For $g_1, g_2 \in S_\varphi$:

$$g_1(x) = g_2(x) \text{ iff } x = g_1^{-1} g_2(x)$$

since S_φ is a group of bijections. Since $g_1^{-1} g_2 \in S_\varphi$, there exists a net $\{t_\alpha\}$ such that $\{f^{t_\alpha}\}$ converges to $g_1^{-1} g_2$ pointwise. In particular $\{f^{t_\alpha}(x)\}$ converges to x in X. By the Hinge Lemma 8.2 it follows that $\{f^{t_\alpha}\}$ converges uniformly to 1_X. Thus $g_1^{-1} g_2 = 1_X$ and $g_1 = g_2$. ■

REMARK. The surjectivity assumption is used to exclude examples like the $\mathbf{Z}_+$ translation action on the one-point compactification of $\mathbf{Z}_+$. The point 0 has a dense orbit, but the action is not transitive. ■

Now we apply these results to consider group compactifications of T.

PROPOSITION 8.19. *Let T be a uniform monoid, S be an Ellis semigroup and G a compact topological group. Following our standing conventions T is assumed abelian, while S is usually not. G need not be abelian, although we write its composition additively.*

a. Let $h : T \to G$ be a continuous monoid homomorphism. Define $\mu_h : T \times G \to G$ by $(t,x) \mapsto h(t)+x$. μ_h is a uniformly reversible, equicontinuous, uniformly rigid, uniform action of T on G.

b. Any closed subsemigroup of G is a subgroup of G.

c. Let $h : S \to G$ be a continuous semigroup homomorphism and let H be a closed co-ideal in S. The images $h(S)$ and $h(H)$ are closed subgroups of G. The kernel $h^{-1}(0)$ is a closed co-ideal in S, and:

$$H = h^{-1}(h(H)) \Leftrightarrow H \supset h^{-1}(0) \tag{8.35}$$

PROOF. (a) A topological group has left invariant and right invariant uniformity. By compactness these are each the unique uniformity on G. Because G is a topological group, the action μ_h is a topological action. By Lemma 1.2 it is a uniform action. By invariance of the uniformity it is an equicontinuous action. It is uniformly rigid and reversible by Proposition 8.17c.

(b) If S is a closed subsemigroup of G and $s \in S$ we define the continuous homomorphism $h : \mathbf{Z}_+ \to G$ by $h(n) = ns$. By (a) the induced action on G is uniformly rigid. Since s is in the subsemigroup S, S is a $+$ invariant subset. Since it is also closed, (8.25) implies that there exists $x_1 \in S$ such that $s + x_1 = s$, i.e., $0 \in S$. Then (8.25) implies there exists $x_2 \in S$ such that $s + x_2 = 0$; i.e., the inverse of s is in S. Thus S is a subgroup.

(c) $h(H)$ and $h(S)$ are closed subsemigroups, so these are subgroups by (b). In particular $h^{-1}(0)$ is nonempty. If $q \in h^{-1}(0)$ and $pq \in h^{-1}(0)$, then:

$$0 = h(pq) = h(p) + h(q) = h(p) + 0$$

Hence $p \in h^{-1}(0)$. Thus $h^{-1}(0)$ is a co-ideal, closed by continuity. Since $0 \in h(H)$, $H = h^{-1}(h(H))$ implies $H \supset h^{-1}(0)$.

Assume, conversely, that $H \supset h^{-1}(0)$. If $p \in S$ and $h(p) \in h(H)$, then because $h(H)$ is a subgroup there exists $q \in H$ such that $h(q) = -h(p)$

$$h(pq) = h(p) + h(q) = 0$$

Then $pq \in h^{-1}(0) \in H$, and if H is a co-ideal, $p \in H$. ■

Now for a uniform monoid T a *character* is a continuous monoid homomorphism $h : T \to \mathbf{R}/\mathbf{Z}$. (We use the quotient group $\mathbf{R}/\mathbf{Z}$ rather than the unit circle in the complex plane because we want to write the group operation additively.) By Proposition 1.1 each such character extends uniquely to the group G_T. Thus if we write T^* for the group of characters (using pointwise addition), we see that T^* is isomorphic to G_T^*.

For any index set I, the product space $(\mathbf{R}/\mathbf{Z})^I$ is a compact abelian topological group using coordinatewise addition. With $I = T^*$ we define the continuous monoid homomorphism:

$$\begin{aligned} \gamma_T &: T \to (\mathbf{R}/\mathbf{Z})^{T^*} \\ \gamma_T(t)(h) &= h(t) \end{aligned} \tag{8.36}$$

Define Γ_T to be the closure of the image of γ_T:

$$\Gamma_T = \overline{\gamma_T(T)} \subset (\mathbf{R}/\mathbf{Z})^{T^*} \tag{8.37}$$

By Proposition 8.19c, Γ_T is a subgroup of the product group, so it is a compact abelian topological group. $\gamma_T : T \to \Gamma_T$ is a dense, continuous monoid homomorphism, called the *Bohr compactification* of T. If T is a locally compact group, then γ_T is injective, though not usually an embedding. However when T is itself a compact (always abelian) group, then injectivity of γ_T implies that it is a homeomorphism and therefore an isomorphism of compact abelian topological groups. This is a special case of the *Pontrjagin Duality Theorem*, which we use but do not prove.

PROPOSITION 8.20. *Let $h : T \to G$ be a continuous monoid homomorphism with T a uniform monoid and G a compact abelian topological group.*

a. There is a unique continuous map $h_\gamma : \Gamma_T \to G$ such that $h = h_\gamma \circ \gamma_T$. h_γ is a group homomorphism.

b. There is a unique continuous map $\beta_u h : \beta_u T \to G$ such that $h = \beta_u h \circ j_u$. $\beta_u h$ is a semigroup homomorphism.

c. There is a unique continuous map $\beta_u \gamma_T : \beta_u T \to \Gamma_T$ such that $\gamma_T = \beta_u \gamma_T \circ j_u$. $\beta_u \gamma_T$ is a surjective semigroup homomorphism. Furthermore:

$$h_\gamma \circ \beta_u \gamma_T = \beta_u h \tag{8.38}$$

PROOF. (a) Because h is a continuous homomorphism, it induces a group homomorphism $h^* : G^* \to T^*$ of character groups, so $(h^*)^* : (\mathbf{R}/\mathbf{Z})^{T^*} \to (\mathbf{R}/\mathbf{Z})^{G^*}$.

Clearly the following diagram of continuous homomorphisms commutes:

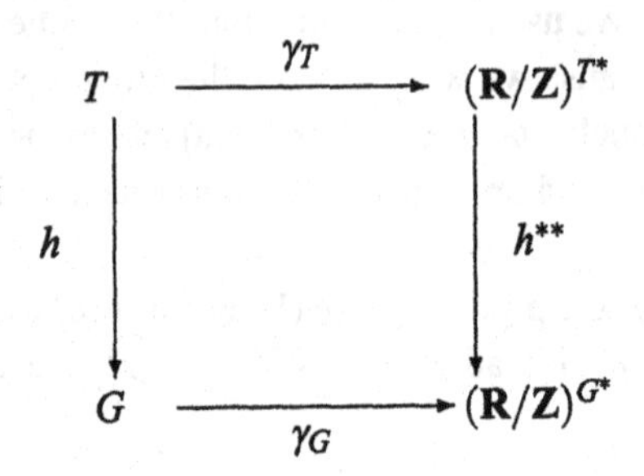

Notice that h^{**} maps Γ_T into Γ_G because $\gamma_T(T)$ is dense in Γ_T.

By Pontrjagin Duality, $\gamma_G : G \to \Gamma_G$ as an isomorphism of compact groups. So we define h_γ on Γ_T to be $(\gamma_G)^{-1} \circ h^{**}$. Uniqueness of h follows because $\gamma_T(T)$ is dense in Γ_T.

(b) and (c) The continuous homomorphism h is uniformly continuous (cf. Lemma 1.1c). Then the continuous extension $\beta_u h$ is defined by Proposition 5.2b; in particular, $\beta_u \gamma_T$ is defined.

$$h_\gamma(\beta_u \gamma_T)(j_u(t)) = h_\gamma(\gamma_T(t)) = h(t) = \beta_u h(j_u(t))$$

By continuity, $h_\gamma(\beta_u \gamma_T)(p) = \beta_u h(p)$ for all p in $\beta_u T$; i.e., (8.38) holds.

$$\beta_u h(pq) = h(p) + h(q)$$

for $p, q \in j_u(T)$, so by the usual two step continuity argument, this equation holds for all $p, q \in \beta_u T$; i.e., $\beta_u h$ and $\beta_u \gamma_T$ are semigroup homomorphisms. ■

If we apply the construction of Proposition 8.19a to the monoid homomorphism $\gamma_T : T \to \Gamma_T$ we obtain a uniformly rigid, reversible, equicontinuous uniform action of T on Γ_T, which we denote $\mu_T : T \times \Gamma_T \to \Gamma_T$. Since $\gamma_T(T)$ is dense in Γ_T, it follows that $(\mu_T, 0)$ is an ambit.

COROLLARY 8.7. *a. Let e be an idempotent of $\beta_u T$ and $J = \beta_u Te$; e.g., let J be a minimal ideal and e any idempotent in J. Let (J, e) denote the ambit obtained by restricting the action $\mu_u : T \times \beta_u T \to \beta_u T$ to J.*

The semigroup homomorphism $\beta_u \gamma_T : \beta_u T \to \Gamma_T$ restricts to a surjective homomorphism of J to Γ_T. $\beta_u \gamma_T$ is an ambit map of (J, e) to $(\mu_T, 0)$.

b. Let $\varphi : T \times X \to X$ be a surjective equicontinuous uniform action with X compact. Let $\mu_\varphi : T \times S_\varphi \to S_\varphi$ be the uniform action obtained by restricting the product action on X^X. S_φ is a compact abelian topological group, and $\Phi^\# : \beta_u T \to S_\varphi$ induces a continuous homomorphism $\varphi^\#_\gamma : \Gamma_T \to S_\varphi$ so that:

$$\Phi^\# = \varphi^\#_\gamma \circ \beta_u \gamma_T \tag{8.39}$$

$\varphi^{\#}_{\gamma}$ is an ambit map of $(\mu_T, 0)$ to $(\mu_\varphi, 1_X)$.

PROOF. (a) $\gamma_T(e) + \gamma_T(e) = \gamma_T(e^2) = \gamma_T(e)$. But the only idempotent in a group is the identity, so $\gamma_T(e) = 0$. Thus $\beta_u \gamma_T$ maps (J, e) to $(\mu_T, 0)$. It is a semigroup homomorphism by Proposition 8.20b, so it is a map of T actions. Any ambit map is surjective.

(b) S_φ is a compact group of homeomorphisms by Theorem 8.5. Applying Proposition 8.20a to $\varphi^{\#} : T \to S_\varphi$, we obtain $\varphi^{\#}_{\gamma} : \Gamma_T \to S_\varphi$, which satisfies (8.39) by (8.38). ■

In particular if (φ, x) is an ambit with φ a surjective equicontinuous action, then by composing the action maps:

$$\begin{gathered}(\mu_T, 0) \xrightarrow{\varphi^{\#}_{\gamma}} (\mu_\varphi, 1_X) \xrightarrow{\mathrm{Ev}_x} (\varphi, x) \\ \Gamma_T \longrightarrow S_\varphi \longrightarrow X\end{gathered} \tag{8.40}$$

we see that $(\mu_T, 0)$ is the universal minimal equicontinuous ambit.

LEMMA 8.3. *Let $g : X \to Y$ be a uniformly continuous map with Y a compact space, and $\beta_u g : \beta_u X \to Y$ the unique continuous map such that $\beta_u g \circ j_u = g$.*

If Y_0 is a closed subset of Y such that $\overline{g(X)} \cap Y_0 \neq \emptyset$, then $g^{-1} u[Y_0]$ is an open filter for X, and:

$$H(g^{-1} u[Y_0]) = (\beta_u g)^{-1}(Y_0) \tag{8.41}$$

PROOF. $u[Y_0]$ is an open filter. The family $g^{-1} u[Y_0]$ is proper because $\overline{g(X)} \cap Y_0 \neq \emptyset$; it is a filter by Proposition 2.4d, and it is open by Proposition 2.13f.

Because g is uniformly continuous, by compactness:

$$(\beta_u g)^{-1}(Y_0) = \cap_U \overline{(\beta_u g)^{-1}(U)}$$

letting U vary over open neighborhoods of Y_0 in Y. By Proposition 5.5a:

$$j_u(X) \cap (\beta_u g)^{-1}(U)$$

is dense in $(\beta_u g)^{-1}(U)$. Hence:

$$\begin{aligned}(\beta_u g)^{-1}(Y_0) &= \cap \overline{j_u(X) \cap (\beta_u g)^{-1}(U)} \\ &= \cap \overline{j_u(g^{-1}(U))} = \cap H(g^{-1}(U)) \\ &= H(g^{-1} u[Y_0])\end{aligned}$$

by (5.40) and (5.33). ■

Using the homomorphism $\gamma_T : T \to \Gamma_T$, we define the open filter $\mathcal{E}_T$:

$$\mathcal{E}_T = \gamma_T^{-1} u[0] \tag{8.42}$$

So Lemma 8.3 implies

$$H(\mathcal{E}_T) = (\beta_u \gamma_T)^{-1}(0) \tag{8.43}$$

which is the co-ideal, the kernel of the semigroup homomorphism $\beta_u\gamma_T : \beta_u T \to \Gamma_T$.

LEMMA 8.4. *Let $h : (\varphi_1, x_1) \to (\varphi_2, x_2)$ be an ambit map with φ_1 a distal action, e.g., a minimal equicontinuous action. If h is sharp, i.e., $h^{-1}(x_2) = \{x_1\}$, then it is an ambit isomorphism.*

PROOF. $h : X_1 \to X_2$ is surjective. By compactness it suffices to prove h is injective. Since x_1 is a distal point, it is a minimal point by Proposition 7.17c. Then the actions φ_1 and φ_2 are minimal. Suppose $x, \tilde{x} \in X_1$ and $h(x) = h(\tilde{x}) = y$. By minimality there exists $p \in \beta_u T$ such that $py = x_2$, then:

$$h(px) = h(p\tilde{x}) = x_2$$

Because h is sharp, $px = p\tilde{x}$. But by Theorem 7.1 $\beta_u T$ acts on X_1 by bijective maps; hence, $x = \tilde{x}$. ∎

PROPOSITION 8.21. *A minimal uniform action $\varphi : T \times X \to X$ with X compact has a maximal equicontinuous factor. That is, there exists $\hat{\varphi} : T \times \hat{X} \to \hat{X}$ a minimal equicontinuous uniform action with $\hat{X}$ compact and a continuous action map $h : \varphi \to \hat{\varphi}$ such that any continuous action map from φ to a minimal equicontinuous action on a compact space factors through h. If for $x \in X$ we let $\hat{x} = h(x)$, then the ambit map $h : (\varphi, x) \to (\varphi, \hat{x})$ is sharp iff x is an $\mathcal{E}_T$ recurrent point. That is:*

$$h^{-1}(\hat{x}) = \{x\} \Leftrightarrow H(\mathcal{E}_T) \subset \mathrm{Iso}_x \tag{8.44}$$

PROOF. For $x \in X$, Iso_x is a closed co-ideal in $\beta_u T$. Then its image under the semigroup homomorphism $\beta_u\gamma_T : \beta_u T \to \Gamma_T$ of Corollary 8.7c is a closed subgroup by Proposition 8.19c. Denote it by K_x. Let $\hat{X}$ denote the compact abelian quotient group Γ_T / K_x with projection π and let $\hat{x}$ be the coset of K_x. We show that the set map $h : X \to \hat{X}$ is well-defined by the following commutative diagram:

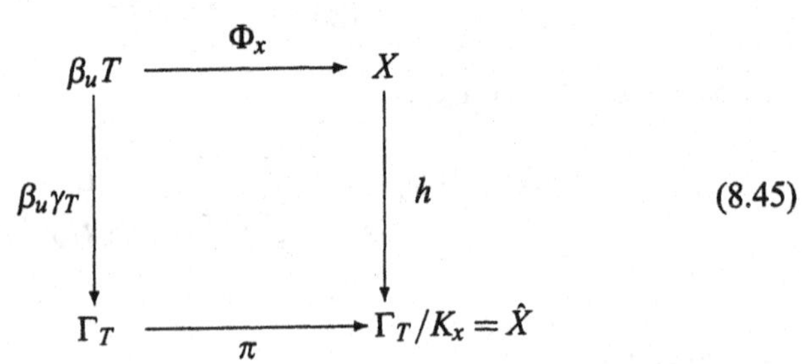

(8.45)

If $p_1, p_2 \in \beta_u T$ with $p_1 x = p_2 x = y$ then by minimality there exists $q \in \beta_u T$ such that $qy = x$. Therefore $qp_1, qp_2 \in \mathrm{Iso}_x$, and in Γ_T:

$$\beta_u \gamma_T(p_1) \equiv -\beta_u \gamma_T(q) \equiv \beta_u \gamma_T(p_2) \pmod{K_x}$$

Thus h is well-defined, and it is continuous by compactness. The homomorphism $\pi \circ \gamma_T : T \to \hat{X}$ induces the equicontinuous action $\hat{\varphi}$ by Proposition 8.19a. Diagram (8.45) consists of action maps. Clearly $h(x) = \hat{x}$; furthermore, $\hat{\Phi}_{\hat{x}} : \beta_u T \to \hat{X}$ is $\pi \circ \beta_u \gamma_T$, since these are both ambit maps from $(\beta_u T, 0)$ to $(\hat{\varphi}, \hat{x})$. Thus $\mathrm{Iso}_{\hat{x}} = h^{-1}(K_x)$. Then by Proposition 8.19c:

$$\mathrm{Iso}_{\hat{x}} = \mathrm{Iso}_x \text{ iff } \mathrm{Iso}_x \supset (\beta_u \gamma_T)^{-1}(0) = H(\mathcal{E}_T)$$

By Proposition 6.7b, (8.44) is true. By (7.4) $\mathrm{Iso}_x = H(\mathcal{N}(x, u[x]))$. Since $\mathcal{E}_T$ is an open filter:

$$H(\mathcal{E}_T) \subset \mathrm{Iso}_x \text{ iff } \mathcal{N}(x, u[x]) \subset \mathcal{E}_T$$

i.e., iff x is $\mathcal{E}_T$ recurrent. In particular if φ was equicontinuous to begin with, then (8.39) implies that $\Phi^{\#}(p) = 1_X$ for every $p \in (\beta_u \gamma_T)^{-1}(0)$, so $H(\mathcal{E}_T) \subset \mathrm{Iso}_x$. Furthermore in that case, $h : (\varphi, x) \to (\hat{\varphi}, \hat{x})$ is an ambit isomorphism by Lemma 8.4.

If $\varphi_1 : T \times X_1 \to X_1$ is an equicontinuous uniform action and $h_1 : \varphi \to \varphi_1$ is a surjective continuous action map, then we let $x_1 = h_1(x)$ so that $h_1 : (\varphi, x) \to (\varphi_1, x_1)$ is an ambit map. Then $\mathrm{Iso}_x \subset \mathrm{Iso}_{x_1}$ and $K_x \subset K_{x_1} \subset \Gamma_T$. Therefore we have a commutative diagram:

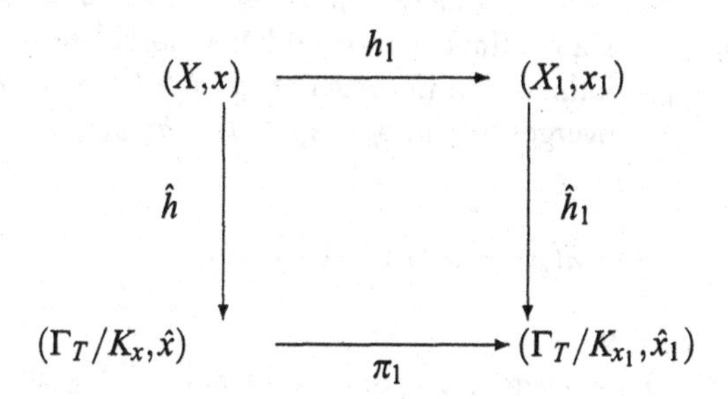

where π_1 is the map of quotient groups. Because φ_1 is equicontinuous, $\hat{h}_1$ is an ambit map isomorphism. The required factorization is

$$h_1 = (\hat{h}^{-1} \circ \pi_1) \circ \hat{h} \tag{8.46}$$

■

PROPOSITION 8.22. *$\mathcal{E}_T$ is an open, semiadditive filter contained in $uk\tau_0\mathcal{P}_+$. For $p \in \beta_u T$, the following conditions are equivalent:*

(1) $p \in H(\mathcal{E}_T)$.

(2) $\tilde{h}(p) = 0$ for every continuous semigroup homomorphism $\tilde{h} : \beta_u T \to G$, where G is a compact topological group.

(3) $px = x$ for every minimal equicontinuous ambit (φ, x).

(4) $\Phi^{\#}(p) = 1_X$ for every surjective, equicontinuous uniform action φ on a compact space.

(5) $\beta_u h(p) = 0$ for every h in the character group T^*.

If $\varphi : T \times X \to X$ is a uniform action with X compact and $x \in X$, then x is $\mathcal{E}_T$ recurrent iff x is recurrent, so that $(\varphi_{\omega\varphi(x)}, x)$ is an ambit, and in addition there exists a sharp ambit map from $(\varphi_{\omega\varphi(x)}, x)$ to a minimal equicontinuous ambit. If x is $\mathcal{E}_T$ recurrent, then it is a distal point. If φ is minimal, then it is equicontinuous iff every point of X is $\mathcal{E}_T$ recurrent.

PROOF. (2) $\Rightarrow$ (5). This is obvious

(5) $\Rightarrow$ (1). The Γ_T is a subgroup of $(\mathbf{R}/\mathbf{Z})^{T^*}$; thus (5) implies that each T^* coordinate of $\beta_u \gamma_T(p)$ is 0. Then $\beta_u \gamma_T(p) = 0$ and $p \in H(\mathcal{E}_T)$ by (8.43).

(1) $\Rightarrow$ (4). $\Phi^{\#}$ factors through $\beta_u \gamma_T$ by (8.39), so $\beta_u \gamma_T(p) = 0$ implies $\Phi^{\#}(p) = 1_X$.

(4) $\Rightarrow$ (3). This is obvious.

(3) $\Rightarrow$ (2). $\tilde{h}(j_u(0))$ is an idempotent in the group G and so is the identity 0. Thus $\tilde{h} \circ j_u : T \to G$ is a continuous monoid homomorphism from which we obtain a surjective equicontinuous uniform action $\varphi : T \times G \to G$ by Proposition 8.19a. When $\{j_u(t_\alpha)\}$ converges to p in $\beta_u T$, $\{f^{t_\alpha}(0) = \tilde{h}(j_u(t_\alpha)) + 0\}$ converges to $\tilde{h}(p)$. That is

$$\tilde{h}(p) = \Phi_0(p) = \Phi^{\#}(p)(0)$$

Thus (3) implies $\tilde{h}(p) = 0$.

Since $\beta_u \gamma_T(e) = 0$ for every idempotent e in $\beta_u T$, (7.53) implies $uk\tau_0\mathcal{P}_+$ contains $K(H(\mathcal{E}_T))$, which is $\mathcal{E}_T$ because the latter is an open filter. Then if x is $\mathcal{E}_T$ recurrent, it is $k\tau_0\mathcal{P}_+$ recurrent and so is $k\tau_0\mathcal{B}_T$ recurrent; i.e., it is distal by Proposition 7.17b. In particular x is a minimal point. The existence of a sharp ambit map to a minimal equicontinuous ambit then follows from Proposition 8.21. On the other hand, if there is a sharp ambit map to a minimal equicontinuous $(\hat{\varphi}, \hat{x})$, then by Proposition 6.7b $\text{Iso}_x = \text{Iso}_{\hat{x}}$. By (3) $H(\mathcal{E}_T) \subset \text{Iso}_{\hat{x}}$; thus x is $\mathcal{E}_T$ recurrent.

Condition (3) implies that for every minimal equicontinuous ambit (φ,x):

$$\mathcal{N}^{\varphi}(x,u[x]) \subset \mathcal{E}_T \tag{8.47}$$

For the action μ_T of T on Γ_T induced by γ_T, we see as in the proof that (3) $\Rightarrow$ (2), $p0$ (action of $p \in \beta_u T$ on the point $0 \in \Gamma_T$) is equal to $\beta_u\gamma_T(p)$. Hence:

$$\mathrm{Iso}_0 = (\beta_u\gamma_T)^{-1}(0) = H(\mathcal{E}_T)$$

That is

$$\mathcal{N}^{\mu_T}(0,u[0]) = \mathcal{E}_T \tag{8.48}$$

by (7.4). Hence $\mathcal{E}_T$ is a semiadditive family by Proposition 7.7.

If φ is minimal and $\hat{\varphi}$ is its maximal equicontinuous factor, then if every point is $\mathcal{E}_T$ recurrent, the action map to $\hat{\varphi}$ is injective by (8.44). Therefore φ is isomorphic to $\hat{\varphi}$. ∎

COROLLARY 8.8. *$F \in \mathcal{E}_T$ iff there exists continuous homomorphism h from T to a finite dimensional torus $(\mathbf{R}/\mathbf{Z})^n$ and a neighborhood U of 0 in $(\mathbf{R}/\mathbf{Z})^n$ such that $F \supset h^{-1}(U)$.*

PROOF. We use (8.42), treating γ_T as a homomorphism of T into $(\mathbf{R}/\mathbf{Z})^{T^*}$. Then we consider a basic neighborhood in the product. ∎

At the end of Chapter 2 we defined $\mathcal{A}_{\mathbf{Z}_+}$ to be the family generated by the infinite subsemigroups of $\mathbf{Z}_+$. If $F \in \mathcal{A}_{\mathbf{Z}}$, then $F \cup \{0\}$ contains the kernel of some character of $\mathbf{Z}_+$. Hence:

$$\mathcal{A}_{\mathbf{Z}_+} \subset \mathcal{E}_{\mathbf{Z}_+} \cdot k\mathcal{B}_{\mathbf{Z}_+} \tag{8.49}$$

It can be shown that a point x is $\mathcal{A}_{\mathbf{Z}_+}$ recurrent for a uniform action $\varphi : \mathbf{Z}_+ \times X \to X$ with X compact iff x is recurrent and $(\varphi_{\omega\varphi(x)}, x)$ has a sharp map to a minimal equicontinuous ambit on a zero-dimensional space.

A uniformly continuous map $f : X \to X$ determines a uniform $\mathbf{Z}_+$ action whose time-one map is f, and conversely. We call the map f transitive, central, etc., if the associated $\mathbf{Z}_+$ action satisfies the corresponding property. For example, f is invertible iff it is a uniform isomorphism on X.

We call x a *periodic point* for f if for some positive integer n, x is a fixed point for f^n. Clearly, a periodic point is $\mathcal{A}_{\mathbf{Z}_+}$ recurrent. We call f *totally transitive* if f^n is transitive for every positive integer n. We conclude with an application of family methods to these concepts.

PROPOSITION 8.23. *Let $f : X \to X$ be a uniformly continuous map.*

a. If $x \in X$ is recurrent for f, then it is recurrent for f^n for all positive integers n.

b. f is totally transitive iff it is $k\mathcal{A}_{\mathbf{Z}_+}$ transitive.

c. If f is weak mixing, it is totally transitive.

d. If f is totally transitive and the periodic points are dense in X, then f is weak mixing.

PROOF. Let $N^f(A,B)$ denote $N^\varphi(A,B)$ where φ is the $\mathbf{Z}_+$ action associated with f.

(a) If x is recurrent, then for every open set U containing x, $N^f(x,U)$ is in a semiadditive family contained in $\mathcal{B}_{\mathbf{Z}_+}$, so it is in $\tau_0\mathcal{B}_{\mathbf{Z}_+}$. But clearly:

$$\mathcal{A}_{\mathbf{Z}_+} \subset k\tau\mathcal{B}_{\mathbf{Z}_+} \subset k\tau_0\mathcal{B}_{\mathbf{Z}_+}$$

Then $\tau_0\mathcal{B}_{\mathbf{Z}_+} \subset k\mathcal{A}_{\mathbf{Z}_+}$. Therefore $N^f(x,U)$ meets every element of $\mathcal{A}_{\mathbf{Z}_+}$.

(b) Both conditions say $N^f(U_1,U_2)$ meets every element of $\mathcal{A}_{\mathbf{Z}_+}$ for all nonempty open subsets U_1,U_2 of X.

(c) $\tau\mathcal{B}_{\mathbf{Z}_+} \subset k\mathcal{A}_{\mathbf{Z}_+}$

(d) If the periodic points are dense, then f is $\mathcal{A}_{\mathbf{Z}_+}$ central, because the $\mathcal{A}_{\mathbf{Z}_+}$ recurrent points are dense. If f is also $k\mathcal{A}_{\mathbf{Z}_+}$ transitive, then it is weak mixing by Proposition 4.10a applied to the full family $k\mathcal{A}_{\mathbf{Z}_+}$. ■

Appendix

Semicontinuous Relations and Almost Open Maps

We begin with some category theory. In a category C an object $*$ is *trivial* if for every object $A \in C$ there is a unique morphism $A \to *$. For trivial objects $*_1$ and $*_2$, the unique morphisms $*_1 \to *_2$ and $*_2 \to *_1$ are inverse isomorphisms. For a family of morphisms $\{f_\alpha : A_\alpha \to A\}$ a *pullback* is a family $\{g_\alpha : B \to A_\alpha\}$ such that $f_\alpha g_\alpha : B \to A$ is independent of α, and for any other such family $\{\overline{g}_\alpha : \overline{B} \to A_\alpha\}$, there is a unique $\overline{g} : \overline{B} \to B$ such that $\overline{g}_\alpha = g_\alpha \overline{g}$ for all α. If the pullback exists, it is unique up to isomorphism. When $*$ is a trivial object, the pullback of the family $\{A_\alpha \to *\}$ is the product and its projections $\{\pi_\alpha : \Pi_\alpha A_\alpha \to A_\alpha\}$. We now consider a category C with trivial objects and finite pullbacks and therefore with finite products. For the pullback of two morphisms $f : A \to B, g : B_1 \to B$ we label the pullback diagram as follows:

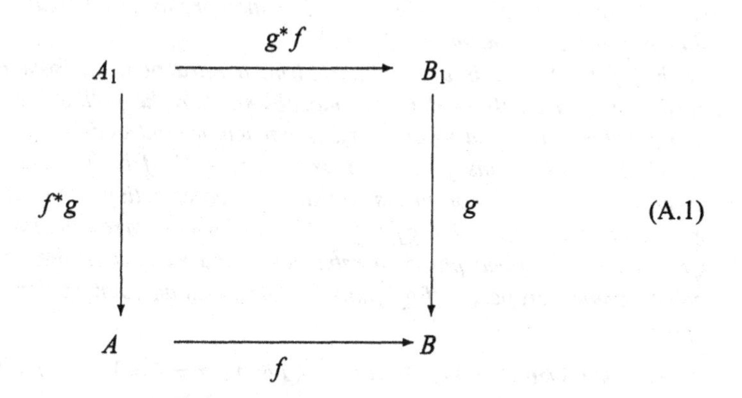

(A.1)

Recall that $h^*(g^*f) = (gh)^*f$. That is, (A.2) is a pullback when each component square is:

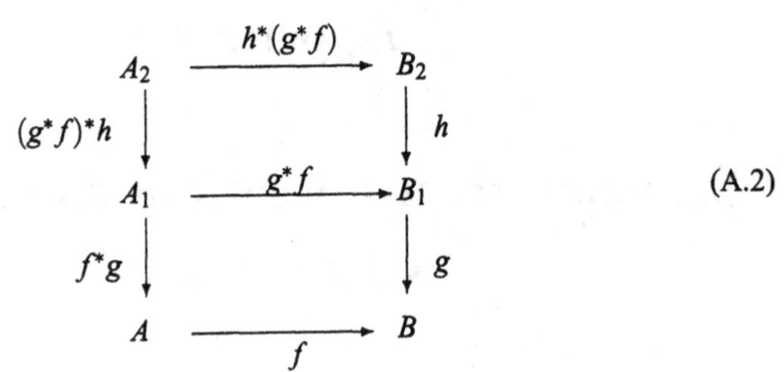

(A.2)

A pair $r : A \to B$ and $\bar{r} : B \to A$ such that $r\bar{r} = 1_B$ is a *retract–coretract pair*; r is a *retraction* (or $\bar{r}$ is a *coretraction*) if an $\bar{r}$ exists (resp. an r exists) such that $r\bar{r} = 1_B$.

$f : A \to B$ is an *epimorphism* (or a *monomorphism*) if for $g_1, g_2 : B \to C$, $g_1 f = g_2 f$ implies $g_1 = g_2$ (resp. for $g_1, g_2 : C \to A$, $fg_1 = fg_2$ implies $g_1 = g_2$). Clearly a retraction is an epimorphism, and a coretraction is a monomorphism.

$f : A \to B$ is a *surjection* if the pullback $g^*f : A_1 \to B_1$ is an epimorphism for any $g : B_1 \to B$. With $g = 1_B$, $f = 1_B^* f$ implies that a surjection is an epimorphism.

$f : A \to B$ *pulls epis* if the pullback $f^*g : A_1 \to A$ is an epimorphism whenever $g : B_1 \to B$ is an epimorphism. Thus in a particular category C, every morphism pulls epis iff every epimorphism is a surjection.

LEMMA A.1. *a. If r is an isomorphism, then r, r^{-1} and r^{-1}, r are both retract–coretract pairs, so r is both a retraction and coretraction. Conversely if $r, \bar{r}$ is a retract–coretract pair and either r is a monomorphism or $\bar{r}$ is an epimorphism, then r is an isomorphism with $\bar{r} = r^{-1}$.*

b. If $f : A \to B$ is a monomorphism, a retraction, an isomorphism, or a surjection and $g : B_1 \to B$ is any morphism, then the pullback $g^*f : A_1 \to B_1$ satisfies the corresponding property. In particular a retraction is a surjection.

c. For morphisms $f : A \to B$ and $g : B \to C$, if both f and g are either monomorphisms, epimorphisms, retractions, coretractions, surjections, or pull epis, then the composition $gf : A \to C$ satisfies the corresponding property. If gf is either an epimorphism, a retraction, or a surjection, then g satisfies the corresponding property. If gf pulls epis and f is an epimorphism, then g pulls epis.

PROOF. (a) Given $r\bar{r} = 1_B$, $r\bar{r}r = r = r1_A$ and $\bar{r}r\bar{r} = \bar{r} = 1_A\bar{r}$. If r is mono or $\bar{r}$ is epi, then $\bar{r}r = 1_A$. Then r is an isomorphism with inverse $\bar{r}$.

(b) If $g_1, g_2 : X \to A_1$ satisfy $(g^*f)g_1 = (g^*f)g_2$, then $g(g^*f)g_i = f(f^*g)g_i$ is independent of $i = 1, 2$. If f is mono, $(f^*g)g_1 = (f^*g)g_2$, so by uniqueness for the pullback $g_1 = g_2$. Thus g^*f is mono when f is.

If $\overline{f} : B \to A$ satisfies $f\overline{f} = 1_B$, then $\overline{f}_1 : B_1 \to A_1$ is uniquely defined so that the following diagram commutes:

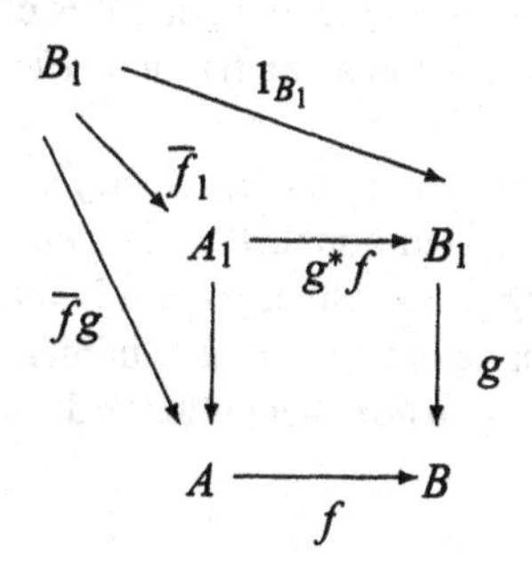

Then $g^*f, \overline{f}_1$ is a retract–coretract pair. If f is iso, then g^*f is also mono. By (a), g^*f is iso. In particular any pullback of a retraction is epi because it is a retraction. Thus a retraction is a surjection.

If f is a surjection and $g : B_1 \to B$, $h : B_2 \to B_1$, then $(gh)^*f$ is epi. This is $h^*(g^*f)$, so g^*f is a surjection.

(c) If f and g are epi and $g_1gf = g_2gf$, then $g_1g = g_2g$, so $g_1 = g_2$. If $f\overline{f} = 1_B$ and $g\overline{g} = 1_C$, then $gf\overline{f}\overline{g} = 1_C$, so gf is a retraction. Duality implies the results for mono and coretraction from which the iso result is clear.

If f and g are surjections and $h : C_1 \to C$, then by (A.2) $h^*(gf) = h^*(g)(g^*h)^*(f)$ which is the composition of two epis because f and g are surjections. Hence $h^*(gf)$ is epi, so gf is a surjection.

If f and g pull epis and $h : C_1 \to C$ is epi, then $(gf)^*h = f^*(g^*h)$ is epi.

On the other hand, if $g_1g = g_2g$, then $g_1gf = g_2gf$, so $g_1 = g_2$ when gf is epi. If $(gf)\overline{r} = 1_C$ then $g(f\overline{r}) = 1_C$, so g is a retraction when gf is.

If gf is a surjection and $h : C_1 \to C$, then $h^*(gf) = h^*(g)(g^*h)^*(f)$ is an epi, so $h^*(g)$ is epi by the result just proved.

If gf pulls epis and $h : C_1 \to C$ is epi, then $(gf)^*(h) = f^*(g^*(h))$ is epi. If f is epi, then $f(gf)^*(h) = g^*(h)(g^*h)^*(f)$ is epi, so $g^*(h)$ is epi. ■

$f : A \to B$ is *minimal* if for $g : A_1 \to A$, fg is epi iff g is epi. In particular $g = 1_A$ epi implies f is epi when it is minimal. Thus f is minimal when f is an epimorphism such that fg epi implies g is epi.

We say that the category *admits minimal surjections* if for every surjection $f : A \to B$ there exists $g : A_1 \to A$ such that fg is a minimal surjection, i.e., a surjection that is minimal.

LEMMA A.2. *a. f is an isomorphism iff f is a minimal retraction.*

b. For $f : A \to B$ and $g : B \to C$, if f and g are both minimal, then $gf : A \to C$ is minimal. Conversely if gf is minimal and f is epi, then f is minimal; if gf is minimal and f is a surjection, then g is minimal.

c. A minimal surjection pulls epis.

PROOF. (a) An isomorphism is a retraction, and it is clearly minimal. If $f\overline{f} = 1_B$ and f is minimal, then $\overline{f}$ is epi and a coretraction. Hence f is an isomorphism by Lemma A.1a.

(b) If f and g are minimal and gfh is epi, then fh is epi, and h is epi because g and f are minimal; hence gf is minimal. If gf is minimal, then g is epi by Lemma A.1c, so fh epi implies gfh is epi. Since gf is minimal, h is epi, so f is minimal.

Now assume f is a surjection and gf is minimal. Let $h : B_1 \to B$ satisfy gh epi. We show that h is epi. Consider the pullback diagram:

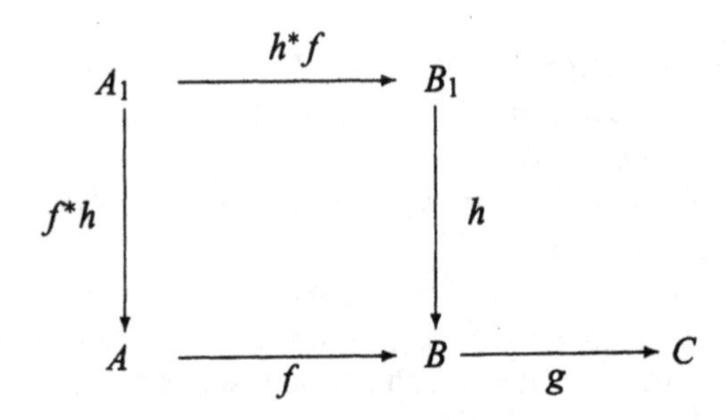

Because f is a surjection, h^*f is epi, so $gf(f^*h)$ is epi. Because gf is minimal, (f^*h) and hence $h(h^*f) = f(f^*h)$ as well are epis. Thus Lemma A.1c implies h is epi.

(c) If $f : A \to B$ is a surjection and $g : B_1 \to B$ is epi, then in (A.1) g^*f is epi because f is a surjection and so the composition $g(g^*f) = f(f^*g)$ is epi. If f is minimal it then follows that f^*g is epi. ■

We let *Spaces* denote the category of uniform spaces and continuous maps and let *Sets* and *Compacts* denote the full subcategories of spaces with discrete uniformity and compact spaces, respectively. Let T *Spaces* denote the category of uniform T actions and continuous action maps, with T *Compacts* the full subcategory of topological T actions on compact spaces. We can identify Spaces with the full subcategory of T Spaces consisting of trivial actions.

LEMMA A.3. *a. Spaces, Sets, Compacts, T Spaces, and T Compacts admit arbitrary pullbacks. The trivial objects are singleton spaces with the trivial T action.*

b. For Sets, $f : A \to B$ is a retraction iff f is a surjection iff f is an epimorphism iff f is a surjective (i.e., onto) map. f is a coretraction iff f is a monomorphism iff f is an injective (i.e., one-to-one) map. f is an isomorphism iff f is a minimal map iff f is bijective. Every map pulls epis. The category admits minimal surjections.

c. *For Compacts and T Compacts, $f : A \to B$ is a surjection iff f is an epimorphism iff f is a surjective map. f is a monomorphism iff f is an injective map. Every map pulls epis. The category admits minimal surjections.*

d. *For Spaces and T Spaces, $f : A \to B$ is an epimorphism iff f is a dense map, i.e., $\overline{f(A)} = B$. f is a surjection if f is a surjective map (and conversely in the Space case). f is a monomorphism iff f is an injective map.*

PROOF. (a) The pullback of $\{f_\alpha : A_\alpha \to A\}$ is the closed subset of the product $\Pi_\alpha A_\alpha$ consisting of those $\{x_\alpha\}$ such that $f_\alpha(x_\alpha)$ is independent of α.

(b) Let C denote the two point space $\{0,1\}$. Map $f(A)$ to 1 and $B\backslash f(A)$ to 0. This characteristic function agrees with the constant function 1 on $f(A)$, so if f is an epimorphism, it must be onto. On the other hand, if f is onto, then by the axiom of choice there exists $\overline{f} : B \to A$ such that $\overline{f}(x) \in f^{-1}(x)$ for every x in B. Hence $f\overline{f} = 1_B$, and f is a retraction. A fortiori $f\overline{f}$ is a minimal surjection, so the category admits minimal surjections. Since every epimorphism is a retraction, and hence a surjection, every map pulls epis. A minimal map is a minimal retraction, so is an isomorphism. A one-to-one map is clearly mono; if $f(x_1) = f(x_2)$ then $* \to x_\alpha$ $(\alpha = 1,2)$ are two maps to A that yield the same map when composed with f. Thus if f is mono, $x_1 = x_2$.

(c) and (d) Clearly if $\overline{f(A)} = B$ and $g_1 f = g_2 f$, then $g_1 = g_2$ on the dense set $f(A)$, so by continuity $g_1 = g_2$. Thus f is an epimorphism. If $\overline{f(A)} = B_0$ is a proper subset of B, then let E be the subalgebra of $\mathcal{B}^u(B)$ consisting of functions constant on B_0. We obtain a dense map $j_E : B \to B_E$, i.e., an epimorphism but with B_E nontrivial because $B_0 \neq B$. On the other hand, $j_E f$ maps A to a singleton x_0, namely, the multiplicative functional on E associating to each $u \in E$ its common value on B_0. Let $g : B \to B_E$ map B to the point x_0. Then $gf = j_E f$, but $g \neq j_E$, so f is not epi. In the T Spaces category, $\overline{f(A)}$ is $+$ invariant, so E is $T+$ invariant. Then B_E obtains an induced uniform T action with j_E a uniformly continuous action map. Hence the subset $\{x_0\} = j_E(\overline{f(A)})$ is $+$ invariant; i.e., x_0 is a fixed point hence g is an action map, too. Thus, when f is not a dense map, it is not an epi in T Spaces, Compacts, or T Compacts either. If f is a closed map, then it is dense iff it is onto. Thus for Compacts and T Compacts, f is epi iff it is onto. In general if $g : B_1 \to B$, then the pullback of f and g is

$$A_1 = \{(x,y) \in A \times B_1 : f(x) = g(y)\}$$

If f is onto, then for any $y \in B_1$ there exists $x \in A$ such that $f(x) = g(y)$. Thus $g^* f : A_1 \to B_1$ is onto. Hence an onto map is a surjection. If $y \in B\backslash f(A)$, then in Spaces $g : * \to B$ mapping to y has an empty pullback, so it is not an epimorphism to $*$. Thus a surjection is onto in Spaces.

For Compacts and T Compacts, every epimorphism is a surjection, so every map pulls epis. Applying Zorn's Lemma as usual to closed subsets, we see that both categories admit minimal surjections. A map is closed, so it is an isomorphism iff it is one-to-one and onto.

Clearly a one-to-one map is mono. For Spaces and Compacts we use $* \to X_\alpha$ $(\alpha = 1,2)$ when $f(x_1) = f(x_2)$ just as in (a). For T Spaces we use $g_\alpha : T \to A$ by $g_\alpha(t) = f^t(x_\alpha)$ $(\alpha = 1,2)$; for T Compacts we use $\beta_u(g_\alpha) : \beta_u T \to \beta_u A \simeq A$, to show that mono implies one-to-one. ■

REMARK. It follows that if an action map f is minimal in the category Spaces, then it is minimal in the category T Spaces. ■

A relation $f : X_1 \to X_2$ is a map iff it satisfies [using relation composition]

$$1_{X_1} \subset f^{-1}f \qquad ff^{-1} \subset 1_{X_2} \tag{A.3}$$

Furthermore for a map f:

$$\begin{aligned} ff^{-1} = 1_{X_2} &\Leftrightarrow f \text{ is surjective} \\ f^{-1}f = 1_{X_1} &\Leftrightarrow f \text{ is injective} \end{aligned} \tag{A.4}$$

For any map f we define the equivalence relation:

$$E_f = (f \times f)^{-1}(1_{X_2}) = \{(x_1, x_2) : f(x_2) = f(x_1)\} = f^{-1} \circ f \tag{A.5}$$

For any relation $F : X_1 \to X_2$ and any relation H on X_2, i.e., $H : X_2 \to X_2$, we define the relation F^*H on X_1 by:

$$F^*H = \{(x_1, x_2) : F(x_2) \subset (H \circ F)(x_1)\} \tag{A.6}$$

LEMMA A.4. *a. For relations H_1, H_2 on X_2 we have*

$$H_1 \subset H_2 \Rightarrow F^*H_1 \subset F^*H_2 \tag{A.7}$$

$$F^*(H_1) \circ F^*(H_2) \subset F^*(H_1 \circ H_2) \tag{A.8}$$

b. For relations H on X_2 and K on X_1:

$$F \circ K \subset H \circ F \Leftrightarrow K \subset F^*H \tag{A.9}$$

For any relation H on X_2:

$$F \circ (F^*H) \subset H \circ F \tag{A.10}$$

with equality when F is a surjective map.

c. $1^*_{X_2}H = H$, and if $G : X_0 \to X_1$ and $F : X_1 \to X_2$ are relations, then:

$$G^*(F^*H) \subset (F \circ G)^*H \tag{A.11}$$

with equality when G is a map.

d. For a map $f : X_1 \to X_2$ and relations $F : X_0 \to X_1$ and $G : X_0 \to X_2$:

$$f \circ F \subset G \Leftrightarrow F \subset f^{-1} \circ G \tag{A.12}$$

e. For a map f, $E_f = f^* 1_{X_2}$:

$$f \circ E_f = f \qquad E_f \circ f^{-1} = f^{-1} \tag{A.13}$$

f. For a map $f : X_1 \to X_2$ and relation H on X_2:

$$\begin{aligned} f^*H &= f^{-1} \circ H \circ f = (f \times f)^{-1}(H) \\ &= \{(x_1, x_2) : f(x_2) \in H(f(x_1))\} \end{aligned} \tag{A.14}$$

So when f is a map:

$$f^*(H^{-1}) = (f^*H)^{-1} \tag{A.15}$$

$$E_f \circ (f^*H) \circ E_f = f^*H \tag{A.16}$$

PROOF. (a) (A.7) is clear. If $(x_1, x_2) \in F^*H_2$ and $(x_2, x_3) \in F^*H_1$, then

$$F(x_2) \subset H_2F(x_1) \qquad F(x_3) \subset H_1F(x_2)$$

Hence $F(x_3) \subset H_1H_2F(x_1)$.

(b) For any set $A \subset X_1$:

$$F(A) \subset HF(x_1) \text{ iff } A \subset (F^*H)(x_1)$$

For (A.9) apply this to $A = K(x_1)$. In particular with $K = F^*H$ we have (A.10). For the surjective map case, we first consider (f).

(f) It is clear that $(x_1, x_2) \in f^*H$ iff $f(x_2) \in H(f(x_1))$. This says $f^*H = (f \times f)^{-1}(H)$ and $f^*H = f^{-1}Hf$, proving (A.14). From this (A.15) is obvious.

When f is a surjective map:

$$f \circ f^*H = ff^{-1}Hf = Hf$$

by (A.4).

(c) $1^*_{X_2}H = H$ is obvious. For the inclusion we apply (A.10):

$$FG(G^*(F^*H)) \subset F(F^*H)G \subset HFG$$

The $G^*F^*H \subset (FG)^*H$ by (A.9). If $G = g$ is a map, we apply (A.10) and (A.3).

$$(Fg)((Fg)^*H)g^{-1} \subset (H(Fg))g^{-1} \subset HF$$

Then by (A.9) we have $g((Fg)^*H)g^{-1} \subset F^*H$. By (A.3) and (A.14):

$$(Fg)^*H \subset g^{-1}g((Fg)^*H)g^{-1}g \subset g^{-1}(F^*H)g = g^*(F^*H)$$

Alternatively observe that $(x_1,x_2) \in (Fg)^*H$ iff $F(g(x_2)) \subset HF(g(x_1))$ and so iff $(g(x_1),g(x_2)) \in F^*H$.

(d) Recall that for a subset A of X_1 and B of X_2:

$$f(A) \subset B \Leftrightarrow A \subset f^{-1}(B) \tag{A.17}$$

For (A.12) apply this and $A = F(x_0)$ and $B = G(x_0)$.

(e) $E_f = f^*1_{X_2}$ by (A.14) and (A.15). By (A.3) we have

$$f = 1_{X_2}f \supset ff^{-1}f \supset f1_{X_1} = f \qquad f^{-1} = f^{-1}1_{X_2} \supset f^{-1}ff^{-1} \supset 1_{X_1}f^{-1} = f^{-1}$$

proving (A.13). Finally by (A.8)

$$(f^*1_{X_2})(f^*H)(f^*1_{X_2}) \subset f^*(H)$$

and the reverse inclusion is obvious, proving (A.16), which completes the proof of (f) as well. ■

For a relation $F : X_1 \to X_2$, $V_2 \in \mathcal{U}_{X_2}$ and $x \in X_1$ we call F V_2 *upper semicontinuous at* x, written V_2 *usc at* x, when $x \in \text{Int}((F^*V_2)(x))$; V_2 *lower semicontinuous at* x, written V_2 *lsc at* x, when $x \in \text{Int}((F^*V_2)^{-1}(x))$; V_2 is *continuous at* x when $(x,x) \in \text{Int}\,F^*V_2$ in $X_1 \times X_1$. F is usc/lsc/continuous at x if it is V_2 usc/V_2 lsc/V_2 continuous at x for all V_2 in $\mathcal{U}_{X_2}$, respectively. F is usc/lsc/continuous if it satisfies the corresponding condition at every x in X_1. We call F *uniformly continuous* if $F^*V_2 \in \mathcal{U}_{X_1}$ for all $V_2 \in \mathcal{U}_{X_2}$.

F is a *closed relation* if it is a closed subset of $X_1 \times X_2$. It is a *pointwise closed* relation if $F(x)$ is a closed subset of X_2 for every $x \in X_1$. Clearly a closed relation is pointwise closed. If $F = f$ is a mapping, then it is pointwise closed. If $f : X_1 \to X_2$ is a continuous map, then the relations f, f^{-1} and $E_f = (f \times f)^{-1}1_{X_2}$ are closed.

For a pointwise closed relation F, there is an associated mapping $F : X_1 \to C(X_2)$ associating to $x \in X_1$ the closed set $F(x) \in C(X_2)$.

LEMMA A.5. *Let $F : X_1 \to X_2$ be a relation.*

a. Let $\tilde{V}_2, V_2 \in \mathcal{U}_{X_2}$ with $\tilde{V}_2 \subset V_2$. If F is $\tilde{V}_2$ usc at x/$\tilde{V}_2$ lsc at x/$\tilde{V}_2$ continuous at x, then F satisfies the corresponding property for V_2.

b. If F is V_2 continuous at x, then F is V_2 usc at x and V_2 lsc at x. If $\tilde{V}_2 \in \mathcal{U}_{X_2}$ with $\tilde{V}_2^2 \subset V_2$ and F is $\tilde{V}_2$ usc at x and $\tilde{V}_2$ lsc at x, then F is V_2 continuous at x.

c. F is continuous at x iff F is usc at x and lsc at x. If F is uniformly continuous, then F is continuous. If F is continuous and X_1 is compact, then F is uniformly continuous.

d. The following are equivalent:

(1) F is uniformly continuous.

(2) For every $V_2 \in \mathcal{U}_{X_2}$ there exists $V_1 \in \mathcal{U}_{X_1}$ such that:

$$F \circ V_1 \subset V_2 \circ F \tag{A.18}$$

(3) For every $V_2 \in \mathcal{U}_{X_2}$, there exists $V_1 \in \mathcal{U}_{X_1}$ such that:

$$V_1 \circ F^{-1} \subset F^{-1} \circ V_2 \tag{A.19}$$

(4) For every $V_2 \in \mathcal{U}_{X_2}$, $V_2 \circ F$ is a uniform neighborhood of F in $X_1 \times X_2$, i.e., $V_2 \circ F \in u[F]$.

e. The inverse relation $F^{-1} : X_2 \to X_1$ is uniformly continuous iff for every $V_1 \in \mathcal{U}_{X_1}$ there exists $V_2 \in \mathcal{U}_{X_2}$ such that:

$$V_2 \circ F \subset F \circ V_1 \tag{A.20}$$

f. If $F = f$ is a mapping then, regarded as a relation, f is usc at x iff f is lsc at x iff f is continuous at x. The relation f is continuous at x/continuous/uniformly continuous iff f satisfies the corresponding map condition.

g. Assume now that F is a pointwise closed relation. F is continuous at x, continuous, or uniformly continuous iff the associated map from X_1 to $C(X_2)$ satisfies the corresponding property. If F is usc, then it is a closed relation. If F is a closed relation and X_2 is compact, then F is usc.

PROOF. (a) is obvious from (A.7). If F is $\tilde{V}_2$ usc at x and $\tilde{V}_2$ lsc at x, then there is an open set O containing x such that:

$$O \subset (F^*\tilde{V}_2)(x) \qquad O \subset (F^*\tilde{V}_2)^{-1}(x)$$

i.e., $O \times \{x\}$ and $\{x\} \times O$ are contained in $F^*\tilde{V}_2$. Then:

$$O \times O \subset (F^*\tilde{V}_2) \circ (F^*\tilde{V}_2)$$

which is contained in F^*V_2 by (A.8) since $\tilde{V}_2^2 \subset V_2$.

F is V_2 continuous iff $1_{X_1} \subset \operatorname{Int}(F^*V_2)$. If X_1 is compact, this implies $F^*V_2 \in \mathcal{U}_{X_1}$. The remaining assertions of (b) and (c) are easy to check.

(d) (1) $\Leftrightarrow$ (2). $F^*V_2 \in \mathcal{U}_{X_1}$ iff there exists $V_1 \in \mathcal{U}_{X_1}$ such that $V_1 \subset F^*V_2$. By (A.9) this is equivalent to $F \circ V_1 \subset V_2 \circ F$.

(2) $\Leftrightarrow$ (3). This is obvious by inversion.

(2) $\Leftrightarrow$ (4). The uniform neighborhoods of F are generated by $\tilde{V}_2 \circ F \circ V_1$s with V_1 and $\tilde{V}_2$ varying over $\mathcal{U}_{X_1}$ and $\mathcal{U}_{X_2}$. Then (4) $\Rightarrow$ (2) is clear. For the converse, choose $\tilde{V}_2$ so that $\tilde{V}_2^2 \subset V_2$ and V_1 so that $F \circ V_1 \subset \tilde{V}_2 \circ F$. Clearly $\tilde{V}_2 \circ F \circ V_1 \subset V_2 \circ F$.

(e) Apply (1) $\Leftrightarrow$ (3) of (d) to F^{-1}.

(f) If $\tilde{V}_2 = \tilde{V}_2^{-1}$ and $\tilde{V}_2 \circ \tilde{V}_2 \subset V_2$, then by (A.15) f is $\tilde{V}_2$ usc at x iff f is $\tilde{V}_2$ lsc at x. By (b) either of these implies f is V_2 continuous at x. So the notions of usc at x/lsc at x/continuous at x agree for a map f. The agreement between the relation and map concepts is easy.

(g) Assume F is usc and $(x,y) \in \overline{F}$. For any $V_2 \in \mathcal{U}_{X_2}$ there is an open set O containing x with $O \subset (F^*V_2)(x)$, i.e.:

$$(O \times X_2) \cap F \subset O \times (V_2(F(x)))$$

Hence:

$$(x,y) \in (O \times X_2) \cap \overline{F} \subset \overline{O} \times (\overline{V_2(F(x))})$$

Thus $y \in \overline{V_2(F(x))}$ for all $V_2 \in \mathcal{U}_{X_2}$. Hence $y \in \overline{F(x)}$, which is $F(x)$ because F is pointwise closed. Thus F is a closed relation.

On the other hand, if F is closed, then for any $x \in X_1$, the intersection of $\overline{F(V_1(x))}$ over all $V_1 \in \mathcal{U}_{X_1}$ is $F(x)$. Given $V_2 \in \mathcal{U}_{X_2}$, then $F(x) \subset \text{Int}\, V_2F(x)$. If X_2 is compact, then $\overline{F(V_1(x))} \subset V_2F(x)$ for some $V_1 \in \mathcal{U}_{X_1}$, so F is usc at x.

Given $V_2 \in \mathcal{U}_{X_2}$ the associated element $C(V_2)$ of $\mathcal{U}_{C(X_2)}$ consists of pairs $(A_1,A_2) \in C(X_1) \times C(X_2)$ such that $A_2 \subset V_2(A_1)$ and $A_1 \subset V_2^{-1}(A_2)$. When $V_2 = V_2^{-1}$, $(F(x_1),F(x_2))$ in $C(X_1) \times C(X_2)$ is in $C(V_2)$ iff $(x_1,x_2) \in (F^*V_2) \cap (F^*V_2)^{-1}$. Thus the equivalence between continuity at x for the relation F and the associated map is clear as are the remaining equivalences. ■

THEOREM A.1. *Let $F : X_1 \to X_2$ be a pointwise closed relation. Assume that either F is usc and $F(x)$ is compact for every $x \in X_1$ or F is lsc and X_2 is compact. For every $V_2 \in \mathcal{U}_{X_2}$ the set of V_2 continuity points, $\{x : (x,x) \in \text{Int}\, F^*V_2\}$, is open and dense in X_1. If in addition X_2 is metrizable and X_1 is a Baire space, then the set of continuity points for F is a residual subset of X_1.*

PROOF. The set of V_2 continuity points is always open. By Lemma A.5b, it suffices to prove that the set of V_2 usc points and the set of V_2 lsc points are each dense. If X_2 is metrizable, the uniformity $\mathcal{U}_{X_2}$ has a countable base. Intersecting over such V_2s we see that the set of continuity points is the countable intersection of dense open sets. Therefore if X_1 is Baire, this intersection is residual.

Assume first that F is usc and pointwise compact. Fix $V_2 \in \mathcal{U}_{X_2}, x_1 \in X_1$, and O_1 an open set containing x_1. We produce $x_0 \in O_1$ at which F is V_2 lsc. Choose $\tilde{V}$, $W \in \mathcal{U}_{X_2}$, symmetric, with $\tilde{V}$ closed and such that $\tilde{V}^2 \subset V_2$ and $W^2 \subset \tilde{V}$. Because F is usc, there is an open set $\tilde{O} \subset O_1$ such that $x \in \tilde{O}$ implies $F(x) \subset W(F(x_1))$. Because $F(x_1)$ is compact, it has a finite subset R such that $F(x_1) \subset W(R)$. Define for $x \in \tilde{O}$:

$$R(x) = R \cap \tilde{V}(F(x)) \tag{A.21}$$

Clearly for $x \in \tilde{O}$:

$$R(x) \subset \tilde{V}(F(x)) \quad \text{and} \quad (R \backslash R(x)) \cap \tilde{V}(F(x)) = \emptyset \tag{A.22}$$

It follows that for $x \in \tilde{O}$:

$$F(x) \cap \tilde{V}(R \backslash R(x)) = \emptyset \qquad F(x) \subset \tilde{V}(R(x)) \tag{A.23}$$

The equation follows from symmetry of $\tilde{V}$, and the inclusion then follows from:

$$F(x) \subset W(F(x_1)) \subset W^2(R) \subset \tilde{V}(R(x)) \cup \tilde{V}(R \backslash R(x))$$

Now choose $x_0 \in \tilde{O}$ so that $R(x_0)$ is minimal in the family $\{R(x) : x \in \tilde{O}\}$ of subsets of the finite set R.

Since $F(x_0)$ is compact and $\tilde{V}(R \backslash R(x_0))$ is the finite union of closed sets, and thus is closed, there exists $W_0 \in \mathcal{U}_{X_2}$ such that:

$$W_0(F(x_0)) \cap \tilde{V}(R \backslash R(x_0)) = \emptyset \tag{A.24}$$

Since F is usc:

$$O_0 = \{x \in \tilde{O} : F(x) \subset W_0(F(x_0))\}$$

is a neighborhood of x_0. For $x \in O_0$, (A.24) implies $R(x) \cap (R \backslash R(x_0)) = \emptyset$, so $R(x) \subset R(x_0)$. By minimality $R(x) = R(x_0)$ for all $x \in O_0$. Hence for all $x \in O_0$:

$$F(x_0) \subset \tilde{V}(R(x_0)) = \tilde{V}(R(x)) \subset \tilde{V}^2(F(x)) \subset V_2(F(x))$$

This means $O_0 \times \{x_0\} \subset F^* V_2$, so F is V_2 lsc at x_0.

If X_2 is compact, F is lsc, $V_2 \in \mathcal{U}_{X_2}$, $x_1 \in X_1$, and O_1 is an open set containing x_1, then choose $\tilde{V}, W \in \mathcal{U}_{X_2}$ symmetric with $\tilde{V}$ open, $\tilde{V}^2 \subset V_2$, and $W^2 \subset \tilde{V}$. Choose R a finite subset of X_2 such that $W(R) = X_2$, and define $R(x)$ for all $x \in X_2$ according to (A.21). Then for all $x \in X_2$, (A.22) and (A.23) hold as before.

Choose $x_0 \in O_1$ so that $R(x_0)$ is maximal in the family $\{R(x) : x \in O_1\}$. Since $\tilde{V}$ is open and $F(x_0)$ is compact, there exists $W_0 \in \mathcal{U}_{X_2}$ symmetric such that for each $y \in R(x_0)$ there exists $z_y \in F(x_0)$ such that $\tilde{V}(y) \supset W_0(z_y)$. Because F is lsc:

$$O_0 = \{x \in O_1 : W_0(F(x)) \supset F(x_0)\}$$

is a neighborhood of x_0. For each $x \in O_0$ and $y \in R(x_0)$, $\tilde{V}(y)$ contains $W_0(z_y)$, which meets $F(x)$. Thus $y \in \tilde{V}(F(x))$. Consequently $R(x_0) \subset R(x)$, so by maximality $R(x_0) = R(x)$ for $x \in O_0$. Hence for all $x \in O_0$:

$$F(x) \subset \tilde{V}(R(x)) = \tilde{V}(R(x_0)) \subset \tilde{V}^2(F(x_0)) \subset V_2(F(x_0))$$

This says $\{x_0\} \times O_0 \subset F^* V_2$, so F is V_2 usc at x_0. ∎

LEMMA A.6. *Let $f: X_1 \to X_2$ be a continuous map.*

a. Let f be a closed map. Assume A is an E_f invariant set, i.e., $A = E_f(A)$. If B is closed and $B \subset A$, then $B_1 = E_f(B)$ is an E_f invariant closed subset of X_1 satisfying $B \subset B_1 \subset A$. If O is open and $A \subset O$ then $O_1 = \{x : E_f(x) \subset O\}$ is an E_f invariant open subset of X_1 satisfying $A \subset O_1 \subset O$. The relations E_f on X_1 and $f^{-1}: X_2 \to X_1$ are usc.

b. If f is a proper map (i.e., point inverses are compact) and f^{-1} is usc, then f is a closed map.

PROOF. (a) If B is closed, then $B_1 = E_f(B) = f^{-1}f(B)$ is closed when f is a continuous, closed map. $B \subset A = E_f(A)$ implies $B \subset E_f(B) \subset E_f(A) = A$. $O_1 = \{x : E_f(x) \subset O\} = X_1 \backslash E_f(X_1 \backslash O)$ which is open when O is open. If A is invariant, then $A_1 = \{x : E_f(x) \subset A\} = A$. Therefore $A \subset O$ implies $A = A_1 \subset O_1 \subset O$. Similarly:

$$\{y \in X_2 : f^{-1}(y) \subset O\} = X_2 \backslash f(X_1 \backslash O) \tag{A.25}$$

is open in X_2. In particular for any $V_1 \in \mathcal{U}_{X_1}$ and $y \in X_2$:

$$y \in \{y_1 \in X_2 : f^{-1}(y_1) \subset \operatorname{Int} V_1(f^{-1}(y))\} \subset ((f^{-1})^* V_1)(y)$$

Then $y \in \operatorname{Int}[((f^{-1})^* V_1)(y)]$ for any $V_1 \in \mathcal{U}_{X_1}$. Thus f^{-1} is usc at y.

By Lemma A.4c, since f is a map:

$$E_f^* V_1 = f^*(f^{-1})^* V_1 = f^{-1}((f^{-1})^* V_1) f \tag{A.26}$$

Hence for any x:

$$(E_f^* V_1)(x) = f^{-1}((f^{-1})^* V_1(f(x)))$$

Then:

$$f(x) \in \operatorname{Int}[(f^{-1})^* V_1(f(x))]$$

implies $x \in \operatorname{Int}[(E_f)^* V_1(x)]$, so E_f is usc at x.

(b) If A is closed in X_1, $y \notin f(A)$, and $f^{-1}(y)$ is compact, then there exists V_1 such that $V_1(f^{-1}(y)) \cap A = \emptyset$. If f^{-1} is usc at y, then there exists O_2 open with $y \in O_2$ and $f^{-1}(O_2) \subset V_1(f^{-1}(y))$. Then $O_2 \cap f(A) = \emptyset$ and $y \notin \overline{f(A)}$. Hence f is a closed map. ■

For a continuous map $f: X_1 \to X_2$ and $A \subset X_1$, we define the relative interior operator:

$$\operatorname{Int}_f A = (\operatorname{Int} A) \cap f^{-1}(\operatorname{Int} f(A)) \tag{A.27}$$

Clearly, $\mathrm{Int}_f A$ is an open subset of $\mathrm{Int}A$. For $x \in X$, we have

$$x \in \mathrm{Int}_f A \Leftrightarrow x \in \mathrm{Int}A \text{ and } f(x) \in \mathrm{Int} f(A) \tag{A.28}$$

For $V_1 \in \mathcal{U}_{X_1}$ and $x \in X_1$, we call f V_1 *open at* x if $x \in \mathrm{Int}_f V_1(x)$. By (A.28) this just says that $f(V_1(x))$ is a neighborhood of $f(x)$.

PROPOSITION A.1. *Let $f : X_1 \to X_2$ be a continuous map.*

a. For $x \in X_1$ the following three conditions are equivalent. When they hold, we call f open at x.

(1) f is V_1 open at x for all $V_1 \in \mathcal{U}_{X_1}$.

(2) If U is a neighborhood of x in X_1, then $f(U)$ is a neighborhood of $f(x)$ in X_2.

(3) For all $A \subset X_1$, $x \in \mathrm{Int}A$ implies $x \in \mathrm{Int}_f A$.

When f is open at x, then the following holds:

(4) $E_f(x) \subset \mathrm{Int}(E_f V_1(x))$ for all $V_1 \in \mathcal{U}_{X_1}$

Conversely, if f is a surjective, closed map then (4) implies f is open at x.

b. The following conditions are equivalent; when they hold we call f open*:*

(1) f is open at x for all $x \in X_1$

(2) If U is open in X_1, then $f(U)$ is open in X_2.

(3) For all $A \subset X_1$, $\mathrm{Int}A = \mathrm{Int}_f A$

c. Let $y \in X_2$ and V_1 be a symmetric element of $\mathcal{U}_{X_1}$. If f^{-1} is V_1 lsc at y, then f is V_1 open at x and E_f is V_1 lsc at x for all $x \in f^{-1}(y)$. If $f^{-1}(y)$ is compact and f is V_1 open at every $x \in f^{-1}(y)$, then f^{-1} is V_1^2 lsc at y. In particular if $f^{-1}(y) = \emptyset$, then f^{-1} is lsc at y.

d. The following conditions are equivalent; when they hold we call f uniformly open:

(1) For every $V_1 \in \mathcal{U}_{X_1}$, there exists $V_2 \in \mathcal{U}_{X_2}$ such that:

$$V_2 \circ f \subset f \circ V_1 \tag{A.29}$$

(2) $f^{-1} : X_2 \to X_1$ is a uniformly continuous relation.

If f is uniformly open and uniformly continuous, then E_f is a uniformly continuous relation on X_1. If f is uniformly open, then it is open. Conversely if X_1 and X_2 are compact and f is open, then it is uniformly open.

PROOF. (a) The equivalence of (1), (2), and (3) is clear. If

$$f(x) \in \mathrm{Int}(f(V_1(x))) \equiv U_2$$

then $E_f(x)$ is contained in the open subset $f^{-1}(U_2)$ of $E_f(V_1(x))$. Conversely if:

$$E_f(x) \subset \mathrm{Int}(E_f(V_1(x))) \equiv U_1$$

and f is a closed map then $X_2 \backslash f(X_1 \backslash U_1)$ is an open set containing $f(x)$. If f is surjective, this set, disjoint from $f(X_1 \backslash U_1)$, is contained in $f(U_1)$.

(b) This is obvious from (a).

(c) If f^{-1} is V_1 lsc at y, then there exists $V_2 \in \mathcal{U}_{X_2}$ such that:

$$V_2(y) \subset ((f^{-1})^* V_1)^{-1}(y)$$

That is $y_1 \in V_2(y)$ implies $f^{-1}(y) \subset V_1(f^{-1}(y_1))$, so: $V_2(y) \subset f(V_1(x))$ for all $x \in f^{-1}(y)$. For $x_1 \in f^{-1}(V_2(y))$ setting $y_1 = f(x_1)$ shows:

$$E_f(x) = f^{-1}(y) \subset V_1(E_f(x_1))$$

So E_f is V_1 lsc at x_1.

If $f^{-1}(y)$ is compact, then there exists $\{x_1, \dots, x_n\}$ in $f^{-1}(y)$ such that $f^{-1}(y) \subset \cup_{i=1}^n V_1(x_i)$. If f is V_1 open at each x_i, then there exists $V_2 \in \mathcal{U}_{X_2}$ such that:

$$V_2(y) \subset \cap_{i=1}^n f(V_1(x_i))$$

If $y_1 \in V_2(y)$, $f^{-1}(y_1)$ meets each $V_1(x_i)$, so it meets $V_1^2(x)$ for every $x \in f^{-1}(y)$. That is

$$V_2(y) \subset ((f^{-1})^*(V_1^2))^{-1}(y)$$

and f^{-1} is V_1^2 lsc at y.

(d) (1) $\Leftrightarrow$ (2) by Lemma A.5e. If f is uniformly open, then it is clearly open. If f also is uniformly continuous, then $E_f = f^{-1} \circ f$ is uniformly continuous by Lemma A.5g. If f is a proper, open map, then by (c) f^{-1} is lsc. If f is a closed map, then by Lemma A.6 f^{-1} is usc. If X_1 is compact, then f is proper and closed. Then f open implies f^{-1} is continuous. If X_2 is compact, then f^{-1} is uniformly continuous. ■

PROPOSITION A.2. *For a continuous map $f: X_1 \to X_2$ the following conditions are equivalent; when they hold, we call f* almost open

(1) *For all $A \subset X_1$, $\mathrm{Int} A \neq \emptyset$ implies $\mathrm{Int} f(A) \neq \emptyset$.*

(2) *If U is open in X_1, then $\mathrm{Int}_f U = U \cap f^{-1}(\mathrm{Int} f(U))$ is dense in U.*

(3) For every $V_1 \in \mathcal{U}_{X_1}$, $\{x : f \text{ is } V_1 \text{ open at } x\}$ is dense in X_1.

(4) For every $V_1 \in \mathcal{U}_{X_1}$, $\{x : (x, f(x)) \in \text{Int}(f \circ V_1)\}$ is open and dense in X_1.

(5) D dense in X_2 implies $f^{-1}(D)$ is dense in X_1.

(6) $f : X_1 \to X_2$ pulls epis in the category Spaces.

In particular if $\{x : f \text{ is open at } x\}$ is dense in X_1, then f is almost open. Conversely if X_1 is a metrizable Baire space and f is almost open, then $\{x : f \text{ is open at } x\}$ is residual in X_1; i.e., it is a dense, G_δ subset of X_1.

PROOF. (1) $\Rightarrow$ (5). If $f^{-1}(D)$ is not dense, then $U = X_1 \setminus \overline{f^{-1}D}$ is a nonempty open set, so $\text{Int} f(U)$ is an open set, nonempty by (1), disjoint from D. Hence D is not dense.

(5) $\Rightarrow$ (2). $D = (\text{Int} f(U)) \cup (X_2 \setminus f(U))$ is dense for any set U. By (5), $f^{-1}(D)$ is dense in X_1. If U is open in X_1, $U \cap f^{-1}(D)$ is dense in U. But $U \cap f^{-1}(D) = \text{Int}_f U$.

(2) $\Rightarrow$ (3). Let $\tilde{V}_1$ be an open, symmetric element of $\mathcal{U}_{X_1}$ such that $\tilde{V}_1^2 \subset V_1$ and let $x \in X$. By (2):

$$\tilde{V}_1(x) \cap f^{-1}(\text{Int} f(\tilde{V}_1(x)))$$

is open and dense in $\tilde{V}_1(x)$. For x_1 in this set:

$$f(V_1(x_1)) \supset f(\tilde{V}_1(x)) \supset \text{Int} f(\tilde{V}_1(x))$$

$f(x_1)$ is in this last set. Since $\tilde{V}_1$ can be chosen arbitrarily small, f is V_1 open at points of a dense set.

(3) $\Rightarrow$ (1). If $V_1^2(x) \subset \text{Int} A$ let $x_1 \in V_1(x)$ at which f is V_1 open.

$$f(A) \supset f(V_1^2(x)) \supset f(V_1(x_1))$$

$f(x_1)$ is the interior of the last set.

(3) $\Leftrightarrow$ (4). For V_1 symmetric in $\mathcal{U}_{X_1}$:

$$\begin{aligned} &\{x : (x, f(x)) \in \text{Int}(f \circ V_1)\} \subset \\ &\{x : f \text{ is } V_1 \text{ open at } x\} \subset \\ &\{x : (x, f(x)) \in \text{Int}(f \circ V_1^2)\} \end{aligned} \tag{A.30}$$

Observe that $U_1 \times U_2 \subset f \circ V_1$ iff $U_2 \subset f(V_1(x_1))$ for all $x_1 \in U_1$. The first inclusion is clear, and the second follows from $f(V_1(x)) \subset f(V_1^2(x_1))$ for all $x_1 \in V_1(x)$. Together these inclusions yield the equivalence.

(5) $\Leftrightarrow$ (6). Let $g : X_2' \to X_2$ be an epimorphism, so $D = g(X_2')$ is dense in X_2. With $X_1' = \{(x_1, x_2') : f(x_1) = g(x_2')\} \subset X_1 \times X_2'$, the pullback $f^*g : X_1' \to X_1$ maps

onto $f^{-1}(D)$. Assuming (5) it follows that f^*g is a dense map and so is epi. Thus f pulls epis. Conversely if $D \subset X_2$ is dense, let $g : D \to X_2$ be the inclusion map. The inclusion of $f^{-1}(D)$ into X_1 is the pullback f^*g. Assuming (6) this inclusion is an epi, so $f^{-1}(D)$ is dense in X_1.

When $\mathcal{U}_{X_1}$ is metrizable and so has a countable base, then (A.30) implies that when f is almost open, the set of points at which f is open is the countable intersection of dense open sets. Thus it is residual when X_1 is Baire. ∎

COROLLARY A.1. *Let $f : X_1 \to X_2$ be a continuous, closed, proper (i.e., point inverses are compact) map. For every $V_1 \in \mathcal{U}_{X_1}$, the set:*

$$\mathrm{Int}\{y \in X_2 : f \text{ is } V_1 \text{ open at each } x \in f^{-1}(y)\} \tag{A.31}$$

is an open and dense subset of X_2. In particular if X_1 is metrizable and X_2 is Baire, then:

$$\{y \in X_1 : f \text{ is open at each } x \in f^{-1}(y)\} \tag{A.32}$$

is a residual subset of X_2.

PROOF. By Lemma A.6, f^{-1} is a usc relation. By Theorem A.1, $\{y : (y,y) \in \mathrm{Int}(f^{-1})^*V_1\}$ is open and dense in X_2, and it is contained in the set of points at which f^{-1} is an lsc relation. By Proposition A.1c, f is V_1 open at every point $x \in X_1$ such that $y = f(x)$ lies in this set. As usual when X_1 is metrizable, we can intersect over a countable base for $\mathcal{U}_{X_1}$ to obtain a residual subset of X_2. ∎

REMARK. If in addition f is almost open then the preimage of the set labeled according to (A.31) is dense in X_1. Conversely density of these sets implies f is almost open. ∎

This result is the category analogue of Sard's Theorem. Call x a *regular point* for f if f is open at x. Otherwise call x a *critical point*. The image of the set of critical points is the set of *critical values*, and its complement the set of *regular values*. Corollary A.1 implies that for a continuous map between compact metric spaces the set of regular values in this sense is residual. Notice that if $f^{-1}(y)$ is empty, then y is a regular value. Compare Oxtoby (1980).

For some purposes a further weakening of the notion of open map is useful. Call a subset A of X *quasi-open* if:

$$A \subset \mathrm{Int}\overline{A} \tag{A.33}$$

For any subset A define the *quasi-interior*:

$$Q\mathrm{Int}A = A \cap \mathrm{Int}\overline{A} \tag{A.34}$$

PROOF. (a) The equivalence of (1), (2), and (3) is clear. If

$$f(x) \in \mathrm{Int}(f(V_1(x))) \equiv U_2$$

then $E_f(x)$ is contained in the open subset $f^{-1}(U_2)$ of $E_f(V_1(x))$. Conversely if:

$$E_f(x) \subset \mathrm{Int}(E_f(V_1(x))) \equiv U_1$$

and f is a closed map then $X_2 \backslash f(X_1 \backslash U_1)$ is an open set containing $f(x)$. If f is surjective, this set, disjoint from $f(X_1 \backslash U_1)$, is contained in $f(U_1)$.

(b) This is obvious from (a).

(c) If f^{-1} is V_1 lsc at y, then there exists $V_2 \in \mathcal{U}_{X_2}$ such that:

$$V_2(y) \subset ((f^{-1})^* V_1)^{-1}(y)$$

That is $y_1 \in V_2(y)$ implies $f^{-1}(y) \subset V_1(f^{-1}(y_1))$, so: $V_2(y) \subset f(V_1(x))$ for all $x \in f^{-1}(y)$. For $x_1 \in f^{-1}(V_2(y))$ setting $y_1 = f(x_1)$ shows:

$$E_f(x) = f^{-1}(y) \subset V_1(E_f(x_1))$$

So E_f is V_1 lsc at x_1.

If $f^{-1}(y)$ is compact, then there exists $\{x_1, \ldots, x_n\}$ in $f^{-1}(y)$ such that $f^{-1}(y) \subset \cup_{i=1}^n V_1(x_i)$. If f is V_1 open at each x_i, then there exists $V_2 \in \mathcal{U}_{X_2}$ such that:

$$V_2(y) \subset \cap_{i=1}^n f(V_1(x_i))$$

If $y_1 \in V_2(y)$, $f^{-1}(y_1)$ meets each $V_1(x_i)$, so it meets $V_1^2(x)$ for every $x \in f^{-1}(y)$. That is

$$V_2(y) \subset ((f^{-1})^*(V_1^2))^{-1}(y)$$

and f^{-1} is V_1^2 lsc at y.

(d) (1) $\Leftrightarrow$ (2) by Lemma A.5e. If f is uniformly open, then it is clearly open. If f also is uniformly continuous, then $E_f = f^{-1} \circ f$ is uniformly continuous by Lemma A.5g. If f is a proper, open map, then by (c) f^{-1} is lsc. If f is a closed map, then by Lemma A.6 f^{-1} is usc. If X_1 is compact, then f is proper and closed. Then f open implies f^{-1} is continuous. If X_2 is compact, then f^{-1} is uniformly continuous.

PROPOSITION A.2. *For a continuous map $f: X_1 \to X_2$ the following conditions are equivalent; when they hold, we call f* almost open

(1) For all $A \subset X_1$, $\mathrm{Int} A \neq \emptyset$ implies $\mathrm{Int} f(A) \neq \emptyset$.

(2) If U is open in X_1, then $\mathrm{Int}_f U = U \cap f^{-1}(\mathrm{Int} f(U))$ is dense in U.

PROPOSITION A.3. *Let $f : X_1 \to X_2$ be a continuous map.*

a. For $x \in X_1$ the following conditions are equivalent; when they hold, we call f quasi-open at x*:*

1. *f is V_1 quasi-open at x for all $V_1 \in \mathcal{U}_{X_1}$.*
2. *If U is a neighborhood of x in X_1, then $\overline{f(U)}$ is a neighborhood of $f(x)$ in X_2.*
3. *For all $A \subset X_1$, $x \in Q\mathrm{Int}A$ implies $x \in Q\mathrm{Int}_f A$*
4. *If U is an open neighborhood of x in X_1, then $Q\mathrm{Int}_f U$ is an open neighborhood of x in X_1 with $f(x)$ in $f(Q\mathrm{Int}_f U) = Q\mathrm{Int}\, f(U)$ quasi-open in X_2.*

b. The following conditions are equivalent; when they hold, we call f quasi-open*:*

1. *f is quasi-open at x for all $x \in X_1$.*
2. *If U is open in X_1, then $f(U)$ is quasi-open in X_2.*
3. *If A is quasi-open in X_1, then $f(A)$ is quasi-open in X_2.*
4. *For all $A \subset X_1$, $Q\mathrm{Int}A = Q\mathrm{Int}_f A$*
5. *For all U open in X_1, $U = Q\mathrm{Int}_f U$*

c. The following conditions are equivalent; when they hold we call f almost quasi-open*:*

1. *For all $A \subset X_1$, $\mathrm{Int}A \neq \emptyset$ implies $\mathrm{Int}\,\overline{f(A)} \neq \emptyset$.*
2. *For all $A \subset X_1$, $Q\mathrm{Int}A \neq \emptyset$ implies $Q\mathrm{Int}\, f(A) \neq \emptyset$.*
3. *If A is quasi-open in X_1, then $Q\mathrm{Int}_f A$ is dense in A.*
4. *For every $V_1 \in \mathcal{U}_{X_1}$, $\{x : f$ is V_1 quasi-open at $x\}$ is dense in X_1.*
5. *For every $V_1 \in \mathcal{U}_{X_1}$, $\mathrm{Int}\{x : f$ is V_1 quasi-open at $x\}$ is open and dense in X_1.*
6. *D open and dense in X_2 implies $f^{-1}(D)$ is dense in X_1.*
7. *B nowhere dense in X_2 implies $f^{-1}(B)$ is nowhere dense in X_1.*

In particular if $\{x : f$ is quasi-open at $x\}$ is dense in X_1, then f is almost quasi-open. Conversely if X_1 is a metrizable Baire space and f is almost quasi-open, then $\{x : f$ is quasi-open at $x\}$ is residual; i.e., it is a dense G_δ subset of X_1.

d. If f is V_1 open at x/open at x/open/almost open, then it is V_1 quasi-open at x/quasi-open at x/quasi-open/almost quasi-open, respectively. If f is a closed map and V_1 is closed, then the converse results are true as well.

PROOF. (a) (3) $\Rightarrow$ (4). Lemma A.7d.

(4) $\Rightarrow$ (2). By (4), $\overline{f(U)}$ contains $\overline{Q\text{Int} f(U)}$, whose interior contains $Q\text{Int} f(U)$ and hence $f(x)$.

(2) $\Rightarrow$ (1). This is obvious.

(1) $\Rightarrow$ (3). Assume V_1 is open in $\mathcal{U}_{X_1}$ with $V_1(x) \subset \overline{A}$. By (1) $\overline{f(V_1(x))}$ contains some open neighborhood U_2 of $f(x)$:

$$U_2 \subset \overline{f(V_1(x))} \subset \overline{f(A)}$$

because $f(A)$ is dense in $\underline{f(\overline{A})}$. Hence $f^{-1}(U_2)$ is an open neighborhood of x, contained in $f^{-1}(\text{Int}\overline{f(A)})$. Since $x \in Q\text{Int}A$, it is in $Q\text{Int}_f A$.

(b) (4) $\Rightarrow$ (5), (3) $\Rightarrow$ (2), and (2) $\Rightarrow$ (1) are obvious. (4) $\Rightarrow$ (3) and (5) $\Rightarrow$ (2) follow from (A.37). (1) $\Rightarrow$ (4) by (1) $\Rightarrow$ (3) of (a).

(c) (3) $\Rightarrow$ (2). Let $A_1 = Q\text{Int}A$, quasi-open and nonempty. By (3) and (A.37) $Q\text{Int} f(A_1) \neq \emptyset$.

(2) $\Rightarrow$ (1). Let $U = \text{Int}A$, open and nonempty. By (2) $Q\text{Int} f(U) \subset \text{Int}\overline{f(A)}$ is nonempty.

(1) $\Rightarrow$ (4). Let $V = V^{-1}$ be open in $\mathcal{U}_{X_1}$, with $V^2 \subset V_1$. For any x, (1) implies $\text{Int}\overline{f(V(x))} \neq \emptyset$. By (A.35) and (A.37) the open set $Q\text{Int}_f V(x)$ is nonempty. For x_1 in this set, $x_1 \in V(x)$, and:

$$f(x_1) \in \text{Int}\overline{f(V(x))} \subset \text{Int}\overline{f(V_1(x_1))}$$

Thus, f is V_1 quasi-open at x_1. Since V can be chosen arbitrarily small, the set described in (4) is dense.

(4) $\Rightarrow$ (5). With $V = V^{-1}$ and $V^2 \subset V_1$ as usual, if f is V quasi-open at x iff for some open neighborhood U of x, $f(U) \subset \text{Int}\overline{f(V(x))}$. If $x_1 \in U \cap V(x)$, then $f(U) \subset \overline{f(V_1(x_1))}$. So f is V_1 quasi-open at every point of a neighborhood of x.

(5) $\Rightarrow$ (3). Given A quasi-open let $U = \text{Int}\overline{A}$ so that $A \subset U \subset \overline{A}$, which implies $\overline{A} = \overline{U}$ and $\overline{f(A)} = \overline{f(U)}$. For $x \in A$ choose V_1 open in $\mathcal{U}_{X_1}$ such that $V_1(x) \subset U$. Since $V_1(x) \cap A$ is quasi-open, (5) implies there exists $x_1 \in V_1(x) \cap A$ such that f is V_1 quasi-open at x_1. Thus $f(x_1)$ is in:

$$\text{Int}\overline{f(V_1(x))} \subset \text{Int}\overline{f(U)} = \text{Int}\overline{f(A)}$$

Since $x_1 \in A = Q\text{Int}A$, it follows that $x_1 \in Q\text{Int}_f A$.

(1) $\Rightarrow$ (7). Let $A = \overline{f^{-1}(\overline{B})}$ closed in X_1 and containing $f^{-1}(B)$. If $\text{Int}A \neq \emptyset$, then (1) implies $\text{Int}\overline{f(A)} \subset \text{Int}\overline{B}$ is nonempty.

(7) $\Rightarrow$ (6). Let $B = \overline{X_2 \backslash D}$.

(6) $\Rightarrow$ (1). If $\text{Int}\overline{f(A)} = \emptyset$, then $D = X_2 \backslash \overline{f(A)}$ is open and dense. By (6) $f^{-1}(D)$ is open and dense and disjoint from A. Then $\text{Int}\overline{A} = \emptyset$.

The Baire argument uses (5) as usual.

(d) The open concepts clearly imply the corresponding quasi-open concepts. If f is a closed map and V_1 is a closed element of $\mathcal{U}_{X_1}$ then:

$$\overline{f(V_1(x))} = f(V_1(x))$$

thus f is V_1 open at x if it is V_1 quasi-open at x. Because the closed elements of $\mathcal{U}_{X_1}$ generate the uniformity, the remaining results follow. ■

REMARK. If for every dense G_δ, B in X_2 $f^{-1}(B)$ is dense in X_1, then f satisfies (6) of (c), so f is almost quasi-open. The converse is true as well if X_1 is Baire. ■

Notice that for families $\mathcal{F}_\alpha$ for X_α $(\alpha = 1,2)$ (2.31) and Proposition 2.4c imply

$$f^{-1}\mathcal{F}_2 \subset \mathcal{F}_1 \Leftrightarrow fk\mathcal{F}_1 \subset k\mathcal{F}_2 \tag{A.39}$$

If $\mathcal{F}$ is the family of dense subsets on any space X, then $k\mathcal{F}$ is the family of sets A with $\text{Int}A \neq \emptyset$, if $\mathcal{F}$ is the family of subsets with a dense interior, then $k\mathcal{F}$ is the family of sets A with $\text{Int}\overline{A} \neq \emptyset$. The equivalence of (1) and (6) in (c), as well as that of (1) and (5) in Proposition A.2, follow from (A.39).

Let $f : X_1 \to X_2$ be a continuous map. For $x \in X_1$ and $V_1 \in X_2$, we call f a V_1 *embedding at* x if there exists $V_2 \in \mathcal{U}_{X_2}$ such that

$$(f^*V_2)(x) = f^{-1}(V_2(f(x))) \subset V_1(x) \tag{A.40}$$

This just says that the preimage of some neighborhood of $f(x)$ is contained in $V_1(x)$. Clearly f is a V_1 embedding at x iff the associated surjective map $f : X_1 \to f(X_1)$ is a V_1 embedding at x. If f is surjective and is a V_1 embedding at x, then f is V_1 open at x because (A.40) then implies $V_2(f(x)) \subset f(V_1(x))$.

PROPOSITION A.4. *Let $f : X_1 \to X_2$ be a continuous map.*

a. For $x \in X_1$ the following conditions are equivalent; when they hold we call f an embedding at x*:*

(1) f is a V_1 embedding at x for all $V_1 \in \mathcal{U}_{X_1}$.

(2) $f^{-1}u[f(x)] = u[x]$

(3) For every $V_1 \in \mathcal{U}_{X_1}$, there exists U_2 a neighborhood of $f(x)$ such that $f^{-1}(U_2) \times f^{-1}(U_2) \subset V_1$.

If f is an embedding at x, then $f^{-1}(f(x)) = E_f(x)$ is $\{x\}$. Conversely if f is a closed map and $E_f(x) = \{x\}$, then f is an embedding at x.

If f is an embedding at x then the surjective map $f : X_1 \to f(X_1)$ is open at x.

b. f is an embedding at x, for all $x \in X_1$ iff f is an embedding, i.e., iff the surjective map $f : X_1 \to f(X_1)$ is a homeomorphism.

c. The following five conditions are equivalent; when they hold we call f an almost embedding*:*

(1) If U_1 is open and nonempty in X_1 then there exists U_2 open in X_2 such that $f^{-1}(U_2)$ is a nonempty subset of U_1.

(2) For all U open in X_1 the open set:

$$U^f \equiv X_1 \backslash f^{-1}(\overline{f(X_1 \backslash U)}) = f^{-1}(\text{Int}(X_2 \backslash f(X_1 \backslash U)))$$

is dense in U.

(3) For every $V_1 \in \mathcal{U}_{X_1}$, $\{x \in X_1 : f$ is a V_1 embedding at $x\}$ is dense in X_1.

(4) For every $V_1 \in \mathcal{U}_{X_1}$, $\{x \in X_1 : f^{-1}(U) \times f^{-1}(U) \subset V_1$ for some neighborhood U of $f(x)\}$ is open and dense in X_1.

(5) For $D \subset X_1$, $f(D)$ dense in $f(X_1)$ implies D dense in X_1.

If f is an almost embedding, then the following conditions hold:

(6) If U is a nonempty open subset of X_1, then there exists U_1 a nonempty open subset of X_1 contained in U, which is E_f invariant, i.e., $f^{-1}f(U_1) = U_1$.

(7) If A is a proper closed subset of X_1, then there exists A_1 a proper closed subset of X_1 containing A, which is E_f invariant, i.e., $f^{-1}f(A_1) = A_1$.

(8) For every $V_1 \in \mathcal{U}_{X_1}$, $\{x \in X_1 : E_f(x) \subset V_1(x)\}$ is dense in X_1.

(9) For every $V_1 \in \mathcal{U}_{X_1}$, $\{x \in X_1 : E_f(x) \times E_f(x) \subset V_1\}$ is dense in X_1.

If f is a closed map, then each of these conditions implies that f is an almost embedding. In particular if f is closed and $\{x \in X_1 : E_f(x) = \{x\}\}$ is dense in X_1, then f is an almost embedding. If X_1 is a metrizable Baire space and f is an almost embedding, then $\{x : E_f(x) = \{x\}\}$ contains $\{x : f$ is an embedding at $x\}$, which is a residual subset of X_1, i.e., a dense G_δ.

d. Assume f is a dense map, i.e., an epimorphism in the category Spaces. The following conditions are then each equivalent to the condition that f is an almost embedding:

(10) For $D \subset X_1$, $f(D)$ is dense in X_2 iff D is dense in X_1.

(11) f is a minimal epimorphism in the category Spaces.

(12) If A is a proper closed subset of X_1, then $\overline{f(A)}$ is a proper closed subset of X_2.

e. We call f an almost homeomorphism *if it is a surjective almost embedding. If f is an almost homeomorphism then for any V_1 closed in $\mathcal{U}_{X_1}$*

$$\{x : f \text{ is } V_1 \text{ open at } x\} = \{x : f \text{ is a } V_1 \text{ embedding at } x\} \tag{A.41}$$

$$\{x : f \text{ is open at } x\} = \{x : f \text{ is an embedding at } x\} \tag{A.42}$$

In particular an almost homeomorphism is almost open, and an open almost homeomorphism is a homeomorphism.

PROOF. (a) (3) obviously implies (1) which is equivalent to (2), i.e., $\{(f^*V_2)(x) : V_2 \in \mathcal{U}_{X_2}\}$ is a base for the neighborhoods of x. If $V = V^{-1}$ and $V^2 \subset V_1$, then $f^{-1}(U) \subset V(x)$ implies $f^{-1}(U) \times f^{-1}(U) \subset V_1$. So (1) $\Rightarrow$ (3). Clearly if f is a V_1 embedding at x, then $E_f(x) \subset V_1(x)$. If f is an embedding at x, $E_f(x) = \{x\}$. If V_1 is open and f is closed, then $f(X_1 \backslash V_1(x))$ is closed in X_2 and disjoint from $f(x)$ if $E_f(x) = \{x\}$. In that case the complement U satisfies $f^{-1}(U) \subset V_1(x)$. The last assertion is already true at the V_1 level as previously discussed.

(b) If f is an embedding at every x, then by (a), $f : X_1 \to f(X_2)$ is a bijective open map. The converse is clear.

(c) (2) $\Rightarrow$ (1) This is obvious.

(1) $\Rightarrow$ (4). If U_1 is open and nonempty, we shrink to get $U_1 \times U_1 \subset V_1$. By (1) there exists $x \in U_1$ and U_2 a neighborhood of $f(x)$ such that $f^{-1}(U_2) \subset U_1$.

(4) $\Rightarrow$ (3). This is obvious.

(3) $\Rightarrow$ (5). Assume $f(D)$ is dense in $f(X_1)$. Given $x_0 \in X_1$ and $W \in \mathcal{U}_{X_1}$, we show that $W(x_0)$ meets D. Choose $V_1 \in \mathcal{U}_{X_1}$ symmetric such that $V_1^2 \subset W$. By (3) there exists $x_1 \in V_1(x_0)$ and $V_2 \in \mathcal{U}_{X_2}$ such that

$$f^{-1}(V_2(f(x_1))) = (f^*V_2)(x_1) \subset V_1(x_1)$$

Because $f(D)$ is dense in $f(X_1)$, it meets $V_2(f(x_1))$. Thus there exists $x_2 \in D$ with $f(x_2) \in V_2(f(x_1))$, so:

$$x_2 \in V_1(x_1) \subset V_1^2(x_0) \subset W(x_0)$$

Hence D is dense in X_1.

(5) $\Rightarrow$ (2). Let $D = U^f \cup X_1 \backslash U$.

$$f(D) = [f(X_1) \cap \text{Int}(X_2 \backslash f(X_1 \backslash U))] \cup f(X_1 \backslash U)$$

which is dense in $f(X_1)$. By (5) D is dense in X_1, so $D \cap U = U^f$ is dense in U.

For the next four conditions observe that (2) $\Rightarrow$ (6) and (6) $\Leftrightarrow$ (7) are obvious. (3) $\Rightarrow$ (8) and (9) $\Rightarrow$ (8) are also obvious. As usual $V = V^{-1}$, $V^2 \subset V_1$, and $E_f(x) \subset V(x)$ imply $E_f(x) \times E_f(x) \subset V_1$, so that (8) $\Rightarrow$ (9).

(8) $\Rightarrow$ (3). If V_1 is open, f is closed, and $E_f(x) \subset V_1(x)$, then $f(x)$ is not in the closed set $f(X_1 \backslash V_1(x))$. Let U be the complement. Then we have $x \in f^{-1}(U)$ and $f^{-1}(U) \subset V_1(x)$.

In any case (6) $\Rightarrow$ (9). For any open set U with $U \times U \subset V_1$ and U_1 such that $f^{-1}f(U_1) = U_1 \subset U$ it follows that $E_f(x) \times E_f(x) \subset V_1$ for all $x \in U_1$. Then (6) implies that the (9) set is dense.

Clearly $E_f(x) = \{x\}$ implies $E_f(x) \subset V_1(x)$ for all $V_1 \in \mathcal{U}_{X_1}$. Then $\{x : E_f(x) = \{x\}\}$ dense implies (8). On the other hand if X_1 is metrizable and Baire, we can intersect the sets of (4) over a countable base of V_1s to obtain the residual set of points at which f is an embedding [cf. (a)].

(d) In general D dense in X_1 implies $f(D)$ is dense in $f(X_1)$. If f is a dense map then $f(D)$ is dense in X_2 iff it is dense in $f(X_1)$. For a dense map (5) $\Leftrightarrow$ (10). (2) $\Rightarrow$ (12) is obvious. On the other hand if U is a nonempty open subset of X_1, then $A = X_1 \backslash U$ is a proper closed subset, so (12) implies that $f(A)$ is a proper subset of X_2. Its complement U_1 is a nonempty open set with $f^{-1}(U_1) \subset U$. But $f(X_1)$ is dense, so U_1 meets $f(X_1)$. Thus $f^{-1}(U_1)$ is nonempty. It follows that (12) $\Rightarrow$ (1).

(10) $\Rightarrow$ (11). f is an epimorphism because it is a dense map. If fg is an epimorphism with $g : X_0 \to X_1$ then $f(g(X_0))$ is dense in X_2. By (10) $g(X_0)$ is dense in X_1, i.e., g is an epimorphism; thus f is minimal.

(11) $\Rightarrow$ (10). Apply (11) to the inclusion map g of D into X_1. Since fg is epi because $f(D)$ is dense in X_2, minimality implies g is epi; i.e., D is dense in X_1.

(e) In general for a surjective continuous map and any $V_1 \in \mathcal{U}_{X_1}$, f is V_1 open at x if f is V_1 embedding at x. If f is an almost embedding, A is closed in X_1, and U is open in X_2, then:

$$f(A) \supset U \Rightarrow A \supset f^{-1}(U) \tag{A.43}$$

If not, then $U_1 = f^{-1}(U) \backslash A$ is a nonempty open subset of X_1, so $U_1 \supset f^{-1}(U_2)$ for some nonempty open subset U_2 of U. But then:

$$U_2 \subset f(U_1) \subset U \subset f(A)$$

A is disjoint from U_1 and so from $f^{-1}(U_2)$. This is a contradiction.

If V_1 is closed and f is V_1 open at x then there exists V_2 open in $\mathcal{U}_{X_2}$ such that:

$$f(V_1(x)) \supset V_2(f(x))$$

By (A.43)

$$V_1(x) \supset f^{-1}(V_2(f(x)))$$

This proves (A.41) which clearly implies (A.42). ■

LEMMA A.8. *Let $f : X_1 \to X_2$ and $g : X_2 \to X_3$ be continuous maps.*

a. Let property P stand for either is open, almost open, quasi-open, or almost quasi-open. If both f and g satisfy P, then gf satisfies P. If gf satisfies P and f is surjective, then g satisfies P.

b. Assume f is surjective. gf is an almost homeomorphism iff both f and g are almost homeomorphisms.

c. Assume g is an almost homeomorphism. If gf is almost open, then f is almost quasi-open. If gf is almost quasi-open and g is closed then f is almost quasi-open.

PROOF. (a) We do the quasi proofs, since the others are similar. Preserving quasi-openness under image ($P =$ quasi-open) and preserving dense and open under preimage ($P =$ almost quasi-open) are satisfied by gf if satisfied by f and g. Now assume f is surjective. If U_2 is open in X_2 then $U_1 = f^{-1}(U_2)$ is open X_1 and $g(U_2) = gf(U_1)$. If gf is quasi-open, then this set is quasi-open, so g is quasi-open. If $U_2 \neq \emptyset$ and gf is almost quasi-open, then this set has a nonempty quasi-interior, so g is almost quasi-open.

(b) Since an almost homeomorphism is a minimal surjection in the category Spaces, these results follow from Lemma A.2b.

(c) Let U_1 be open and nonempty in X_1. If gf is almost open, then let $U_3 = \mathrm{Int}\, gf(U_1)$. If gf is almost quasi-open and g is closed, then let

$$U_3 = \mathrm{Int}\, \overline{gf(U_1)} = \mathrm{Int}\, g(\overline{f(U_1)})$$

In either case U_3 is open, nonempty, and satisfies $g(\overline{f(U_1)}) \supset U_3$. By (A.43) $\overline{f(U_1)} \supset g^{-1}(U_3)$ so f is almost quasi-open. ■

We collect the examples of these sorts of maps that are most important for our purposes.

LEMMA A.9. *For $X_0 \subset X$ let $i : X_0 \to X$ be the inclusion map; i is an embedding. If X_0 is open, i.e., $\mathrm{Int}X_0 = X_0$, then i is open. If $\overline{\mathrm{Int}X_0} \supset X_0$, then i is almost open. If X_0 is quasi-open, i.e., $\mathrm{Int}\overline{X_0} \supset X_0$, then i is quasi-open.*

PROOF. The first two results are clear. Now let $U = \mathrm{Int}X_0$. If $U \subset X_0 \subset \overline{U}$, then i is open at the points of U which are dense in X_0. Then i is almost open. If instead X_0 is quasi-open, then any open subset of X_0 is of the form $X_0 \cap U$ for some U open in X. Such a set is quasi-open in X. Thus the inclusion takes open sets to quasi-open sets, so it is quasi-open. ■

PROPOSITION A.5. *Let $h : (\varphi_1, x_1) \to (\varphi_2, x_2)$ be a sharp, continuous map of ambits. If the actions φ_1 and φ_2 are reversible, then h is a closed, almost homeomorphism. h is a minimal map; i.e., h is minimal for the categories Spaces, Compacts, T Spaces, and T Compacts.*

PROOF. Because h is sharp $h^{-1}(x_2) = \{x_1\}$. Because the actions are reversible:

$$h^{-1}(f_2^t(x_2)) = \{f_1^t(x_2)\}$$

for all $t \in T$. Since the map is closed by compactness and the orbit of x_1 is dense in X_1, h is an almost embedding. It is surjective and so is an almost homeomorphism. The map h is then minimal by Proposition A.4d. ■

REMARK. In the case of minimal systems, such maps h are called *highly proximal extensions* by Auslander and Glasner (1977). ■

The minimal mappings of Chapter 5 (cf. Propositions 5.15 and 5.16) are minimal morphisms of the category T Compacts. We now extend these minimal morphisms to the noncompact case.

PROPOSITION A.6. *Let $\varphi_\alpha : T \times X_\alpha \to X_\alpha$ be uniform actions $(\alpha = 1,2)$ and let $h : \varphi_1 \to \varphi_2$ be a continuous action map. The following two conditions are equivalent; when they hold, we call h a* minimal action map*:*

(1) h is a minimal morphism in the category T Spaces.

(2) If D is a + invariant subset of X_1, then D is dense in X_1 iff $h(D)$ is dense in X_2.

Assume that h is a minimal action map. Then h is a dense map; i.e., $h(X_1)$ is dense in X_2, and:

$$\text{Trans}_{\varphi_1} = h^{-1}(\text{Trans}_{\varphi_2}) \tag{A.44}$$

If h is a minimal surjection with T separable, X_2 Baire, and φ_2 invertible, then h is almost quasi-open.

PROOF. The equivalence is an easy exercise, since an epimorphism in the category is a dense map. In particular h is dense because a minimal is epi. For any $t \in T$ the set $f^{T_t}(x)$ is + invariant and x is a transitive point iff this set is dense for all t. Then (A.44) follows.

Now assume T_0 is a countable dense submonoid of T, X_2 is Baire, and U_2 is a dense open subset of X_2. Define

$$D_2 = \cap_{t \in T_0} f_2^{-t}(U_2) \tag{A.45}$$

Because the action is invertible, each $f_2^{-t}(U_2)$ is dense, and so, because X_2 is Baire, D_2 is dense. Because h is surjective:

$$h(D) = D_2 \tag{A.46}$$

where $D \equiv h^{-1}(D_2)$. For $x \in D$, $f^{T_0}(x) \subset D$. Then $\overline{D}$ is a + invariant subset of X_1. By (A.46) $h(\overline{D})$ is dense in X_2, so by minimality, $\overline{D}$ is dense, i.e., $\overline{D} = X_1$. Since $D \subset h^{-1}(U_2)$, it follows that $h^{-1}(U_2)$ is dense, so h is an almost quasi-open map. ■

REMARK. That φ_2 is invertible was used only to show each $f_2^{-t}(U_2)$ is dense. It is sufficient for each f_2^t to be almost quasi-open. ■

As a special case we obtain the following theorem of Auslander.

THEOREM A.2. *Let $h : \varphi_1 \to \varphi_2$ be a minimal action map of actions on compact spaces. If T is separable and φ_2 is invertible, then h is an almost open map.*

PROOF. By compactness, h is closed and surjective. Then almost quasi-open implies almost open. ■

We recall some well-known results about open maps.

LEMMA A.10. *Let $f : X_1 \to X_2$ be a continuous, open map. If $A \subset X_2$, then:*

$$f^{-1}(\overline{A}) = \overline{f^{-1}(A)} \tag{A.47}$$

Furthermore the restriction $f|f^{-1}(A) : f^{-1}(A) \to A$ is an open map.

PROOF. $f^{-1}(\overline{A})$ is a closed set containing $f^{-1}(A)$ and hence its closure. If U is a neighborhood of x with $f(x) \in \overline{A}$, then $f(U)$ is a neighborhood of $f(x)$, so it meets A. Hence every such U meets $f^{-1}(A)$, proving (A.47).

If O is open in $f^{-1}(A)$, then $O = f^{-1}(A) \cap U$ for some open subset U of X. Then:

$$f(O) = A \cap f(U)$$

which is open in A. ■

We conclude with some results specific to the compact case.

If $f : X_1 \to X_2$ is continuous and X_1 is compact, then for every V open in $\mathcal{U}_{X_1}$ the set:

$$S_f^V = \{x \in X_1 : E_f(x) \times E_f(x) \subset V\} \tag{A.48}$$

is open by compactness. Since f is a closed mapping, Proposition A.4c implies that f is an almost embedding iff S_f^V is dense in X_1 for all V in $\mathcal{U}_{X_1}$.

PROPOSITION A.7. *Let $f : X_1 \to X_2$ be a continuous surjection with X_1, X_2 compact. The following conditions are equivalent; when they hold, we say that* there is a unique minimal subset for f:

(1) *There is a unique closed subset X_0 of X_1 such that the restriction $f|X_0 : X_0 \to X_2$ is an almost homeomorphism.*

(2) *If A_1, A_2 are closed subsets of X_1 such that $f(A_1) = X_2 = f(A_2)$, then:*

$$f(A_1 \cap A_2) = X$$

(3) *For every $V \in \mathcal{U}_{X_1}$, $f(S_f^V)$ is a dense subset of X_2.*

PROOF. (2) $\Rightarrow$ (1). Condition (2) says that $\mathcal{F} = \{A \subset X_1 : A \text{ closed and } f(A) = X_2\}$ forms a filterbase of closed sets. Let $X_0 = \cap\mathcal{F}$. For each $y \in X_2$ $\{f^{-1}(y) \cap A : A \in \mathcal{F}\}$ is a filterbase of closed sets, so it has a nonempty intersection, which is $f^{-1}(y) \cap X_0$. Hence $X_0 \in \mathcal{F}$, and it is clearly the unique minimal member of the family.

(3) $\Rightarrow$ (2). By compactness $f(\overline{S_f^V}) = \overline{f(S_f^V)}$ which is X_2 by (3). Hence $\overline{S_f^V}$ is in the filterbase $\mathcal{F}$ previously described. Since:

$$S_f^{V_1 \cap V_2} \subset S_f^{V_1} \cap S_f^{V_2}$$

the collection $\{\overline{S_f^V} : V \in \mathcal{U}_{X_1}\}$ is a filterbase. Let $\tilde{X}_0$ denote its intersection; as before $\tilde{X}_0 \in \mathcal{F}$. If $A \in \mathcal{F}$ and $x \in S_f^V$, then $f(x) \in f(A)$ and $E_f(x) \times E_f(x) \subset V$ imply $x \in V(x_1)$ for some $x_1 \in A$, i.e., $V(A) \supset S_f^V$. Hence for all $V \in \mathcal{U}_{X_1}$, $\overline{V(A)} \supset \tilde{X}_0$. Intersecting over $V \in \mathcal{U}_{X_1}$ we have $A \supset \tilde{X}_0$ because A is closed. Consequently $\cap\mathcal{F} \supset \tilde{X}_0$; since $\tilde{X}_0 \in \mathcal{F}$, $\cap\mathcal{F} = \tilde{X}_0$. In particular (2) holds. Furthermore the unique minimal subset $X_0 = \cap\mathcal{F}$ satisfies

$$X_0 = \cap\{\overline{S_f^V} : V \in \mathcal{U}_{X_1}\} \tag{A.49}$$

(1) $\Rightarrow$ (3). Assume (3) is false. For some $V \in \mathcal{U}_{X_1}$ there exists a nonempty open subset U of X_2 that is disjoint from $f(S_f^V)$. Choose $W = W^{-1}$, closed in $\mathcal{U}_{X_1}$ such that $W^2 \subset V$. By Corollary A.1 we can choose a point $y \in U$ such that f is W open at every point x of $f^{-1}(y)$. If for every pair $x_1, x_2 \in f^{-1}(y)$, $W(x_1)$ meets $W(x_2)$, then:

$$f^{-1}(y) \times f^{-1}(y) \subset W \circ W \subset V$$

contradicting $y \notin f(S_f^V)$. Thus there exist $x_1, x_2 \in f^{-1}(y)$ with $W(x_1) \cap W(x_2) = \emptyset$. Because f is W open at x_1 and x_2, there exists $\tilde{U}$ an open set containing y such that $\tilde{U} \subset f(W(x_\alpha))$ for $\alpha = 1, 2$. Now for $\alpha = 1, 2$ define

$$A_\alpha = f^{-1}(X_2 \backslash \tilde{U}) \cup W(x_\alpha) \tag{A.50}$$

Since W is closed and $\tilde{U}$ is open, A_α is closed. Since f is surjective, $f(A_\alpha) = X_2$ $(\alpha = 1,2)$, i.e., $A_1, A_2 \in \mathcal{F}$. But:

$$A_1 \cap A_2 = f^{-1}(X_2 \backslash \tilde{U})$$

which maps to the proper closed subset $X_2 \backslash \tilde{U}$ under f. By Zorn's Lemma we can choose minimal elements in $\mathcal{F}$ $\tilde{A}_\alpha \subset A_\alpha$ $(\alpha = 1,2)$; i.e., the category Compacts admits minimal surjections by Lemma A.3c. $\tilde{A}_1 \neq \tilde{A}_2$ because each maps onto X_2 and the intersection does not. It follows that (1) is false. ■

REMARK. Observe that:

$$f(S_f^V) \equiv \{y \in X_2 : f^{-1}(y) \times f^{-1}(y) \subset V\} \tag{A.51}$$

■

LEMMA A.11. *Assume that the following is a pullback diagram in the category Compacts:*

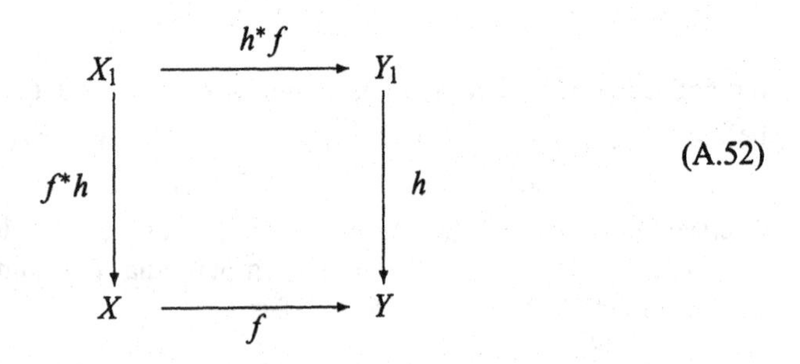

(A.52)

Assume that there is a unique minimal subset Y_0 for h. If f is an almost open surjection, then there is a unique minimal subset X_0 for f^*h, and:

$$h^*f(X_0) = Y_0 \tag{A.53}$$

If, in addition, the pullback map h^*f is open, then:

$$X_0 = (h^*f)^{-1}(Y_0) \tag{A.54}$$

PROOF. Without loss of generality we can assume

$$X_1 = \{(x,y) \in X \times Y_1 : f(x) = h(y)\}$$

with f^*h and h^*f restrictions of the projections. For $\tilde{V} \in \mathcal{U}_X$ and $V \in \mathcal{U}_{Y_1}$ we write $\tilde{V} \times V$ for:

$$(X_1 \times X_1) \cap \{(x_1, y_1, x_2, y_2) : (x_1, x_2) \in \tilde{V} \text{ and } (y_1, y_2) \in V\}$$

Observe that for $(x,y) \in X_1$:

$$E_{f^*h}(x,y) = \{x\} \times E_h(y) \tag{A.55}$$

So for all $\tilde{V} \in \mathcal{U}_X$ and $V \in \mathcal{U}_{Y_1}$:

$$S_{f^*h}^{\tilde{V}\times V} = (h^*f)^{-1}(S_h^V) \tag{A.56}$$

$$f^*h(S_{f^*h}^{\tilde{V}\times V}) = f^{-1}(h(S_h^V)) \tag{A.57}$$

By Proposition A.7, $h(S_h^V)$ is dense in Y for all $V \in \mathcal{U}_Y$. If f is almost open, each $f^*h(S_{f^*h}^{\tilde{V}\times V})$ is dense in X by (A.57). Because the $\tilde{V} \times V$s generate the uniformity $\mathcal{U}_{X_1}$ it follows from Proposition A.7 that f^*h has a unique minimal subset X_0. By compactness $h^*f(X_0)$ is closed in Y_1 and:

$$h(h^*f(X_0)) = f(f^*h(X_0)) = f(X) = Y$$

because f is surjective. Hence $Y_0 \subset (h^*f)(X_0)$. On the other hand since $h(Y_0) = Y$ for every $x \in X$ there exists $y \in Y_0$ such that $h(y) = f(x)$. Then $(x,y) \in h^{-1}(Y_0)$ with $f^*h(x,y) = x$. Hence:

$$X_0 \subset (h^*f)^{-1}(Y_0)$$

proving (A.53).

If h^*f is an open map, then applying (A.47) to (A.56)

$$\overline{S_{f^*h}^{\tilde{V}\times V}} = (h^*f)^{-1}(\overline{S_h^V}) \qquad (h^*f \text{ open}) \tag{A.58}$$

If $y \in Y_0$, $(h^*f)^{-1}(y)$ meets each $\overline{S_{f^*h}^{\tilde{V}\times V}}$. By compactness and (A.49) it follows that $(h^*f)^{-1}(y)$ meets X_0, proving (A.54). ■

We now generalize a useful construction from Glasner (1996).

PROPOSITION A.8. *Let Y and Z be compact spaces and F a usc relation from Y to Z, i.e., $F : Y \to C(Z)$. Define Y_1 to be the closure in $Y \times C(Z)$ of $\{(y, F(y)) : y \in Y\}$ and let $h_1 : Y_1 \to Y$ and $F_1 : Y_1 \to C(Z)$ denote the restrictions of the projections to Y_1. F_1 is a continuous relation from Y_1 to Z, and there is a unique minimal subset Y_0 for h_1.*

Suppose now that F_0 is a pointwise closed lsc relation from Y to Z, and F is the closure of F_0, regarded as subsets of $Y \times Z$. Define Y_2 to be the subset of $Y \times C(Z)$ consisting of:

$$\{(y,K) : F_0(y) \subset K \subset F(y)\}$$

Y_2 is a closed subset of $Y \times C(Z)$ containing Y_1. Letting $h_2 : Y_2 \to Y$ and $F_2 : Y_2 \to C(Z)$ denote the restrictions of the projections to Y_2, F_2 is a continuous relation from Y_2 to $C(Z)$ and Y_0 is the unique minimal subset for h_2.

PROOF. We prove the two cases together. First we show that Y_2 is closed. If $\{(y_\alpha, K_\alpha)\}$ is a net in Y_2 converging to $(y,K) \in Y \times C(Z)$ and $V \in \mathcal{U}_Z$, then for sufficiently large α, $V(K_\alpha) \supset K$ and $V(K) \supset K_\alpha$ since $\{K_\alpha\}$ converges to K, and:

$$V(F(y)) \supset F(y_\alpha) \supset K_\alpha \qquad V(K_\alpha) \supset V(F_0(y_\alpha)) \supset F_0(y)$$

since F is usc and F_0 is lsc. Hence $V^2(F(y)) \supset K$ and $V^2(K) \supset F_0(y)$. Intersecting over all V in $\mathcal{U}_2$, we have $(y,K) \in Y_2$, since $F(y)$ and K are closed. Notice that $\{(y,F(y)) : y \in Y\} \subset Y_2$, so the closure Y_1 is contained in Y_2. The relations F_1 and F_2 are clearly continuous.

Now suppose we are given $V \in \mathcal{U}_Z$. Because F is usc, Theorem A.1 implies that:

$$C_F^V = \{y \in Y : (y,y) \in \text{Int}(F^*V)\} \tag{A.59}$$

is open and dense in Y. For $y \in C_F^V$ we can choose an open set U in Y containing y such that $U \times U \subset F^*V$, i.e.:

$$F(y_1) \subset V(F(y_2)) \text{ for all } y_1, y_2 \in U \tag{A.60}$$

If V is assumed closed, then it follows that $K_1 \subset V(K_2)$ for all $(y_1,K_1), (y_2,K_2) \in Y_1$ such that $y_1, y_2 \in U$. In particular for any $W \in \mathcal{U}_Y$ and any closed $V \in \mathcal{U}_Z$:

$$C_F^V \subset h_1(S_{h_1}^{W \times C(V)}) \tag{A.61}$$

Then by Proposition A.7 there is a unique minimal set for h_1.

Similarly for $y \in C_{F_0}^V$ and $U \times U \subset F_0^*V$ the analogue of (A.60) for F_0 implies that when V is closed:

$$F(y_1) \subset V(F_0(y_2)) \text{ for all } y_1, y_2 \in U \tag{A.62}$$

because F is the closure of F_0 and so each point of $F(y_1)$ is the limit of points in $F_0(y_\alpha)$ for some net $\{y_\alpha\}$ converging to y_1 and eventually in U. It follows that $K_1 \subset V(K_2)$ for all $(y_1,K_1), (y_2,K_2) \in Y_2$ such that $y_1, y_2 \in U$. As before there is a unique minimal set for h_2. Because $Y_1 \subset Y_2$ and h_1 is the restriction of h_2 to Y_1, the minimal set Y_0 for h_1 is the unique minimal set for Y_2. ■

THEOREM A.3. *Let $f : X \to Y$ be a continuous surjection with X and Y compact. f is almost open if and only if there exist compact spaces X_0 and Y_0, almost homeomorphisms $g : X_0 \to X$ and $h : Y_0 \to Y$, and a continuous open surjection*

$f_0 : X_0 \to Y_0$ such that the following diagram commutes:

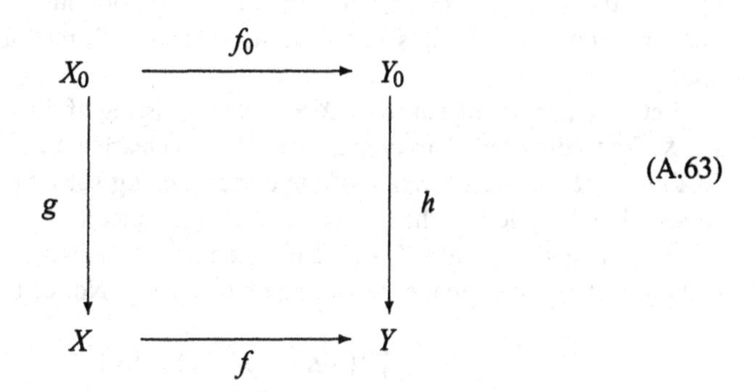

(A.63)

PROOF. If such a factorization exists, then $fg = hf_0$ is almost open by Proposition A.4e and Lemma A.8a. Since g is surjective, f then is almost open by Lemma A.8a.

Now assume f is almost open. By compactness the relation f^{-1} from Y to X is usc (cf. Lemma A.6a). We apply Proposition A.8 to define Y_1 to be the closure in $Y \times C(X)$ of the set $\{(y, f^{-1}(y)) : y \in Y\}$. Consider the map:

$$f \times 1_{C(X)} : X \times C(X) \to Y \times C(X)$$

Let $X_1 = (f \times 1_{C(X)})^{-1}(Y_1)$, so that:

$$X_1 = \{(x, K) : (f(x), K) \in Y_1\}$$

Let $f_1 : X_1 \to Y_1$ denote the restriction of $f \times 1_{C(X)}$ and let $g_1 : X_1 \to X$ be the restriction of the projection. It is easy to check that:

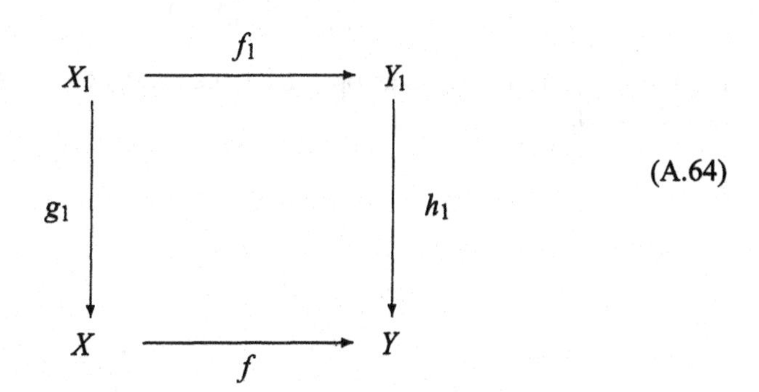

(A.64)

is a pullback diagram. By Proposition A.8 and Lemma A.11, there are unique minimal subsets X_0 and Y_0 for g_1 and h_1 respectively and $f_1(X_0) = Y_0$. This is where we use the fact that f is almost open. Let g, h and f_0 denote the restrictions to the subsets X_0 and Y_0. Then diagram (A.63) commutes, g and h are almost homeomorphisms, and f_0 is a continuous surjection. It remains to show that f_0 is open.

Let EL_X denote the subset of $X \times C(X)$ consisting of the pairs (x, K) such that $x \in K$. Regarding the identity map of $C(X)$ as a clearly continuous a relation from $C(X)$ to X, EL_X is the inverse of the corresponding subset of $C(X) \times X$, so it is closed. Let $X_{1/2}$ denote the subset $X_1 \cap (EL_X)$. Since $(x, E_f(x)) \in X_{1/2}$ for every $x \in X$, g_1 maps $X_{1/2}$ onto X, so the unique minimal subset X_0 for g_1 is contained in $X_{1/2}$. Let $f_{1/2}$ denote the restriction of f_1 to $X_{1/2}$. Notice that for $(y, K) \in Y_1$:

$$\begin{aligned} f_1^{-1}(y,K) &= f^{-1}(y) \times \{K\} \\ f_{1/2}^{-1}(y,K) &= K \times \{K\} \end{aligned} \tag{A.65}$$

It follows that the relation $f_{1/2}^{-1}$ from Y_1 to X_1 is continuous. By Proposition A.1c $f_{1/2} : X_{1/2} \to Y_1$ is an open map.

Now for $V \in \mathcal{U}_{Y_1}$, (A.56) implies

$$(f_{1/2})^{-1}(S_{h_1}^V) \subset S_{g_1}^{\tilde{V} \times V}$$

Applying (A.47) we have

$$\begin{aligned} (f_{1/2})^{-1}(Y_0) &\subset (f_{1/2})^{-1}(\overline{S_{h_1}^V}) \\ &= \overline{(f_{1/2})^{-1}(S_{h_1}^V)} \subset \overline{S_{g_1}^{\tilde{V} \times V}} \end{aligned}$$

Intersecting over the $\tilde{V} \times V$s we have by (A.49) that $(f_{1/2})^{-1}(Y_0) \subset X_0$. But the reverse inclusion follows from $f_0(X_0) = Y_0$. Hence:

$$(f_{1/2})^{-1}(Y_0) = X_0 \tag{A.66}$$

Because $f_{1/2} : X_{1/2} \to Y_1$ is an open map, the restriction $f_0 : X_0 \to Y_0$ is open by Lemma A.10. ■

References

1. E. Akin, *The General Topology of Dynamical Systems*, Amer. Math. Soc., Providence, RI (1993).
2. ———, "On chain continuity," *Discrete and Cont. Dynam. Sys.* **2**, 111–120 (1996).
3. ———, "Dynamical systems: the topological foundations," in: *Six Lectures on Dynamical Systems*, (B. Aulbach and F. Colonius, eds.), pp. 1–43, World Scientific, Singapore (1996).
4. E. Akin, J. Auslander, and K. Berg, "When is a transitive map chaotic?" in: *Conference in Ergodic Theory and Probability* (V. Bergelson, K. March, and J. Rosenblatt, eds.), pp. 25–40, Walter de Gruyter, Berlin (1996).
5. J-P. Aubin and H. Frankowska, *Set-Valued Analysis*, Birkhäuser, Boston (1990).
6. J. Auslander, "Filter stability in dynamical systems," *SIAM J. Math. Anal.* **8**, 573–579 (1977).
7. ———, *Minimal Flows and Their Extensions*, North-Holland, Amsterdam (1988).
8. J. Auslander and H. Furstenberg, "Product recurrence and distal points," *Trans. Amer. Math. Soc.* **343**, 221–232 (1994).
9. J. Auslander and E. S. Glasner, "Distal and highly proximal minimal flows," *Indiana Univ. Math. J.* **26**, 731–749 (1977).
10. J. Auslander and F. Hahn, "Point transitive flows, algebras of functions and the Bebutov system," *Fund. Math.* **60**, 117–137 (1967).
11. J. Auslander and P. Siebert, "Prolongations and generalized Liapunov functions," in: *International Symposium on Nonlinear Differential Equations and Nonlinear Mechanics*, pp. 454–462, Academic Press, New York (1963).
12. J. Auslander and J. Yorke, "Interval maps, factors of maps and chaos," *Tôhoku Mat. J.* **32**, 177–188 (1980).
13. N. Dunford and J. Schwartz, *Linear Operators Part II: Spectral Theory, Self Adjoint Operators in Hilbert Space*, John Wiley and Sons, New York (1963).
14. R. Ellis, "Locally compact transformation groups," *Duke Math. J.* **24**, 119–125 (1957).
15. ———, "Distal transformation groups," *Pacific Math. J.* **8**, 401–405 (1958).
16. ———, "A semigroup associated with a transformation group," *Trans. Amer. Math. Soc.* **94**, 272–281 (1960).
17. ———, "Universal minimal sets," *Proc. Amer. Math. Soc.* **11**, 540–543 (1960).
18. ———, "Point transitive transformation groups," *Trans. Amer. Math. Soc.* **101**, 384–395 (1961).

19. ———, "Group-like extensions of minimal sets," *Trans. Amer. Math. Soc.* **127**, 125–135 (1967).

20. ———, "The structure of group-like extensions of minimal sets," *Trans. Amer. Math. Soc.* **134**, 261–287 (1968).

21. ———, *Lectures on Topological Dynamics*, W. A. Benjamin, Inc., New York (1969).

22. R. Ellis and H. Keynes, "Bohr compactification and a result of Folner," *Israel J. of Math.* **12**, 314–330 (1972).

23. H. Furstenberg, "Disjointness in ergodic theory, minimal sets and a problem in Diophantine approximations," *Math. Sys. Th.* **1**, 1–49 (1967).

24. ———, *Recurrence in Ergodic Theory and Combinatorial Number Theory*, Princeton Univ. Press, Princeton (1981).

25. H. Furstenberg and B. Weiss, "Topological dynamics and combinatorial number theory," *J. Anal. Math.* **34**, 61–85 (1978).

26. E. S. Glasner, *Proximal Flows*, Lecture Notes in Math No. 517, Springer-Verlag, Berlin (1976).

27. ———, "Compressibility properties in topological dynamics," *Amer. J. Math.* **97**, 148–171 (1975).

28. ———, "Divisible properties and the Stone–Čech compactification," *Canad. J. Math.* **32**, 993–1007 (1980).

29. ———, *Homogeneous flows*, 1996, to appear.

30. E. S. Glasner and D. Maon, "Rigidity in topological dynamics," *Ergod. Th. and Dynam. Sys.* **9**, 309–320 (1989).

31. E. S. Glasner and B. Weiss, "Interpolation sets for subalgebras of $l^\infty(\mathbf{Z})$," *Israel J. of Math.* **44**, 345–360 (1983).

32. ———, "Sensitive dependence on initial conditions," *Nonlinearity* **6**, 1067–1075 (1993).

33. W. Gottschalk, "Some general dynamical notions," in: *Recent Advances in Topological Dynamics* (A. Beck, ed.), Lecture Notes in Math. No. 318, Springer-Verlag, Berlin (1973).

34. W. Gottschalk and G. Hedlund, *Topological Dynamics*, Amer. Math. Soc., Providence, RI (1955).

35. K. Haddad, "New limiting notions of the IP type in the enveloping semigroup," *Ergod. Th. and Dynam. Sys.* **16**, 719–733 (1996).

36. N. Hindman, "Summable ultrafilters and finite sums," *Contemp. Math.* **65**, 263–274 (1987).

37. ———, "The semigroup βN and its application to number theory," in: *The Analytical and Topological Theory of Semigroups* (K. H. Hofmann, J. D. Lawson, and J. S. Pym, eds.) pp. 347–360, Walter de Gruyter, Berlin (1990).

38. Y. Katznelson, *An Introduction to Harmonic Analysis*, 1968, reprint, Dover, New York (1976).

39. Y. Katznelson and B. Weiss, "When all points are recurrent/generic," in: *Ergodic Theory and Dynamical Systems 1, College Park 1979–1980*, (A. Katok, ed.), pp. 195–210, Progress in Math. 10, Birkhäuser Verlag, Boston (1981).

40. J. Kelley, *General Topology*, D. Van Nostrand, New York (1955).

41. V. V. Nemytskii and V. V. Stepanov, *Qualitative Theory of Differential Equations*, Princeton Univ. Press, Princeton (1960).

42. J. Oxtoby, *Measure and Category* (2nd ed.), Springer-Verlag, Berlin (1980).

43. D. E. Penazzi, *Groups and Relations in Topological Dynamics*, Ph. D. Thesis, University of Minnesota (1993).

44. T. Ura, "Sur les courbes définies par les equations differentielles dans l'espace à m dimensions," *Ann. Sci. Ecole Norm. Sup.* **70**, 287–360 (1953).

45. J. de Vries, *Elements of Topological Dynamics*, Kluwer, Dordrecht, Holland (1993).

Index